Automatic Differentiation of Algorithms

Springer

New York
Berlin
Heidelberg
Barcelona
Hong Kong
London
Milan
Paris
Singapore
Tokyo

George Corliss   Christèle Faure
Andreas Griewank   Laurent Hascoët
Uwe Naumann
Editors

Automatic Differentiation of Algorithms
From Simulation to Optimization

With 94 Illustrations

 Springer

INCLUDES
CD-ROM

George Corliss
Department of Electrical and Computer
  Engineering
Marquette University
P.O. Box 1881
1515 W. Wisconsin Ave.
Milwaukee, WI 53233, USA
George.Corliss@Marquette.edu

Laurent Hascoët
INRIA, projet Tropics
Route des Lucioles 06902
Sophia Antipolis, France
Laurent.Hascoet@sophia.inria.fr

Christèle Faure
PolySpace Technologies
28 Rue Estienne d'Orves 92120
Montrouge, France
Christele.Faure@polyspace.com

Uwe Naumann
Department of Computer Science
University of Hertfordshire
AL10 9AB Hatfield
Herts, United Kingdom
U.1.Naumann.@herts.ac.uk

Andreas Griewank
Institute of Scientific Computing
Technical University
Dresden, Mommsenstr. 13
01062 Dresden, Germany
griewank@math.tu-dresden.de

Library of Congress Cataloging-in-Publication Data
Automatic differentiation of algorithms : from simulation to optimization / editors
George Corliss . . . [et al.].
    p. cm.
    " . . . Selected papers from the Third International Conference on Automatic
Differentiation, . . . June 2000, . . . Côte d'Azur."
    Includes bibliographical references and index.
    ISBN 0-387-95305-1 (alk. paper)
    1. Differential calculus—Data processing—Congresses.   I. Corliss, George F.
II. International Conference on Automatic Differentiation (3rd : 2000 : Côte d'Azur, France)
QA304 .A78 2001
515'.33—dc21   2001042959

Printed on acid-free paper.

Production managed by Jenny Wolkowicki; manufacturing supervised by Jacqui Ashri.
Photocomposed pages prepared from the authors' LATEX files.
Printed and bound by Sheridan Books, Inc., Ann Arbor, MI.
Printed in the United States of America.

9 8 7 6 5 4 3 2 1

ISBN 0-387-95305-1   SPIN 10839134

Springer-Verlag   New York   Berlin   Heidelberg
A member of BertelsmannSpringer Science+Business Media GmbH

Preface

This volume comprises selected papers from the Third International Conference on Automatic Differentiation, which took place in June 2000 near the old harbor of Nice, Côte d'Azur. Like the earlier meetings in Breckenridge, Colorado (1991) and Santa Fe, New Mexico (1996), it was supported by SIAM, but this time the organization was in the most capable hands of INRIA, Sophia Antipolis.

Goals

As demonstrated by many case studies in this volume, a growing number of scientific and engineering computing applications require derivatives. This need for accurate sensitivity information arises when parameters of nonlinear models need to be adjusted to fit data or when one wishes to optimize performance characteristics by varying design and control variables. Differencing is still a very popular way to approximate derivatives. When exact (truncation-error free) sensitivity values are required, computer algebra, hand-coding and automatic differentiation (AD, also sometimes called computational or algorithmic differentiation) are the only choices.

As many studies have found, derivative-free numerical methods yield results at acceptable costs only if the number of degrees of freedom is rather low. This applies even more for fully symbolic modeling, which is sometimes employed, for example, in multi-body dynamics. The limitation to a small number of free variables severely restricts the modeling fidelity, and it strongly reduces the chance of finding solutions whose structure deviates from conventional designs. Hand-coding, on the other hand, usually yields efficient and precise derivatives, but at a very high development cost. For the same precision, AD requires *far* less human time, enabling more flexibility in model development. For example, the automatic generation of (discrete) adjoints in the reverse mode is possible using TAMC [211] or Odyssée [186].

Since the first meeting some ten years ago, the sophistication of AD tools has continually grown. Many software packages for scientific computation and many other applications now use embedded AD. The techniques of AD have become reliable and efficient enough that users are satisfied and need not concern themselves about the issue of differentiation. Very often the evaluation of all problem-specific nonlinear functions takes only a fraction of a second on modern work stations. On the overall run time, computation of the gradients is negligible. Only the full accuracy of derivative values is important and greatly enhances robustness. In the modeling systems GAMS, AMPL, and even Microsoft Excel, the application of AD variants now often happens "under the hood" without the user being aware of it. Similarly the ODE/DAE packages ATOMFT [118] and more recently DAEpack [481] perform extensive higher order derivative calculations in an AD fashion. Other more conventional AD software tools such as ADIFOR and ADOL-C are tied into larger numerical environments such as PETSc [26], NEOS [389], ADIFFSQP [359], DASPK [354], and ADMIT [141], a Matlab interface, specifically for computing sparse Jacobians and Hessians and exploiting structure for M-files. The Maple package `gencode` now contains the commands GRADIENT, JACOBIAN, and HESSIAN, which extend Maple procedures to make them evaluate derivatives with respect to the specified independent parameters, either as floating point numbers or as expressions in other problem parameters.

However, for cutting-edge problems in large-scale applications, the use of AD is not yet fully automatic. To maintain some nearly optimal degree of efficiency, AD must still be combined with hand-coding or user directives, while AD tool developers are trying to close the gap. Hand-coded derivatives can take advantage of structural properties that may be quite obvious to the code developer, but are still very hard to detect automatically in the source program. In any case, we require more significant efforts in tool design and user education. We hope that the present volume will further these twin causes with the ultimate goal of turning simulation software into optimization software. In many engineering fields, this transition has begun after a great deal of effort on developing and implementing ever more complex mathematical models of the physical reality. The term *optimization* should be understood in the broad sense of systematic exploration of system response ranges rather than the down hill minimization of a scalar objective. Multi-criteria optimization, factor-analysis, and even global optimization can all greatly benefit from accurate but affordable derivative values; often they are impossible without such sensitivity information.

Organization

In Parts 1 to 4 of this volume, we consider the application and adaptation of AD to algebraic nonlinear least squares, discretized PDE, and ODE/DAE problems. On all these problem classes, black-box application of automatic differentiation typically yields reasonable derivative values, but sometimes with degraded accuracy and often with unnecessarily high run-time and storage costs. Then only a combination of fully automatic code translation at low levels with user-directed algorithmic transformations at higher levels can yield optimal results.

In Parts 5 to 8, we consider the development of AD techniques and tools with respect to parallelism, problem structure, and space-time trade-offs.

It becomes clear that through methodological advances in these areas, there is plenty of scope for significant improvement in efficiency and user convenience. The last two parts, 9 and 10, are concerned with the evaluation of higher derivatives, the estimation of truncation errors, and the tight inclusion of function values. The volume concludes with a common bibliography.

Challenges

Future development of algorithms and software tools for AD should emphasize further improvements in robustness, user-friendliness, and efficiency. AD-specific (pre)compile-time optimizations require static analyses, performed on suitable high-level abstractions of the program. In this, convergence with research themes in compilation or parallelization is evident. Challenges include fixed point iterations with a variable number of steps and other adaptive program branches, automatic sparsity detection, selective storage and retrieval or recalculation of intermediate values, especially in the calculation of (discrete) adjoints in the reverse mode, and derivative accumulation based on elimination techniques in computational graphs.

Furthermore, a quasi-standard intermediate representation of scientific programs will provide a common platform for the development of AD algorithms. Solutions for the exploitation of sparsity, for the minimization of memory requirements, or for decreasing the number of arithmetic operations could thus become available to a larger number of end-users. The composition of all these block-level optimal derivatives has to be investigated in a general way to unify the diverse specific strategies. In such a platform, new derivation algorithms could be invented and compared on industrial programs with a minimal development effort.

Dedication

The editors dedicate this volume to Alan Carle and to the late Fred Howes. Fred, as program manager for the Applied Mathematical Sciences program in the Department of Energy's Mathematical, Information and Computational Sciences division, was an early supporter of the AD research at Argonne National Laboratory. Alan, the implementor of ADIFOR, could not participate in Nice because of serious health problems that persist. We wish him and his family well.

Marquette University	George Corliss
PolySpace Technologies	Christèle Faure
Technical University Dresden	Andreas Griewank
INRIA Sophia-Antipolis	Laurent Hascoët
University of Hertfordshire	Uwe Naumann

May 2001

Contents

VII   Exploiting Structure and Sparsity   245

VIII   Space-Time Tradeoffs in the Reverse Mode   281

Contributors

Jason Abate: Texas Institute for Computational and Applied Mathematics, University of Texas at Austin, Austin, TX 78712 USA
abate@mcs.anl. gov

Alvard Arazyan: Department of Mathematics, University of Louisiana at Lafayette, P.O. Box 4-1010, Lafayette, LA 70504-1010 USA
aya3602@louisiana.edu

Pierre Aubert: Echo Interactive, 15, Parc des Hautes Technologies, 06250 Mougins France
Pierre.Aubert@insa-lyon.fr

Michael Bartholomew-Biggs: Faculty of Engineering and Information Sciences, University of Hertfordshire, College Lane, AL10 9AB Hatfield, Herts United Kingdom
M.Bartholomew-Biggs@herts.ac.uk

Jean-Daniel Beley: CADOE, Novacite Alpha, B.P. 2131 43, boulevard du 11 Novembre, 69603 Villeurbanne Cedex France
jean-daniel.beley@cadoe.com

Adel Ben-Haj-Yedder: ENPC/Cermics, 6/8 avenue Blaise Pascal, 77455 Champs sur Marne France
benhaj@cermics.enpc.fr

Steve Benson: Argonne National Laboratory, Bldg. 221, Rm. C-236 9700 S. Cass Ave., Argonne, IL 60439 USA
benson@mcs.anl.gov

Théo Berclaz: Department of Physical Chemistry, Universite de Geneve, Sciences II, 30 Quai Ernest Ansermet, CH-1211 Geneve-4 Switzerland
berclaz@sc2a.unige.ch

Martin Berz: National Superconducting Cyclotron Laboratory, Michigan State University, East Lansing, MI 48824 USA
berz@pilot.msu.edu

Christian H. Bischof: Institute for Scientific Computing, Aachen University of Technology, Seffenter Weg 23, 52074 Aachen Germany
bischof@rz.rwth-aachen.de

François Bodin: IRISA-INRIA, Campus Universitaire de Beaulieu, 35052 Rennes France
Francois.Bodin@irisa.fr

Thierry Braconnier: Department of Mathematics, 15 MLH, The University of Iowa, Iowa City, IA 52242-1419 USA
tbraconn@univ-reunion.fr

H. Martin Bücker: Computing Center, Aachen University of Technology, Seffenter Weg 23, 52074 Aachen Germany
buecker@sc.rwth-aachen.de

Jean-Baptiste Caillau: ENSEEIHT-IRIT UMR CNRS 5505, 2 rue Camichel, 31071 Toulouse France
caillau@enseeiht.fr

Eric Cances: ENPC/Cermics, 6/8 avenue Blaise Pascal, 77455 Champs sur Marne France
cances@cermics.enpc.fr

Bernard Cappelaere: IRD - UMR HydroSciences Montpellier, B.P. 5045, 34032 Montpellier Cedex 1 France
Bernard.Cappelaere@mpl.ird.fr

Alan Carle: Computational and Applied Mathematics, Rice University, 6100 Main Street MS-134, Houston, TX 77005-1892 USA
carle@rice.edu

Daniele Casanova: Benetton Formula Ltd., Whiteways Technical Centre, OX7 4EE Enstone Chipping Norton United Kingdom
Daniele.Casanova@benettonformula.com

Isabelle Charpentier: INRIA Rhone-Alpes, projet Idopt, B.P. 53, 38041 Grenoble Cedex 9 France
Isabelle.Charpentier@imag.fr

Bruce Christianson: Faculty of Engineering and Information Sciences, University of Hertfordshire, College Lane, AL10 9AB Hatfield, Herts United Kingdom
B.Christianson@hertfordshire.ac.uk

George F. Corliss: Marquette University, P.O. Box 1881, Haggerty Engineering 296, Milwaukee, WI 53201-1881 USA
George.Corliss@Marquette.edu

Josse De Baerdemaeker: Faculty of Agricultural and Applied Biological Sciences, Catholic University Leuven, Kasteelpark Arenberg 30, B-3001 Leuven Belgium
josse.debaerdemaeker@agr.kuleuven.ac.be

Nicolas Di Césaré: 24, rue Censier, 75005 Paris France
nicolas.dicesare@ann.jussieu.fr

Florian Dignath: Institute B of Mechanics, University of Stuttgart, Pfaffenwaldring 9, 70550 Stuttgart Germany
fd@mechb.uni-stuttgart.de

Peter Eberhard: Institute of Applied Mechanics, Friedrich Alexander University Erlangen Nuremberg, Egerlandstraße 5, 91058 Erlangen Germany
eberhard@ltm.uni-erlangen.de

David Elizondo: IRD - UMR HydroSciences Montpellier, B.P. 5045, 34032 Montpellier Cedex 1 France
david@delta.ft.uam.es

Trevor P. Evans: DERA, Ively Road, GU14 0LX Farnborough Hampshire United Kingdom
TPEVANS@dera.gov.uk

Yuri G. Evtushenko: Computing Centre Russian Academy of Sciences, Vavilov str. 40, 117967 GSP-1 Moscow Russia
evt@ccas.ru

Mike Fagan: Computational and Applied Mathematics, Rice University, 6100 Main Street MS-134, Houston, TX 77005-1892 USA
mfagan@rice.edu

Christèle Faure: PolySpace Technologies, 28 Rue Estienne d'Orves, 92120 Montrouge, France
Christele.Faure@polyspace.com

Stefka Fidanova: Universite Libre de Bruxelles, 50 Av. Roosevelt, 1050 Bruxelles Belgium

Mark Final: Faculty of Engineering and Information Sciences, University of Hertfordshire, College Lane, AL10 9AB Hatfield, Herts United Kingdom
M.Final@herts.ac.uk

Harley Flanders: University of Michigan, Dept. of Mathematics, East Hall, Ann Arbor, MI 48109 USA
harley@umich.edu

Shaun A. Forth: Cranfield University (RMCS Shrivenham), Shrivenham, SN6 8LA Swindon, Wiltshire United Kingdom
S.A.Forth@rmcs.cranfield.ac.uk

Axel Fritz: Institute B of Mechanics, University of Stuttgart, Pfaffenwaldring 9, 70550 Stuttgart Germany
af@mechb.uni-stuttgart.de

Stephane Garreau: CADOE, Novacite Alpha, B.P. 2131 43, boulevard du 11 Novembre, 69603 Villeurbanne Cedex France
sgarreau@enst.fr

Michel Geoffroy: Department of Physical Chemistry, Universite de Geneve, Sciences II, 30 Quai Ernest Ansermet, CH-1211 Geneve-4 Switzerland
geoffroy@sc2a.unige.ch

Ralf Giering: FastOpt, Kieselstr. 1a, 22929 Hamfelde Germany
giering@dkrz.de

Michael Giles: Oxford University Computing Laboratory, Wolfson Building Parks Road, OX1 5NH Oxford United Kingdom
Mike.Giles@comlab.ox.ac.uk

Mark S. Gockenbach: Department of Mathematical Sciences, Michigan Technological University, Houghton, MI 49931 USA
msgocken@mtu.edu

Andreas Griewank: Institute of Scientific Computing, Technical University Dresden,, Mommsenstr. 13, 01062 Dresden Germany
griewank@math.tu-dresden.de

Lisa Grignon: Mathematics Department, University of North Carolina, Chapel Hill, NC, USA

José Grimm: INRIA, projet Miaou, Route des Lucioles, 06902 Sophia Antipolis France
Jose.Grimm@sophia.inria.fr

Gundolf Haase: Johannes Kepler University Linz, Institute for Analysis and Computational Mathematics, Altenberger Straße 69, A-4040 Linz Austria
ghaase@numa.uni-linz.ac.at

Steve Hague: The Numerical Algorithms Group, Jordan Hill, OX2 8DR Oxford United Kingdom
steve.hague@nag.co.uk

Laurent Hascoët: INRIA, projet Tropics, Route des Lucioles, 06902 Sophia Antipolis France
Laurent.Hascoet@sophia.inria.fr

Christophe Held: INRIA, projet Tropics, Route des Lucioles, 06902 Sophia Antipolis France
Christophe.Held@sophia.inria.fr

Jens Hoefkens: National Superconducting Cyclotron Laboratory, Shaw Lane, Michigan State University, East Lansing, MI 48824 USA
hoefkens@msu.edu

Shahadat Hossain: Mathematics and Computer Science Department, University of Lethbridge, 4401 University Drive, T1K 3M4 Lethbridge, Alberta Canada
hossain@cs.uleth.ca

Paul D. Hovland: Argonne National Laboratory, Bldg. 221, Rm. C-236 9700 S. Cass Ave., Argonne, IL 60439 USA
hovland@mcs.anl.gov

Mark Huiskes: National Research Institute for Mathematics and Computer Science, Kruislaan 413, 1098 SJ Amsterdam, The Netherlands
Mark.Huiskes@cwi.nl

Noel Jakse: Laboratoire de Theorie de la Matiere Condensee, Universite de Metz, 1 Boulevard F. D. Arago, 57078 Metz cedex 3 France

Thomas Kaminski: FastOpt, Kieselstr. 1a, 22929 Hamfelde Germany
kaminski@dkrz.de

R. Baker Kearfott: Department of Mathematics, University of Louisiana at Lafayette, P.O. Box 4-1010, Lafayette, LA 70504-1010 USA
rbk@louisiana.edu

David E. Keyes: Mathematics and Statistics Department Old Dominion University, BAL 500, Norfolk, VA 23321-0077 USA
keyes@icase.edu

Jong G. Kim: Argonne National Laboratory, 9700 S. Cass Avenue MCS Division - Bldg. 221/C216, Argonne, IL 60439 USA
kim@mcs.anl.gov

Wolfram Klein: Siemens AG, Corporate Technology, Otto-Hahn Ring 6, 81730 Muenchen Germany
Wolfram.Klein@mchp.siemens.de

Ulrich Langer: Johannes Kepler University Linz, Institute for Analysis and Computational Mathematics, Altenberger Strasse 69, A-4040 Linz Austria
ulanger@numa.uni-linz.ac.at

Philippe Langlois: LIP - ENS Lyon - INRIA Rhone-Alpes, projet Arenaire, 46, allee d'Italie, 96364 Lyon cedex 07 France
Philippe.Langlois@ens-lyon.fr

Claude Le Bris: ENPC/Cermics, 6/8 avenue Blaise Pascal, 77455 Champs sur Marne France
lebris@cermics.enpc.fr

Steven L. Lee: Center for Applied Scientific Computing, Lawrence Livermore National Laboratory, Box 808, L-560, Livermore, CA 94551 USA
slee@llnl.gov

Ewald Lindner: Johannes Kepler University Linz, Institute for Analysis and Computational Mathematics, Altenberger Strasse 69, A-4040 Linz Austria
lindner@numa.uni-linz.ac.at

Kyoko Makino: Department of Physics, 1110 West Green Street, University of Illinois at Urbana-Champaign, Urbana, IL 61801-3080 USA
makino@uiuc.edu

Marco Mancini: Department of Electronics, Computer Science and Systems, University of Calabria, Arcavacata di Rende, 87030 Rende - Cosenza Italy
mancini@parcolab.unical.it

Mohamed Masmoudi: MIP - Université Paul Sabatier, 118 route de Narbonne, 31062 Toulouse Cedex France
masmoudi@cix.cict.fr

Lois C. McInnes: Argonne National Laboratory, Bldg. 221, Rm. C-236 9700 S. Cass Ave., Argonne, IL 60439 USA
mcinnes@mcs.anl.gov

Dieter an Mey: Computing and Communication Center, Aachen University of Technology, Seffenter Weg 23, 52074 Aachen Germany
anmey@rz.rwth-aachen.de

Christo Mitev: Treppenmeister GmbH, Emminger Strasse 38, 71131 Jettingen Germany

Antoine Monsifrot: IRISA-INRIA, Campus Universitaire de Beaulieu, 35052 Rennes France
Antoine.Monsifrot@irisa.fr

Jorge J. Moré: Argonne National Laboratory, 9700 South Cass Avenue, Argonne, IL 60439 USA
more@mcs.anl.gov

Wolfram Mühlhuber: Universitaet Linz Spezialforschungsbereich SFB F013, Freistaedterstr. 313, 4040 Linz Austria
wmuehlhu@sfb013.uni-linz.ac.at

Uwe Naumann: University of Hertfordshire, AL10 9AB Hatfield, Herts United Kingdom
U.1.Naumann@herts.ac.uk

Joseph Noailles: ENSEEIHT-IRIT, UMR CNRS 5505, 2, rue Camichel, 31071 Toulouse France
jnoaille@enseeiht.fr

Boyana Norris: Argonne National Laboratory, 9700 South Cass Avenue, Argonne, IL 60439 USA
norris@mcs.anl.gov

John D. Pryce: Cranfield University (RMCS Shrivenham), Shrivenham, SN6 8LA Swindon, Wiltshire United Kingdom
J.D.Pryce@rmcs.cranfield.ac.uk

Herman Ramon: Faculty of Agricultural and Applied Biological Sciences, Catholic University Leuven, Kasteelpark Arenberg 30, B-3001 Leuven Belgium
Herman.Ramon@agr.kuleuven.ac.be

Kurt J. Reinschke: Institut für Regelungs- und Steuerungstheorie TU Dresden, Mommsenstr. 13, 01062 Dresden Germany
kr@erss11.et.tu-dresden.de

Daniel R. Reynolds: MS-134 Rice University, 6100 Main Street MS-134, Houston, TX 77005-1892 USA
reynoldd@caam.rice.edu

Klaus Röbenack: Institut für Regelungs- und Steuerungstheorie TU Dresden, Mommsenstr. 13, 01062 Dresden Germany
klaus@roebenack.de

Widodo Samyono: Mathematics and Statistics Department, Old Dominion University, BAL 500, Norfolk, VA 23321-0077 USA
wsamyono@lions.odu.edu

Robin S. Sharp: School of Mechanical Engineering, Cranfield University, MK43 0AL Cranfield, Bedford United Kingdom
R.S.Sharp@cranfield.ac.uk

Edgar Soulié: Service de chimie moleculaire CEA-DSM-DRECAM, CEA Saclay, 91191 Gif-sur-Yvette Cedex France
bsoulie@drecam.cea.fr

Trond Steihaug: University of Bergen, Department of Informatics, Postbox 7800, N-5020 Bergen Norway
Trond.Steihaug@ii.uib.no

William W. Symes: MS-134 Rice University, 6100 Main Street MS-134, Houston, TX 77005-1892 USA
symes@caam.rice.edu

Pat Symonds: Benetton Formula Ltd., Whiteways Technical Centre, OX7 4EE Enstone Chipping Norton United Kingdom

Mohamed Tadjouddine: Cranfield University (RMCS Shrivenham), Shrivenham, SN6 8LA Swindon, Wiltshire United Kingdom
M.Tadjouddine@rmcs.cranfield.ac.uk

Frederic Thevenon: CADOE, Novacite Alpha, B.P. 2131 43, boulevard du 11 Novembre, 69603 Villeurbanne Cedex France

Engelbert Tijskens: Faculty of Agricultural and Applied Biological Sciences, Catholic University Leuven, Kasteelpark Arenberg 30, B-3001 Leuven Belgium
engelbert.tijskens@agr.kuleuven.ac.be

Fabrice Veersé: INRIA Rhone-Alpes, projet Idopt, B.P. 53, 38041 Grenoble Cedex 9 France
Fabrice.Veerse@imag.fr

Olaf Vogel: Institute of Scientific Computing, Technical University Dresden, Mommsenstr. 13, 01062 Dresden Germany
vogel@math.tu-dresden.de

Andrea Walther: Institute of Scientific Computing, Technical University Dresden, Mommsenstr. 13, 01062 Dresden Germany
awalther@math.tu-dresden.de

E.S. Zasuhina: Computing Centre Russian Academy of Sciences, Vavilov str. 40, 117967 GSP-1 Moscow Russia

V.I. Zubov: Computing Centre Russian Academy of Sciences, Vavilov str. 40, 117967 GSP-1 Moscow Russia

Part I

Invited Contributions

1

Differentiation Methods for Industrial Strength Problems

Wolfram Klein, Andreas Griewank and Andrea Walther

ABSTRACT The importance of simulation has been growing in industrial production for many years. Because of reduced product cycles, new and more complicated computer models have to be developed more quickly. The correct description and implementation of the interaction between different components of the entire simulation model as well as the nonlinear behaviour of these components lead in many cases to the need for derivative information. In this chapter, we use software packages for automatic differentiation (AD) in three real world simulation systems typical for a wide range of tasks that have to be solved in numerous industrial applications. We consider difficulties arising from particular aspects of the modelling such as the integration of ordinary differential equations or fixed-point iterations for the solution of equations. Furthermore, we discuss challenges caused by technical software issues such as inhomogeneous source codes written in different languages or table look-ups.

Several results concerning the use of tools such as ADIFOR, Odyssée, and ADOL-C are presented. We discuss the benefits and the difficulties of current AD techniques applied to real industrial codes. Finally, we outline possible future developments.

1.1 Introduction

1.1.1 Scope and purpose

In this introductory article we use three industrial problems to illustrate the state of the art in automatic differentiation (AD) theory and software from (and for) a user's point of view. This survey includes comparisons with divided difference approximations and fully symbolic differentiation by the computer algebra system Maple. Readers looking for a primer on automatic differentiation might consult Part I of the recent book [238] or the introductory chapters of the previous two proceedings volumes [239] and [57] written by Iri [300] and by Rall and Corliss [433], respectively. Here we restrict ourselves to a verbal description of the forward and reverse modes of automatic differentiation and their essential characteristics as they have evolved over the last four decades.

The key motivation for considering the collection of techniques and tools called automatic differentiation is the growing need to calculate accurate sensitivity information for complex industrial and scientific computer models. Here and throughout this volume "only" local sensitivity information in the form of first and higher partial derivatives are considered. These may be evaluated at arbitrary arguments and can thus be used to quantify input-output dependence in a more global sense as is done in nonlocal sensitivity analysis (see [446] for example).

Gradients, Jacobians, Hessians, and other derivative arrays are useful if not indispensable for many computational purposes, including the numerical integration of stiff differential equations, the optimization of objective functions, and the solution of discretized partial differential equations, which are considered in §1.2, 1.3, and 1.4, respectively. Most numerical packages for these and similar problem classes still rely on inaccurate divided difference approximations unless the user is prepared to provide additional code for the needed derivative values. Here preparedness is required both in terms of the technical ability to derive such code by hand and the willingness to invest the necessary human effort. The latter is typically rather high and also hard to predict because an extended debugging phase is probably required.

1.1.2   Basic differentiation modes

The big advantage of chain-rule based derivative codes compared to divided differences is that they yield truncation error free derivative values at a cost that is either about the same or significantly lower, depending on the kind of derivative object required. More specifically, if the derivative code implements the reverse mode, then gradients of scalar functions are obtained with two to three times the operations count of the function itself. However, the need to in some sense "run the original code backwards" usually results in a significant increase in memory requirement and entails some other complications (see e.g., [185, 214]). The reverse mode can be thought of as a discrete analog of the adjoint equations method familiar in the control of ordinary differential equation systems. In contrast, the forward, or direct, mode is much easier to implement and has about the same run time and memory characteristics as divided differences. It is still the method of choice for the differentiation of vector functions whose number of components m is comparable to or significantly larger than its number of variables n. In other words n and m are the number of independent and dependent variables selected by the user. In the future the results of current research into hybrid methods (see e.g. [399]) and their implementation may yield Jacobians at much reduced costs. Already sparsity in the Jacobian can be exploited by matrix compression (see e.g. [287]).

In many applications the key problem parameters n and m are not as fixed as they might appear at first. For example, scalar objective functions such as total energies or error functionals are often calculated as sums of many in some sense "local" contributions, each of which depends only on a few of the independent variables. Interpreting these subobjectives as dependent variables, we obtain a problem whose Jacobian is sparse and probably can be evaluated quite cheaply by matrix compression in the forward mode. The next generation of AD tools should be able to detect and exploit this kind of partial separability [248] automatically, so that the user will only notice the improved efficiency.

If not only first but also second derivatives are required and m is much smaller than n, then a combination of the forward and reverse modes described for example in [175] is the most attractive solution. Higher derivative tensors and their restrictions to certain subspaces can be computed on the basis of multivariate or univariate Taylor polynomial expansions as suggested in [48] and [249], respectively. Automatic differentiation yields accurate derivatives of order ten or more even for codes simulating particle accelerators [47] and other industrial strength computer models. In contrast it must be noted that the approximation of second and higher derivatives by divided differences is a dicey proposition since at least two thirds of the available number of significant digits are lost.

1.1.3   Software and tool issues

Perhaps the most obvious software issues are whether source code for the given function evaluation procedure is available and in which language(s) it is written. Most existing AD tools are geared to Fortran or C and their various extensions. Fortran AD tools typically generate source codes with extra instructions for the required derivative calculations, whereas C tools mostly use the overloading facilities of C++ to make the compiler insert these extra instructions directly into the object code (see e.g., [477]). Generally speaking, the main advantage of the first technique is user convenience and run time performance, whereas the latter approach offers more flexibility in terms of the source structure and the kind of derivative object desired. More specifically, overloading easily covers all C++ constructs and allows the evaluation of derivatives of any order. In contrast upgrades of the source transformation tools from Fortran 77 to Fortran 90 and from first to second derivatives continue to require substantial software development and maintenance efforts. This burden on the tool developer may pay off nicely in fast derivative code for the user, provided the original code is well structured as we will see in §1.2.

There are many ways in which derivatives can be calculated given a procedure for the evaluation of the underlying function. Apart from actual mathematical variations concerning the application of the chain rule, there are many computer science issues that may have a strong bearing on the efficiency of the derivative code. For example, certain algorithmic examinations and actual computations can be performed either at run time, at compile time, or even earlier during a preprocessing phase. Also, the values of subexpressions that occur repeatedly may be either saved and later retrieved or reevaluated several times. In particular, this choice between storage and recalculation must be made for the very many intermediate quantities that are needed during the backward sweep of the reverse mode (see e.g. [214, 495]). Its handling also represents a major difference between automatic differentiation and fully symbolic differentiation as carried out by most computer algebra packages.

In principle, anything that can be done by some kind of software tool for the generation of derivative codes can also be achieved by "hand coding" and vice versa. Of course the tool can perform compiler-type optimizations that are far too messy for a pencil and paper attack, while the user may exploit insight into the nature of the function that cannot be deduced by any kind of general code analysis. The latter observation applies in particular if the function code itself is highly adaptive so that its control flow is strongly dependent on its input data, for example the geometry of the domain in a partial differential equations problem. On the other hand, for a straight-line code where dimensions and loop limits are fixed and no branching occurs, one can expect the differences between automatically and hand generated derivative code to be minor. For a much larger class of codes one can expect significant differences in performance but still virtually identical numerical derivative values (see e.g., [263]).

1.1.4   Potential trouble spots

Significant discrepancies in the results are possible and may require some user attention if the evaluation procedure contains either one of the following aspects:

- Table look-ups
- Quadratures
- Fixed point iterations
- Numerical integrators

On the level of the computer programs these mathematical constructs materialize in the form of `if-then-else` statements. Even though these branches may very well destroy differentiability or even continuity in the strict mathematical sense, one can still strive to obtain some kind of useful approximate derivative. This situation will be discussed on the industrial example in §1.4. Unfortunately, there is as yet no agreement on how the users should be alerted to these potential problems and how they should be enabled to steer the tool to perform the differentiation in a sensible way. It was found in [74] that the appropriate handling of a quadrature subroutine was critical for the accuracy and efficiency of the overall differentiation.

If an evaluation procedure relies on modules that are written in various languages, the user may have to differentiate them separately and splice the resulting component Jacobians together according to the chain rule. If some modules are proprietary or their source code is not available for other

reasons, their Jacobians can only be approximated by divided differences. It is to be hoped that eventually software providers will supply their numerical routines enhanced with corresponding derivative procedures.

The potential troubles of AD in differentiating complicated mathematical algorithms as well as technical aspects of applying AD tools will be discussed in the following sections. The development and differentiation of simulation models for MOS-transistors as described in §1.2 encompass a comparison of traditional Fortran AD techniques with computer algebra systems and with advanced AD techniques realized internally in a VHDL compiler. Nonlinear optimization problems as the one studied originally in [329] and discussed here in §1.3 are typical applications of AD techniques. Here, a complex but homogeneous software structure written in C enables the comparison of different AD tools and divided differences with respect to run time, memory costs, and the time for the application of AD techniques itself.

The model of a power plant considered in §1.4 represents a typical industrial simulation problem, where single components such as heating surfaces are implemented using fixed-point iterations, ODE systems or table look-ups. The differentiation of these units in combination with an inhomogeneous and very complex software structure is still a real challenge for automatic differentiation.

1.2   Application of AD in Circuit Simulation

The situation studied here is typical for a large class of problems arising in industrial simulation systems, where Jacobians of complex systems are accumulated as sums of relatively small component Jacobians. The overall Jacobians may be required for the solution of stationary equations or the numerical iteration of stiff dynamical systems by implicit schemes.

In the field of circuit simulation, models for semiconductor elements such as MOS transistors are of great interest. Because the terminal current and charge functions are nonlinear with respect to the input voltages, the derivatives of these transistor functions are required.

The application of AD techniques to transistor models within the Siemens AG started several years ago [325, 326]. Here we review some of the results concerning the applicability, execution time, and memory costs of automatically generated derivatives of various AD tools from the report [326]. Currently, the transistor model description starts on the next higher level of modelling, i.e. on the functional specification itself. Therefore, we also discuss the use of Maple V and the current state of functional modelling evaluation and differentiation realized in a VHDL-AMS compiler.

Due to reduced product cycles in the manufacturing of integrated circuits, new simulation models for larger and more complicated technologies need to be developed in even shorter times. Compact analytical models of

MOS transistors yield a nonlinear dependence of the terminal current and charge functions at drain, gate, source, and bulk on the three branch voltages V_{GS}, V_{DS} and V_{BS} between gate and source, drain and source as well as between bulk and source. The combination of many such transistor models to entire circuit simulators results in a nonlinear and nonautonomous set of differential algebraic equations. This system is mathematically treated by an implicit discretization in time that requires a Newton iteration at each time-step. The derivatives of the transistor model functions are needed with respect to the input voltages for this Newtonian process.

1.2.1 AD tools working with Fortran source code

Several AD tools for the application of AD techniques to Fortran source code such as ADIFOR [78] or Odyssée [186] were applied to simulation models of different MOS-transistor models. In this section, a brief discussion of the application and corresponding results shown in [326] is given.

Application of ADIFOR to MOS transistors

The ADIFOR tool for automatic differentiation in Fortran enables the automatic generation of derivative code from user-supplied Fortran 77 code [78]. It was developed jointly by Argonne National Laboratory and Rice University. ADIFOR Version 2.0 was released in spring 1995, adding exception handling and support for automatic exploitation of sparsity in derivative objects. The results shown in [326] and discussed here were made using ADIFOR V2.0. A newer version incorporating also the reverse mode is available at [71].

Several versions of ADIFOR were applied to different MOS transistor models used in the circuit simulation tools within Siemens. The CPU time for several transistor models with respect to different input circuits was determined for derivatives automatically generated by ADIFOR V2.0 and compared with hand-coded derivatives. The corresponding ratio of the CPU times was computed.

The comparison of the ratio of CPU times for different MOS transistor models showed that the CPU time needed for the execution of automatically generated derivatives was in general slightly larger than the time for hand-coded derivatives. This ratio varies by a factor of about 1.13 to 1.6 depending on the specific model and application [326]. These timing discrepancies could be attributed to the following effects, which still arise in one form or another in many AD tools:

- A large number of assignments to intermediate variables within one formula is generated.

- The derivatives of standard functions are determined in a very correct manner with branches for all possible exceptions. For example, the test for $x > 0$ is contained in the differentiation of $y = \sqrt{x}$.

- There are many "trivial" operations, e.g. multiplications with 1 or 0 and additions of 0 resulting from differentiation of constants or independent variables within the automatically generated code. This could be avoided using differentiation "by hand" or a more extensive dependence analysis by the AD tool.

- The detection of common subexpressions within different functions and/or derivatives is very hard to do automatically. It might be achieved through peep hole optimization of the compiler on the original code but may fail on the enhanced code, where the original statements are spread further apart.

An additional optimization of the functional description of the current and charge functions of the transistor models with respect to the understanding of the results of the previous attempt was done in a further step. These changes of the input formulas yielded a significant improvement of the run time ratio of automatically generated to hand-coded derivatives to a value of about 1.01 to 1.22.

On the other hand, the generation of hand-coded derivatives usually requires a very significant human effort. This rather error-prone process takes 2-4 weeks depending on the complexity of the transistor model. It can be drastically accelerated by the use of ADIFOR to 2-4 days.

Odyssée

The main difference between ADIFOR V2.0 and Odyssée [186] developed at INRIA Sophia Antipolis concerns the mode of differentiation. ADIFOR V2.0 implements a globally forward mode combined with a statement level reverse mode. This differentiation technique can be also used for sparse Jacobian matrices. Odyssée provides a global reverse mode in addition to the forward mode.

When applied to different MOS transistors, both AD tools Odyssée and ADIFOR yield very similar results with respect to the execution speed of the function and derivatives and to code size and memory requirements [326]. The fact that forward and reverse mode are here similarly efficient is no real surprise since n and m are of comparable and rather small size.

ADOL-F

Compared to the AD tools ADIFOR or Odyssée, ADOL-F [459] has a totally different philosophy. Based on a Fortran 90 input file for a function, the concept of operator overloading is used to prepare for differentiation in either forward or reverse mode. For that purpose, the input file has to be modified slightly as described in more detail in §1.3.2.

Similar to previous results, the accuracy of the derivatives varies only in the last digit. The execution time is only slightly larger than that of ADIFOR or Odyssée [326].

1.2.2 Maple V

The application of Maple V to a MOS transistor model consists essentially of four steps [327]:

Step 1: The internal kernel of Maple V is enlarged. Here, the functionality of piecewise defined functions must be generalized. The evaluation, differentiation, and optimization of piecewise defined functions as well as the generation of corresponding Fortran codes requires additionally defined, user-supplied functions.

Step 2: The definition of the four terminal functions of the transistor model forms the essential step during the model development in Maple V. Using the dynamically defined piecewise continuous functions, the transistor functions are implemented in a mathematical notation. The symbolic manipulation of these formulas in Maple V combines all intermediate functions leading to only one, but very large functional expression for the corresponding transistor functions.

Step 3: Eight terminal functions are symbolically differentiated with respect to three independent variables using the Maple V differentiation statement combined with user defined functions of Step 1.

Step 4: Automatically generate an optimized Fortran code for functions and derivatives, again using the Maple V functionality for generating program codes, which is able to detect common subexpressions of functions and derivatives.

Unfortunately, Maple V is rather slow. The differentiation and code generation for both functions and derivatives take about 1.5 hrs. For a small transistor model, the differentiation of about 350 lines Maple V input generates approximately 1.500 lines of Fortran output. Larger models cause memory and run time problems [327]. Furthermore, the run time of the generated Fortran code is about a 1.2 to 1.8 factor slower than the code developed "by hand" [327].

1.2.3 Comparative conclusion

According to the results reported in the previous two subsections, automatic and symbolic differentiation have the following common characteristics.

1. The derivative values obtained are truncation error free and agree within working accuracy.

2. The derivative code obtained is 5% to 80% slower than the one coded "by hand".

3. The convenient, correct, and fast generation of derivative code provides more modelling flexibility.

4. The time for the development of new transistor models can be essentially reduced.

Rather than viewing automatic and symbolic differentiation as alternatives, one should realize that the two are special cases in a range of options, which differ less from a mathematical than from a computer science point of view. The practical difference between current implementations is that Maple and other computer algebra systems involve a very significant preprocessing effort that precludes the treatment of really large and complex models. Another possibility in the range of options discussed above is the compiler integrated differentiation system considered in the next subsection.

1.2.4   VHDL-AMS compiler

A future trend in circuit simulation is the generalization and standardization of simulation models for circuit elements or building blocks such as operational amplifiers or voltage controlled oscillators, which are available either in the public domain or can be obtained from several simulation systems of different companies. The development VHDL-AMS is an effort to standardize an extension of the "very high description language" VHDL-1076 in order to support the description and the simulation of "analog and mixed-signal" (AMS) circuits and systems. The resulting language VHDL-1076.1 is informally known as VHDL-AMS. Up to now, VHDL-1076 is only suitable for the modelling and simulation of discrete systems, but designers wish to work with an uniform description language, including continuous models based on differential algebraic equations (DAEs). This requires the handling of continuous models based on DAEs, where the corresponding equations are solved by a dedicated simulation kernel. Furthermore, it must be possible to handle initial conditions, piecewise-defined behaviour, and discontinuities. This requires the ability to automatically differentiate an internal model description within the compiler.

There exist several compilers for the handling of VHDL-1076 models, and the source code of some compilers is even available. A first version of a VHDL-1076.1 compiler, written in Java, was available within the Infineon circuit simulation group [156].

For the additional implementation of a forward mode of automatic differentiation, the dependent and independent variables as well as the required formulas are marked in the internal Java representation. The realization of the forward mode of automatic differentiation is achieved through the generation of a new formula node. The corresponding differentiated tree is constructed relative to the operations of the original function tree, where subformulas of the original tree are added and replaced by appropriate differentiated subformulas.

The compiler enables the generation of Fortran or C source code for the function tree and/or the differentiated function tree using a fixed model interface. This generated source code can be compiled and linked to a given circuit simulation system.

The VHDL-AMS compiler including the additional differentiation module was applied to various models coming from real industrial applications. The results obtained have shown that the differentiation module works well and does not lead to any problems concerning run time or memory costs, even for larger models.

The essential point of this compiler integrated automatic differentiation is the enhanced flexibility in the development of new transistor models. The extreme simplification and acceleration of the process of modelling, the absence of any need for additional user insight or input for differentiation, and the safe knowledge of correct and accurate generated derivatives dominates discussions about run time and memory costs. Therefore, a compiler integrated automatic differentiation at least for specialized modelling languages might be a trend for the future.

1.3   Application of AD to a Cooling System

The situation studied in this section is typical for a large class of problems where the operating performance of certain engineering systems is to be optimized by suitably adjusting some control parameters and/or design variables. Here, the model is integrated over a certain period of time as is also the case for the important task of 4D data assimilation in weather forecasting [472]. These calculations form the simulation of the process under consideration. In an additional step, an optimization stage is performed to fit some model parameters. For that purpose, the gradient of the considered function is calculated by automatic differentiation using either the forward or the reverse mode. Alternatively the needed derivative values are approximated by divided differences for comparison of the numerical results and the run time behaviour.

1.3.1   Problem description

When steel has been mechanically treated its temperature is about $900°C$. In order to preserve material features of the produced steel tape and to get a homogeneous, tight micro-structure within the steel, the tape must be cooled down to about $600°C$ by spraying it with water. The decrease of temperature within the steel tape and therefore the quality of the final produced steel depends critically on the control of the cooling system as well as on the properties of the specific steel.

Figure 1.1 shows the physical layout of a cooling system. We consider a cooling system consisting of several rollers along a path of 73 m. The rollers are responsible for the transport of the steel tape. The cooling of the steel is mainly achieved by sprinkling it with water. Several groups of water valves, combining some nozzles of different sizes, are placed both

Cooling system

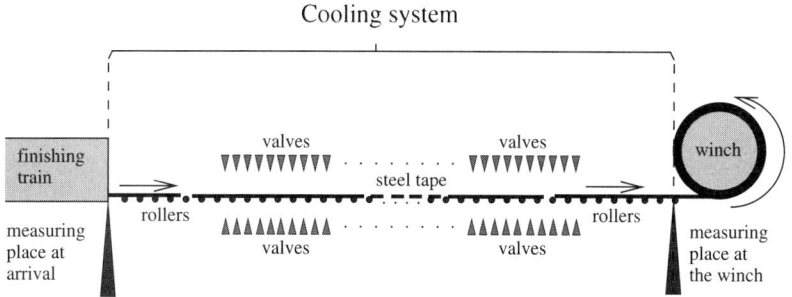

FIGURE 1.1. Physical structure of the cooling system

above and under the rollers. The temperature of the steel can be measured only at the beginning and at the end of the cooling system.

There are several possibilities for modelling a cooling system. Here, a one-dimensional, time-dependent Fourier equation for the heat transfer is used. The simple model neglects two of the three dimensions of the steel tape, namely width and length, and takes only the space coordinate of the steel tape thickness into account. More specifically, we have:

$$\frac{\partial T(t, x)}{\partial t} = \kappa(T) \frac{\partial^2 T(t, x)}{\partial x^2} \tag{1.1}$$

with the variables

$T(t, x)$ temperature of the tape of steel

t coordinate of time, i.e., position within the cooling system

x coordinate of space within the thickness of the steel tape

$\kappa(T)$ temperature conduction.

In addition to the partial differential equation, there are several boundary conditions concerning the heat conduction $\rho(T)$ of steel at both upper and lower surfaces of the steel tape.

This specific application is controlled by a set of 33 independent parameters $p = (p_1, \ldots, p_{33})$. For example, one has to adapt parameters for the steel properties temperature conduction and heat conduction. Also, some layout details of the cooling system are described by the independent variables. Given these parameters, the differential equation (1.1) is solved using data from several different steel tapes. In order to improve the given model of the system, i.e. in order to adapt parameters such as heat conduction for the special steel, the difference between the computed temperature and the observed temperature at the end of the cooling system has to be minimized. This difference is measured by a functional of the form

$$f(p) = \sum_{i=0}^{m} g\left(T(t_{final}, x_i, p)_{computed} - T(t_{final}, x_i, p)_{observed}\right).$$

The partial derivatives $\partial f/\partial p_j$, $j = 1, \ldots, 33$, are used to adapt the model of the cooling system in an optimization stage. Based on the finally calculated parameters, the process of cooling can be computed in advance. Using this resulting model, the steering of the real cooling process can be improved.

An implicit one-step method was used to integrate the spatial discretization of the differential equation (1.1) with respect to time. The corresponding software program consists of about 1500 lines of C code. There is a modular and hierarchical structure within the software program. Several if-then-else branches and loops are used. Therefore, the basic structure of the software is nonlinear. This leads to some challenges in applying AD tools to that real-life software program for the computation of the needed derivative information.

1.3.2   Application of AD

The AD tool ADOL-C [242] using the concept of operator overloading was applied to the model of the cooling system. The key ingredient of AD by overloading is the concept of an *active variable*. All program variables that may at some time during the program execution hold differentiable quantities must be declared to be of an active type. For the new data type the standard operations such as + or * and elementary functions such as sin() or exp() are redefined. This overloading causes the storage of the data needed to apply the chain rule to the whole sequence of operations. The required information is either stored on a sequential data structure or in a component of the active variables depending on the software tool applied.

The AD tool ADOL-C uses the new type adouble for all active variables. Hence, for the application of ADOL-C all 33 independent variables must be defined as adoubles. Subsequently, the data type of all intermediate variables, which are depending on the 33 independent variables, must also be changed to adouble. These modifications serve to generate a sequential data structure called *tape* during the evaluation of the cooling system model. The tape contains all information needed to calculate the derivatives. Using this sequential record interpreter-like functions of ADOL-C execute the forward and reverse modes of AD. These problem independent functions allow us to freely choose the order of the derivatives at run time. In addition, there are modules that compute standard derivative structures such as gradients, Jacobians or Hessians.

In the AD tool FADBAD [41] the "tape" is not a linear array or file but a dynamic data structure that more directly represents the computational graph. In either case the generation of the tape is likely to take longer than the subsequent derivative calculations. Since the structure of the tape reflects the actual control flow during the function evaluation, it must be recreated at least partially whenever program branching changes the computational graph. Such retaping is generally not required in source transformation AD tools.

1.3.3   Comparison with divided differences

The forward and reverse modes of automatic differentiation using ADOL-C are applied to the model of the cooling system. For the computation of the desired gradients, the initialization of the independent variables is done using values from practical applications in steel manufacturing. Additionally, the AD tool FADBAD also based on operator overloading was applied to differentiate the cooling system model. The numerical values of the gradients generated by both software tools are identical.

Since the derivatives are also computed by divided differences, comparisons between the values yielded by both approaches can be made [329]. I.e. the quality of the gradient approximated by difference quotients can be analyzed using the AD results. This relative error is independent of the choice of the corresponding AD tool because both software packages compute the same results as mentioned above. As expected, the choice of the increment size in the divided difference method strongly influences its result. The additional information supplied by the AD values allows an optimal choice of different increment sizes for all 33 independent variables and the determination of the relative error of the divided differences for the various increments.

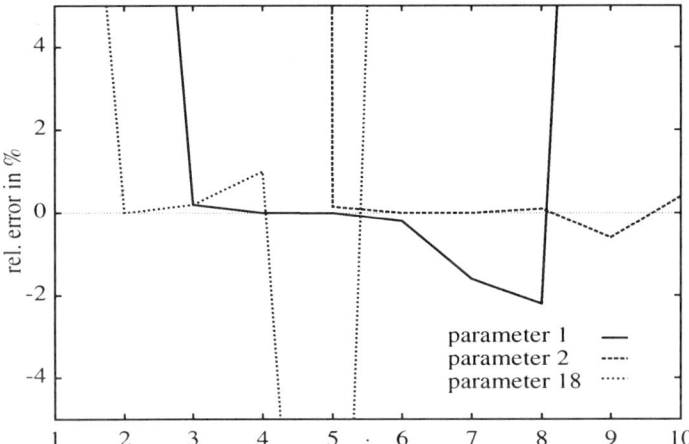

FIGURE 1.2. Relative error for parameters 1, 2, and 18 and increment size 10^{-i}

Figure 1.2 shows the typical behaviour of the relative error for a variation of the increment size. Here, three independent variables and a fixed input dataset of different kinds of steel are taken into account. The increment sizes 10^{-i}, $i = 1, \ldots, 10$, are chosen taking the absolute value of the parameters into account. On the left-hand side of Figure 1.2, the relative error $> 5\%$ is caused by the large increment sizes selected. On the right-hand side, the gradient determined by a rather small increment size is influenced by rounding errors, which leads again to an increase of the relative error. The range of suitable increment sizes is rather small for the parameters considered in Figure 1.2. Moreover, the optimal increment size for the three parameters is totally different. Hence, an optimal choice of the increment size for the divided differences is possible only with the knowledge of the correct values of the gradient. Without this knowledge, it is impossible to judge the quality of the numerical derivatives for a complex model such as the cooling system.

In the present study we could not isolate or modify the code for the numerical ODE integration. As observed by Bock et al. [87] and shown by Bischof and Eberhard [173] the "deactivation" of step size calculations and other adaptive elements may reduce numerical noise and improve efficiency. By "deactivation" we mean the deliberate truncation of certain dependencies that are known to have little bearing on the derivative values or to adversely effect their accuracy. For similar comparisons of AD and divided differences consider the contributions [123, 158, 199, 323, 465].

1.3.4   Run time and storage observations

Depending on the particular steel tape and on the chosen cooling strategy, the function value can be computed in 0.1 to 0.4 seconds. The determination of the gradient using one sided divided differences needs about 6.5 – 24 seconds, again depending on the steel tape.

Using the forward mode of AD one finds that the determination of the gradient using AD is up to 100% slower than divided differences. The calculation of the gradient applying the reverse mode of AD is up to 40% faster than the determination of the gradient by divided differences. However, the reverse mode needs more memory compared to the forward mode. In order to store the generated tape for the cooling system mode, one needs 3 - 4 MByte. This amount of memory can be drastically reduced if one applies the checkpointing technique (see e.g. [255]). Instead of generating the complete tape, several intermediate states are stored as checkpoints. Then the data required are saved piecewise by restarting the forward integration of the model repeatedly from the suitably placed checkpoints. Using this approach a logarithmic increase in run time and memory needed can be achieved as shown in [235].

1.3.5   Summary

The AD tool ADOL-C enables the application of automatic differentiation to the rather complex industrial problem of a cooling system written in C. The required changes of the source code described in §1.3.2 were done in about 4 days, based on knowledge of C as well as in the application of ADOL-C, whereas the internal structure of the cooling system software was unknown. The 40% gain in run time we observed can be expected to be even bigger on systems with more than 33 independent variables, where the much lower operations count of the reverse mode amortizes its general overhead compared to differencing and the forward mode. Such experiences are reported in the paper [295], where also second and even third order derivatives need to be calculated in the context of nonlinear regression applied to a fishery model.

The application of automatic differentiation to our real-world problem leads to the detection of serious problems concerning the application of divided differences. Therefore, a totally new simulation system was developed having the desired differentiation in mind. Hence, the application of AD to the given model of the cooling system causes an improvement of the simulation process beyond the adaptation of control parameters.

1.4 Application of AD to Power Plant Models

The following section deals with numerical calculations for the stationary thermodynamic behaviour of power-plant steam generators. Especially in part-load calculations this requires highly sophisticated mathematical models. The simulation system [360] is a large nonlinear system of equations, which is solved by a Newton-like process. Hence one needs the derivatives of the functions describing different components with respect to several model parameters.

The aim of this section is to show the application of different AD tools to a very complex real world simulation system. Moreover, the resulting difficulties using AD tools will be described.

1.4.1 Mathematical concepts

A reliable computer code for the determination of the stationary thermodynamic behaviour of heat recovery steam generators has to be quite sophisticated. The small temperature differences between flue gas and water/steam in this type of boilers call for a very high degree of accuracy. The models under consideration cover a wide range of applications. Users have to cope with many different plant configurations and are usually short of time. Therefore the computer code must be extremely stable and easy to handle.

The basic mathematical concept consists of three different approaches:

1. The physical models of components of the steam generator have to be supported by mathematical algorithms.

2. The connections between these components in the boiler model lead to a nonlinear system of equations.

3. A matching number of boundary conditions is necessary.

In contrast to many other codes that solve the system of equations sequentially, the approach considered here relieves the user of having to supply additional information when cyclic flows occur. Furthermore, the conditions at flow cycles can change during the calculation. Each component is described by several nonlinear functions, including ODEs and DAEs, for the physical/technical relation between input and output parameters. The entire simulation model is therefore given by a set of coupled functions forming a nonlinear system of equations [10]. The solution of that system using a Newton-like algorithm requires the gradient of the functions and the Jacobian matrix of the entire system with respect to model parameters of singular components. Hence, the aim of the application of automatic differentiation is to reach a higher stability and an acceleration of the computer code.

The calculation model of a steam generator consists of

1. a number of components, e.g. heating surfaces, drums, pipes, valves, pumps, flow splitter and mixer,

2. corresponding connections between these components, e.g. mass flow rate, pressure or enthalpy,

3. a matching number of boundary conditions, e.g. prescriptions of pressure, temperature, mass flow rate at specific locations, and

4. control units, i.e. groups of boundary conditions between which the status of application given by "apply" or "do not apply" can switch.

The components may have different modes of calculation: For example a heating surface has three modes: thermodynamic design, geometric design, and part load calculation. From an industrial point of view, the design modes are needed to specify contracts for the planning of new power plants, whereas the calculation mode runs for the simulation of the functional behaviour of existing plants. From a mathematical point of view, the number of equations within the components for the different modes may vary. For example, the thermodynamic design mode of a heating surface has one equation less than the part load mode. This is caused by the fact that no momentum conservation equation is solved in the thermodynamic design mode, whereas in the part load mode a representative heating tube is integrated along its length leading to the corresponding momentum conservation equation. The integration is always done in the direction of the water flow because otherwise, due to boiling, the characteristics may diverge.

Different methods are used to determine the Jacobian matrix. Simple linear functions are solved in a preprocessing step. For trivial nonlinear functional dependencies between function and model parameters, the derivatives are determined analytically and hand coded into the source code. Other functional dependencies are given as ODEs, e.g., the components

"heating surface," and lead to the problem of determining a suitable increment of numerical differentiation relative to the step width of the needed
integration algorithm. Therefore, the application of automatic differentiation to these problems is a real challenge.

1.4.2 AD applied to a C++ software code

The automatic differentiation of the program of the power plant simulation
tool was done using the AD tools ADOL-C [242] and FADBAD [41]. The
number of dependent and independent variables is given by the specific
component and its mode of calculation. In power plant simulation, a single
component is defined by fewer than 5 dependent and 5 independent variables (mass flow, pressure, enthalpy, e.g.). The collection of the non-trivial
equations of all components of the plant topology alone leads to a sparse
square Jacobian with dimension of about 500.

The success of applying AD systems is strongly influenced by the C++
class concept used. In the case of relatively simple classes, the application of
the concept of operator overloading as used in the AD tools ADOL-C and
FADBAD to the required functions is straightforward. Hence, the provided
predefined functions of these packages can be used for the differentiation
of the code.

In the case of large and complex C++ classes, the application of the
concept of operator overloading is much more complicated. All functions,
which are often widely scattered over the entire source code, have to be
analyzed carefully, whether operator overloading is applicable to all subfunctions within the classes. In that case, all active variables of all needed
C++ (sub-)classes must be determined and retyped as described in §1.3.2.
The brute force approach of redefining all doubles as adoubles, which works
quite reliably in NEOS [401], was not tried here.

To avoid the troublesome implementation of the concept of operator
overloading to all (sub-) classes, it may even be preferable in terms of convenience and efficiency to develop derivative versions of member functions
by hand. One potential advantage of this approach is better locality in
data layout and access. The application of automatic differentiation leads
to problems if the concept of operator overloading is not applicable to some
function that is called from within the considered C++ class. Then the corresponding derivatives must be determined in some other way, for example
by divided differences.

Because of the points raised in the last two paragraphs, the application
of automatic differentiation especially to the component "heating surface"
is extremely complicated. Here, the internally used heat flux is determined
as a solution of a one-dimensional Newtonian process. Within this nonlin-
ear process, further parameters α_{inside} and $\alpha_{outside}$ denoting heat transfer
parameters for flue gas at the outside of the pipe and the inside of the pipe,
respectively, must be computed. These functions are given by very complex

formulas, which depend on further functions of the independent variables such as behaviour of temperature, Prandl formula, or steam concentration. In a final step, integration algorithms have to be called for the solution of the ODE system. Here, the application of ADOL-C for the differentiation leads to the generation of a new tape for each time the integration step is performed. In total, the concept of operator overloading must be applied to all these very complex subfunctions, to a solution algorithm of a one-dimensional nonlinear equation, and to integration algorithms.

1.4.3   Dealing with nonsmoothness

In general, the output of an algebraic equation solver or a numerical ODE integrator is not a differentiable or even a continuous function of its inputs. Hence, the evaluation of derivative values as well as their usage for locally approximating the function in question is a little doubtful under such (rather typical) circumstances. This limitation applies not only to AD but also any other method for evaluating or approximating derivatives.

Especially table look-ups, which are quite often contained in simulation models, may destroy differentiability. For example, the water steam table (Figure 1.3) is a basic function used by several components within the simulation system. The function $\rho(p, h)$ represents a complex dependency between water steam concentration, enthalpy h and the corresponding water pressure p. Besides the function $\rho(p, h)$ itself, the derivatives $\frac{\partial \rho}{\partial p}(p, h)$ and $\frac{\partial \rho}{\partial h}(p, h)$ are already used in the simulation code. Therefore, another differentiation reaching second order derivatives is needed for overall sensitivity studies and optimization calculations.

If the computation of these derivatives is not very accurate, the special functional behaviour of this water steam table as shown in Figure 1.3 leads to complications during the application of certain fixed-point iterations. Therefore, the water steam table $\rho(p, h)$ is a function, which must be handled carefully with respect to its steep gradients and its non-differentiability in certain regions.

Several attempts were made over the last years for a good mathematical representation of the water steam table. There exist some versions of linear interpolation schemes for the function values originally given simply as an array of values. The differentiation of these schemes leads at the phase boundary shown in Figure 1.3 to a certain smoothing of the theoretically sharp edge and therefore to a rather good natured behaviour of fixed-point or integration algorithms. On the other hand, the inaccuracy of linear interpolation yields additional (convergence) problems concerning the assignment of function values to the correct region of the table. A bicubic spline interpolation of the water steam table may result in a certain over/undershooting of the mentioned phase boundary and therefore to potential problems in fixed-point iterations. However, a correct representation of the steepness as well as the almost constant behaviour of the gradients near the strong phase boundary might lead to problems of overshooting in Newton algorithms into regions of undefined arguments for the water steam table. This shows some of the problems which might appear using even one of the most basic routines of the entire power steam simulation.

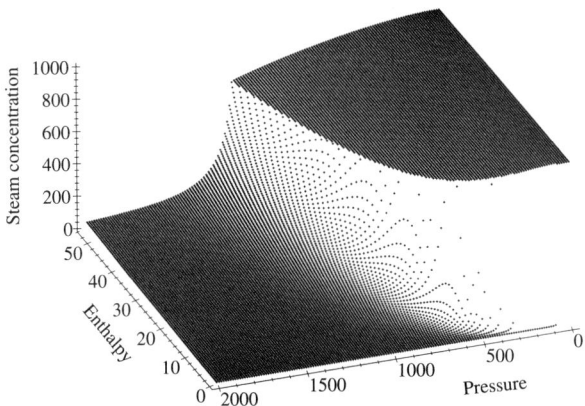

FIGURE 1.3. Water steam table

Similar problems with table look-ups occur in many real world simulations. For example, the optimization of race car control and design [116] involves a table yielding the engine torque as a function of throttle position and revolution count. Moreover, gear shifting generates a genuine discontinuity in acceleration so that the resulting velocities and positions coordinates are only C^0 and C^1, respectively. There is some theory (see e.g., Chapter 11 in [238]) regarding the existence and computability of one sided directional derivatives of functions defined by programs with branches or nonsmooth elemental functions such as $\rho(p, h)$ or simply abs, max, and min. Moreover, it has been shown for contractive fixed-point iterations that the derivatives of the iterates converge asymptotically to the derivative of the solution being approximated. However, the derivative values typically lag behind the iterates themselves, so that it is advisable to tighten the stopping criteria by requiring smallness not only of the residual but also its derivatives. Unfortunately, these modifications must still be performed by the programmer, as currently no AD tool implements them automatically.

The papers on optimal design in PDEs in Part III of this volume all face the problem of computing adjoints for some variant of the Navier Stokes equation, which can only be solved by rather slowly converging fixed point iterations. As discussed in [218] and Section 11.4 of [238], mechanically applying the reverse mode to these problems is definitely not the right thing to do, as one can iteratively solve the adjoint equations without undue increase in storage and a comparable asymptotic rate of convergence.

1.4.4   Industrial challenges to AD

The application of automatic differentiation to the problem of power plant simulation shows essentially two problems: i) Technical software aspects — AD tools have to be used for a mixture of Fortran, C, and C++ source code, which is partially self written, from the public domain, or perhaps the source code is not available. ii) Mathematical concerns — the application of automatic differentiation to iterative fixed-point algorithms, integral equations, and ODE solvers has to be analyzed very carefully concerning the evaluation and use of derivative values.

1.5   Summary

In this chapter, we applied AD tools such as ADIFOR, Odyssée, ADOL-C, and computer algebra tools such as Maple V to real world industrial applications coming from circuit simulation, steel fabrication, and steam power plants. These applications are used every day within the Siemens AG. There are stringent requirements concerning the stability and efficiency of the simulation systems. The progress in the technical systems and therefore the further development of the corresponding simulation systems leads to the need of an efficient development of new simulation models. This report has shown that the benefits that can be achieved by the use of automatic differentiation depend critically on the complexity of the specific computer program as well as of the underlying mathematical model and its algorithmic realization.

In circuit simulation the application of AD tools improves the development of new transistor models impressively. The accurate gradient information is computed with only a slight increase in the run time of the models compared to hand-coded derivatives, resulting in the successful integration of automatic differentiation into a VHDL-AMS compiler, which will be an important trend for the future.

Furthermore, the technique of automatic differentiation can be integrated successfully into the simulation of a cooling system. Using AD, we detected significant problems concerning the accuracy of divided differences used in the old simulation system. Therefore, a totally new simulation system was developed including an improved approach of hybrid differentiation techniques.

The combination of several programming languages and the combination of several mathematical algorithms as used in a power plant simulation system is still a great challenge for automatic differentiation. Here, AD tools can be applied only partially requiring a great effort for the realization.

The results presented in this chapter show that existing AD tools are available for the successful differentiation of moderate complex industrial problems. Accurate derivative information can be provided within an acceptable run time. Also, we describe difficulties that arise if the underlying computer program is written in different languages and/or use potential trouble spots such as fixed-point iterations or table look-ups. This illustrates and underlines the need of further research concerning both the theory and the software of AD. Future trends may include the development of one AD tool for the differentiation of an internal standard representation with front and back ends for different programming languages or the integration of AD into a C or Fortran compiler.

A difficulty that cannot be resolved by AD tool development is the increasing reliance by industrial users on commercial software and other precompiled subroutines. Quite a few systems such as GAMS, AMPL, and Windows Excel already use automatic differentiation internally. However, we know of no example where sensitivity information is provided directly to a calling program, and there is certainly no standard for such additional data transfer. It would seem that the seeding and harvesting mechanism embodied in the vector modes of ADIFOR and TAMC [212] represent a good standard for the dense case. Naturally, sparse dependencies are very important and must also be handled eventually.

Acknowledgments: The authors are grateful to Dr. U. Feldmann and Dr. C. Hammer (Siemens Infineon) for the application of techniques of AD to the circuit simulator TITAN for several years. For the possibility to apply AD to the power plant simulator, the authors would like to thank Dr. G. Löbel and R. Altpeter, Siemens KWU.

2

Automatic Differentiation Tools in Optimization Software

Jorge J. Moré

ABSTRACT We discuss the role of automatic differentiation tools in optimization software. We emphasize issues that are important to large-scale optimization and that have proved useful in the installation of nonlinear solvers in the NEOS Server. Our discussion centers on the computation of the gradient and Hessian matrix for partially separable functions and shows that the gradient and Hessian matrix can be computed with guaranteed bounds in time and memory requirements.

2.1 Introduction

Despite advances in automatic differentiation algorithms and software, researchers disagree on the value of incorporating automatic differentiation tools in optimization software. There are various reasons for this state of affairs. An important reason seems to be that little published experience exists on the effect of automatic differentiation tools on realistic problems, and thus users worry that automatic differentiation tools are not applicable to their problems or are too expensive in terms of time or memory. Whatever the reasons, few optimization codes incorporate automatic differentiation tools.

Without question, incorporating automatic differentiation tools into optimization is not only useful but, in many cases, essential in order to promote the widespread use of state-of-the-art optimization software. For example, a Newton method for the solution of large bound-constrained problems

$$\min \left\{ f(x) : x_l \leq x \leq x_u \right\},$$

where $f : \mathbb{R}^n \mapsto \mathbb{R}$ and x_l and x_u define the bounds on the variables, requires that the user provide procedures for evaluating the function $f(x)$ and also the gradient $\nabla f(x)$, the sparsity pattern of the Hessian matrix $\nabla^2 f(x)$, and the Hessian matrix $\nabla^2 f(x)$. The demands on the user increase for the constrained optimization problem

$$\min \left\{ f(x) : x_l \leq x \leq x_u, \ c_l \leq c(x) \leq c_u \right\},$$

where $c : \mathbb{R}^n \mapsto \mathbb{R}^m$ are the nonlinear constraints. In this case the user must also provide the sparsity pattern and the Jacobian matrix $c'(x)$ of the constraints. In some cases the user may even be asked to provide the Hessian matrix of the Lagrangian

$$L(x, u) = f(x) + \langle u, c(x) \rangle \qquad (2.1)$$

of the optimization problem. The time and effort required to obtain this information and verify their correctness can be large even for simple problems. Clearly, any help in simplifying this effort would promote the use of the software.

In spite of the advantages offered by automatic differentiation tools, relatively little effort has been made to interface optimization software with automatic differentiation tools. Dixon [160, 161] was an early proponent of the integration of automatic differentiation with optimization, but to our knowledge Liu and Tits [359] were the first to provide interfaces between a general nonlinear constrained optimization solver (FSQP) and automatic differentiation tools (ADIFOR).

Modeling languages for optimization (e.g., AMPL [11] and GAMS [203]) provide environments for solving optimization problems that deserve emulation. These environments package the ability to calculate derivatives, together with state-of-the-art optimization solvers and a language that facilitates modeling, to yield an extremely attractive problem-solving environment.

The NEOS Server for Optimization [401] is another problem-solving environment that integrates automatic differentiation tools and state-of-the-art optimization solvers. Users choose a solver and submit problems via the Web, email (neos@mcs.anl.gov), or a Java-enabled submission tool. When a submission arrives, NEOS parses the submission data and relays that data to a computer associated with the solver. Once results are obtained, they are sent to NEOS, which returns the results to the user. Submissions specified in Fortran are processed by ADIFOR [78, 79], while C submissions are handled by ADOL-C [242]. Since the initial release in 1995, the NEOS Server has continued to add nonlinear optimization solvers with an emphasis on large-scale problems, and the current version contains more than a dozen different nonlinear optimization solvers.

Users of a typical computing environment would like to solve optimization problems while only requiring that the user provide a specification of the problem; all other quantities required by the software (for example, gradients, Hessians, and sparsity patterns) would be generated automatically. Optimization modeling languages and the NEOS Server provide this ability, but as noted above, users of nonlinear optimization solvers are usually asked to provide derivative information.

Our goal in this chapter is to discuss techniques for using automatic differentiation tools in large-scale optimization software. We highlight issues that are relevant to solvers in the NEOS Server. For recent work on the interface between automatic differentiation tools and large-scale solvers, see [1, 322]. We pay particular attention to the computation of second-order (Hessian) information since there is evidence that the use of second-order information is crucial to the solution of large-scale problems, while a major concern is the cost of obtaining such second-order information. See [2, 213, 238] for related work.

Currently, most optimization software for large-scale problems uses only first-order derivatives. Of the nonlinear solvers available in the NEOS Server, only LANCELOT, LOQO, and TRON accept second-order information. We expect this situation to change, however, as automatic differentiation tools improve and provide second-order information with the same reliability and efficiency as are currently available for first-order information.

2.2 Partially Separable Functions

We consider the computation of the gradient and Hessian matrix of a partially separable function, that is, a function $f : \mathbb{R}^n \mapsto \mathbb{R}$ of the form

$$f(x) = \sum_{k=1}^{m} f_k(x), \qquad (2.2)$$

where the component functions $f_k : \mathbb{R}^n \mapsto \mathbb{R}$ are such that the *extended* function

$$f_E(x) = \begin{pmatrix} f_1(x) \\ \vdots \\ f_m(x) \end{pmatrix}$$

has a sparse Jacobian matrix. Our techniques are geared to the solution of large-scale optimization problems. For an extensive treatment of techniques for computing derivatives of general and partially separable functions with automatic differentiation tools, we recommend the recent book by Griewank [238].

Griewank and Toint [248] introduced partially separable functions, showing that if the Hessian matrix $\nabla^2 f(x)$ is sparse, then $f : \mathbb{R}^n \mapsto \mathbb{R}$ is partially separable. Partially separable functions also arise in systems of nonlinear equations and nonlinear least squares problems. For example, if each component of the mapping $r : \mathbb{R}^n \mapsto \mathbb{R}^m$ is partially separable, then

$$f(x) = \tfrac{1}{2} \|r(x)\|^2$$

is also partially separable. As another example, consider the constrained optimization problem

$$\min \left\{ f(x) : x_l \leq x \leq x_u, \; c_l \leq c(x) \leq c_u \right\},$$

where $c : \mathbb{R}^n \mapsto \mathbb{R}^m$ specifies the constraints. For this problem, the Lagrangian function $L(\cdot, u)$ defined by (2.1) is partially separable if f and all the components of the mapping c are partially separable. For specific examples note that the functions f and c in the parameter estimation and optimal control optimization problems in the COPS [163] collection are partially separable.

We are interested in computing the gradient and the Hessian of a partially separable function with guaranteed bounds in terms of both computing time and memory requirements. We require that the computing time be bounded by a multiple of the computing time of the function, that is,

$$T\{\nabla f(x)\} \le \Omega_{T,G} T\{f(x)\}, \qquad T\{\nabla^2 f(x)\} \le \Omega_{T,H} T\{f(x)\}, \qquad (2.3)$$

for constants $\Omega_{T,G}$ and $\Omega_{T,H}$, where $T\{\cdot\}$ is computing time. We also require that

$$M\{\nabla f(x)\} \le \Omega_{M,G} M\{f(x)\}, \qquad M\{\nabla^2 f(x)\} \le \Omega_{M,H} M\{f(x)\} \qquad (2.4)$$

for constants $\Omega_{M,G}$ and $\Omega_{M,H}$, where $M\{\cdot\}$ is memory.

These are important requirements for large-scale problems. In particular, if the constants in these expressions are small and independent of the structure of the extended function f_E, then the computational requirements of an iteration of Newton's method are comparable with those of a limited-memory Newton's method.

The constants in (2.3) and (2.4) can be bounded in terms of a measure of the sparsity of the extended function. We use ρ_M, where

$$\rho_M \equiv \max\{\rho_i\},$$

and ρ_i is the number of nonzeros in the ith row of $f_E{}'(x)$. We can also view ρ_M as the largest number of variables in any of the component functions.

Decompositions (2.2) with the number m of element functions of order n, and with ρ_M small and independent of n, are preferred. Since the number of nonzeros in the Hessian $\nabla^2 f(x)$ is no more than $m\rho_M{}^2$, decompositions with these properties are guaranteed to have sparse Hessian matrices. Discretizations of parameter estimation and optimal control problems, for example, have these properties because in these problems each element function represents the contributions from an interval or an element in the discretization.

One of the aims of this chapter is to present numerical evidence that we can compute the gradient $\nabla f(x)$ and the Hessian matrix $\nabla^2 f(x)$ of a partially separable function with

$$\Omega_{T,G} \le \kappa_1 \rho_M, \qquad \Omega_{T,H} \le \kappa_2 \rho_M^2, \qquad (2.5)$$

where κ_1 and κ_2 are constants of modest size and independent of f_E. We normalize $\Omega_{T,G}$ by ρ_M because the techniques in §2.3 require at least ρ_M functions evaluations to estimate the gradient. Similarly, the number of gradient evaluations needed to estimate the Hessian matrix by the techniques in §2.4 is at least ρ_M. Thus, these techniques require at least ρ_M^2 function evaluations to estimate the Hessian matrix.

2.3 Computing Gradients

We now outline the techniques that we use for computing the gradients of partially separable functions. For additional information on the techniques in this section, see [69, 73].

Computing the gradient of a partially separable function so that the bounds (2.3) and (2.4) are satisfied is based on the observation, due to Andreas Griewank, that if $f : \mathbb{R}^n \to \mathbb{R}$ is partially separable, then

$$f(x) = f_E(x)^T e,$$

where $e \in \mathbb{R}^m$ is the vector of all ones, and hence

$$\nabla f(x) = f_E'(x)^T e. \tag{2.6}$$

We can then compute the gradient by computing the Jacobian matrix $f_E'(x)$.

At first sight the approach based on (2.6) does not look promising, since we need to compute a Jacobian matrix and then obtain the gradient from a matrix-vector product. However, the key observation is that the Jacobian matrix is sparse, while the gradient is dense. Thus, we can use sparse techniques for the computation of the extended Jacobian.

We could also use the reverse approach of automatic differentiation to compute the gradient of f. The reverse approach works directly on f and does not require the partial separability structure of f. Moreover, for the reverse approach, (2.3) holds with $\Omega_{T,G}$ small and independent of ρ_M. Theoretically $\Omega_{T,G} \leq 5$, but practical implementations may not satisfy this bound. However, the memory requirements of the reverse approach depend on the number of floating point operations needed to compute f, and thus (2.4) can be violated. A careful comparison between the reverse approach and the techniques described below would be of interest.

In this section we consider two methods for computing the gradient of a partially separable function via (2.6). In the *compressed* AD approach, automatic differentiation tools are used to compute a compressed form of the Jacobian matrix of the extended function f_E, while in the *sparse* AD approach, automatic differentiation tools are used to compute a sparse representation of the Jacobian matrix of the extended function.

In the compressed AD approach we assume that the sparsity pattern of the Jacobian matrix $f_E'(x)$ is known. Given the sparsity pattern, we partition the columns of the Jacobian matrix into groups of *structurally orthogonal* columns, that is, columns that do not have a nonzero in the same row position. Given a partitioning of the columns into p groups of structurally orthogonal columns, we determine the Jacobian matrix by computing the *compressed Jacobian* matrix $f_E'(x)V$, where $V \in \mathbb{R}^{n \times p}$. There is a column of V for each group, and $v_{i,j} \neq 0$ only if the ith column of $f_E'(x)$ is in the jth group. Software for this partitioning problem [133] defines the groups with an array ngrp that sets the group for each column.

The extended Jacobian can be determined from the compressed Jacobian matrix $f_E'(x)V$ by noting that if column j is in group k, then

$$\langle e_i, f_E'(x)V e_k \rangle = v_{i,j} \partial_{i,j} f_E(x).$$

Thus $\partial_{i,j} f_E(x)$ can be recovered directly from the compressed Jacobian matrix.

We note that for many sparsity patterns, the number of groups p needed to determine $A \in \mathbb{R}^{m \times n}$ with a partitioning of the columns is small and independent of n. In all cases there is a lower bound of $p \geq \rho_M$. We also know [136] that if a matrix A can be permuted to a matrix with bandwidth $band(A)$, then $p \leq band(A)$.

The sparse AD approach uses a sparse data representation, usually in conjunction with dynamic memory allocation, to carry out all intermediate derivative computations. At present, the SparsLinC library in ADIFOR [78] is the only automatic differentiation tool with this capability. The main advantage of the sparse AD approach over the compressed AD approach is that no knowledge of the sparsity pattern is required. On the other hand, the sparse AD approach is almost always slower, and can be significantly slower on vector machines.

In an optimization setting, a hybrid approach [91] is the best approach. With this strategy, the sparse AD approach is used to obtain the sparsity pattern of the Jacobian matrix of the extended function at the starting point. See §2.4 for additional information on techniques for computing the sparsity pattern of the extended function. Once the sparsity pattern is determined, the compressed AD approach is used on all other iterations. The hybrid approach is currently the best approach to compute gradients of partially separable functions, and is used in all solvers installed on the NEOS Server.

We conclude this section with some recent results on using the sparse AD approach to compute the gradients of partially separable functions drawn from the MINPACK-2 [22] collection of test problems. We selected ten problems; the first five problems are finite element formulations of variational problems, while the last five problems are systems of nonlinear equations derived from collocation or difference formulations of systems of differential equations.

Table 2.1 provides the value of ρ_M for the ten problems in our performance results. For each of the problems we used three values of n, usually $n \in \{1/4, 1, 4\} \cdot 10^4$, to observe the trend in performance as the number of variables increases. The results were essentially independent of the number of variables, so our results are indicative of the performance that can be expected in large-scale problems.

We want to show that the bounds (2.5) for $\Omega_{T,G}$ holds for these problems. For these results we used the sparse approach to compute the Jacobian matrix $f_E'(x)$ of the extended function, and then computed the gradient of

TABLE 2.1. Data for MINPACK-2 test problems

	PJB	MSA	ODC	SSC	GL2	FIC	SFD	IER	SFI	FDC
ρ_M	5	4	4	4	5	9	14	17	5	13

f with (2.6). For each problem we computed the ratio κ_1, where

$$T\{\nabla f(x)\} = \kappa_1 \, \rho_M \, \max T\{f(x)\}.$$

Table 2.2 presents the quartiles for κ_1 obtained on a Pentium 3 (500 MHz clock, 128 MB of memory) with the Linux operating system.

TABLE 2.2. Quartiles for κ_1 on the MINPACK-2 problems

min	q_1	q_2	q_3	max
1.3	2.8	4.5	5.3	7.8

The results in Table 2.2 show that the bound (2.5) for $\Omega_{T,G}$ holds for the MINPACK-2 problems, with κ_1 small.

   These results are consistent with the results in [73], where it was shown that $\kappa_1 \in [3, 15]$ on a SPARC-10 for another set of test problems drawn from the MINPACK-2 collection. Note that in [73] the ratio κ_1 was computed with ρ_M replaced by the number of columns p in the matrix V. Since $p \geq \rho_M$, the ratios in Table 2.2 would decrease if we replaced ρ_M by p. The advantage of using ρ_M is that the ratio κ_1 is then dependent only on the structure of the function.

2.4 Computing Hessian Matrices

We have already shown how automatic differentiation tools can be used to compute the gradient of a partially separable function. We now discuss the tools that are needed to compute the Hessian of a partially separable so that the requirements (2.3) on computing time and (2.4) on memory are satisfied.

   The techniques that we propose require the sparsity pattern of the Hessian matrix and that the Hessian-vector products $\nabla^2 f(x)v$ be available. In our numerical results we approximate the Hessian-vector product with a difference of gradient values, but in future work we expect to compute Hessian-vector products with ADIFOR.

   We now show how to compute the sparsity pattern of the Hessian matrix from the sparsity pattern of $f_E'(x)$. We define the sparsity pattern of a matrix-valued mapping $A : \mathbb{R}^n \mapsto \mathbb{R}^{n \times n}$ in a neighborhood $N(x_0)$ of a point x_0 by

$$\mathcal{S}\{A(x_0)\} \equiv \left\{ (i,j) : a_{i,j}(x) \not\equiv 0, \ x \in N(x_0) \right\}. \tag{2.7}$$

We are interested in the sparsity pattern of the extended Jacobian and the Hessian matrix of a partially separable function $f : \mathbb{R}^n \mapsto \mathbb{R}$ in a region $\mathcal{D}$ of the form

$$\mathcal{D} = \{x \in \mathbb{R}^n : x_l \le x \le x_u\}.$$

Given $x \in \mathcal{D}$, we evaluate the sparsity pattern $\mathcal{S}\{f_E'(x)\}$ by computing $f_E'(\bar{x}_0)$, where $\bar{x}_0$ is a random, small perturbation of x_0, for example,

$$\bar{x}_0 = (1 + \varepsilon)x_0 + \varepsilon, \qquad |\varepsilon| \in [10^{-6}, 10^{-4}].$$

Then we can reliably let $\mathcal{S}\{f_E'(x)\}$ be the set of (i, j) such that $\partial_{i,j} f_E(\bar{x}_0) \ne 0$. We should not obtain the sparsity pattern of the Jacobian matrix by evaluating f_E' at the starting point x_0 of the optimization process because this point is invariably special, and thus the sparsity pattern of the Jacobian matrix is unlikely to be representative.

The technique that we have outlined for determining the sparsity pattern is used by the solvers in the NEOS Server and has proved to be quite reliable. The sign of ε must be chosen so that $\bar{x}_0 \in \mathcal{D}$, and special care must be taken to handle the case when x_l and x_u agree in some component.

Given the sparsity pattern of the Jacobian matrix of the extended function, we determine the sparsity pattern for the Hessian $\nabla^2 f(x)$ of the partially separable function f via

$$\mathcal{S}\{\nabla^2 f(x)\} \subset \mathcal{S}\{f_E'(x)^T f_E'(x)\}. \tag{2.8}$$

Note that (2.8) is valid only in terms of the definition (2.7) for a sparsity pattern. For example, if $f : \mathbb{R}^2 \mapsto \mathbb{R}$ is defined by

$$f(x) = \phi(\xi_1 \xi_2)$$

for a function ϕ such that $\phi'(0) \ne 0$, then $\partial_{1,2} f(0) \ne 0$, but $\partial_1 f(0) = \partial_2 f(0) = 0$. However, (2.8) holds because $\partial_2 f(x) \not\equiv 0$ and $\partial_1 f(x) \not\equiv 0$ in a neighborhood of the origin.

In most cases equality holds in (2.8). This happens, in particular, if f does not depend linearly on the variables, and

$$\bigcup_{k=1}^{m} \mathcal{S}\{\nabla^2 f_k(x)\} \subset \mathcal{S}\{\nabla^2 f(x)\}. \tag{2.9}$$

If f depends linearly on some variables, say,

$$f(x) = \xi_1 + \phi(\xi_2, \ldots, \xi_n),$$

then equality does not hold in (2.8). Assumption (2.9) implies that there is no cancellation in the computation of the Hessian $\nabla^2 f(x)$. This assumption can fail in some cases, for example, when $f_1 \equiv -f_2$, but holds in most cases.

Since we are able to estimate the sparsity pattern of the Hessian matrix via (2.8), we could use the compressed AD approach described in §2.3 to compute the Hessian matrix from a compressed Hessian $\nabla^2 f(x)V$. However, these techniques ignore the symmetry of the Hessian matrix and thus may require an unnecessarily large number of columns p in the matrix V. For example, an arrowhead matrix requires $p = n$ if symmetry is ignored, but $p = 2$ otherwise.

Powell and Toint [423] were the first to show that symmetry can be used to reduce the number p of columns in the matrix V. They proposed two methods for determining a symmetric matrix A from a compressed matrix AV. In the direct method the unknowns in A are determined directly from the elements in the compressed matrix AV. In this method unknowns are determined independently of each other. In the substitution method the unknowns are determined in a given order, either directly or as a linear combination of elements that have been previously determined.

These definitions of direct and substitution methods are precise but do not readily yield algorithms for determining symmetric matrices. Coleman and Moré [137] and Coleman and Cai [132] extended [423] by interpreting the problem of determining symmetric matrices in terms of special graph coloring problems. This work led to new algorithms and a deeper understanding of the estimation problem.

Software for the symmetric graph coloring problem is available [134] for both direct and substitution methods. Numerical results in [137] suggest that a direct method yields a 20% improvement over methods that disregard symmetry, and that the substitution method yields about a 30% reduction over the direct method.

⋄ Evaluate $f_E(x_0)$ and obtain $m = \text{size} f_E(x_0)$.
⋄ Compute $\text{nnz}\{f_E'(x_0)\}$.
⋄ Allocate space for $f_E'(x_0)$.
⋄ Compute the sparsity pattern $\mathcal{S}\{f_E'(x_0)\}$.
⋄ Compute $\text{nnz}\{f_E'(x_0)^T f_E'(x_0)\}$.
⋄ Allocate space for $\nabla^2 f(x_0)$
⋄ Compute $\nabla^2 f(x_0)$ from the compressed Hessian matrix $\nabla^2 f(x_0)V$.

Algorithm 2.4.1: Computing the Hessian matrix for a partially separable function.

We use Algorithm 2.4.1 to compute the Hessian matrix from a user-supplied extended function f_E. This algorithm uses static memory allocation so that it is first necessary to determine the number of nonzeros in $f_E'(x_0)$ by computing $f_E'(x_0)$ by rows, but not storing the entries. Once this is done, we allocate space for $f_E'(x_0)$ and compute $f_E'(x_0)$ and the sparsity pattern. Another interesting aspect of Algorithm 2.4.1 is that we compute the number of nonzeros in $f_E'(x_0)^T f_E'(x_0)$ directly from the sparsity pattern of $f_E'(x_0)$. In view of (2.8), we then have an accurate idea of the amount of memory needed to store the Hessian matrix. The final step is to compute the Hessian matrix from the the compressed Hessian matrix $\nabla^2 f(x_0)V$ by either a direct or a substitution method.

We consider both direct and substitution methods to determine the Hessian matrix from the compressed Hessian. In both cases we are interested in the ratio κ_2, where

$$T\left\{\nabla^2 f(x)\right\} = \kappa_2 \rho_M^2 T\{f(x)\},$$

since this provides a measure of the cost of evaluating the Hessian matrix relative to the cost of the function. The κ_2 quartiles for both direct and substitution methods on the MINPACK-2 problems used in §2.3 appear in Table 2.3.

TABLE 2.3. Quartiles for κ_2 on MINPACK-2 problems

Method	min	q_1	q_2	q_3	max
Direct	1.6	5.1	11.2	15.2	46.4
Substitution	1.5	4.1	9.0	12.5	30.2

Direct and substitution methods usually require more than ρ_M gradient evaluations to determine the Hessian matrix, and thus the increase in the value of κ_2 relative to κ_1 in Table 2.2 was expected. Still, it is reassuring that the median value of κ_2 is reasonably small. The largest values of κ_2 are due to one of the problems; if this problem is eliminated, then the maximal value drops by at least a factor of two. In general, problems with the longest computing times yield the smallest values of κ_2 since these problems tend to mask the overhead in the automatic differentiation tools and in determining the Hessian matrix. Moreover, these results are based on using a gradient evaluation that relies on sparse automatic differentiation tools; the use of the hybrid approach mentioned in §2.3 should reduce κ_2 substantially.

Acknowledgments: Paul Hovland merits special mention for sharing his considerable knowledge of automatic differentiation tools. Liz Dolan used a preliminary implementation of the techniques in this chapter to install TRON on the NEOS Server, and in the process sharpened these techniques. Gail Pieper provided the final touches on the chapter with her careful editing.

This work was supported by the Mathematical, Information, and Computational Sciences Division subprogram of the Office of Advanced Scientific Computing, U.S. Department of Energy, under Contract W-31-109-Eng-38, and by the the National Science Foundation (Information Technology Research) grant CCR-0082807.

3

Using Automatic Differentiation for Second-Order Matrix-free Methods in PDE-constrained Optimization

David E. Keyes, Paul D. Hovland, Lois C. McInnes and Widodo Samyono

ABSTRACT Classical methods of constrained optimization are often based on the assumptions that projection onto the constraint manifold is routine, but accessing second-derivative information is not. Both assumptions need revision for the application of optimization to systems constrained by partial differential equations, in the contemporary limit of millions of state variables and in the parallel setting. Large-scale PDE solvers are complex pieces of software that exploit detailed knowledge of architecture and application and cannot easily be modified to fit the interface requirements of a black box optimizer. Furthermore, in view of the expense of PDE analyses, optimization methods not using second derivatives may require too many iterations to be practical. For general problems, automatic differentiation is likely to be the most convenient means of exploiting second derivatives. We delineate a role for automatic differentiation in matrix-free optimization formulations involving Newton's method, in which little more storage is required than that for the analysis code alone.

3.1 Introduction

Years of two-sided (from architecture up, from applications down) algorithms research has made it possible to solve partial differential equation (PDE) problems implicitly with reasonable scalability. PDEs are equality constraints on the state variables in many optimization problems. Hardly auxiliary, the PDE system may contain millions of degrees of freedom. In problems of shape optimization and control, the number of optimization parameters is typically much smaller than the number of state variables. In problems of parameter identification, the number of parameters to be optimized may be comparable to the number of state variables, but few general-purpose optimization frameworks have been demonstrated at the scale required for three-dimensional problems. We therefore propose that large-scale PDE-constrained optimization codes usually should be constructed around the data structures and functional capabilities of the PDE solver.

Optimization is easily incorporated through the Lagrange saddle-point formulation into a Newton-like parallel PDE framework that accommodates substructuring. Newton's method is a common element in the most rapidly convergent solvers and optimizers. Furthermore, a PDE solver that is not part of an optimization framework is probably short of what the client really wants. Hence, for both algorithmic and teleological reasons, analysis and optimization belong together.

We focus in §3.2 on the Newton-Krylov-Schwarz (NKS) family of parallel implicit root finders, and we give an example of pseudo-transient globalization of NKS (ΨNKS) in a large-scale parallel context, aerodynamics. The first-order optimality conditions of equality-constrained optimization using the Lagrangian are presented in §3.3, which introduces a parallel optimization framework called LNKS (Lagrange-Newton-Krylov-Schur or Lagrange-Newton-Krylov-Schwarz). In §3.4 we sketch a prototype parameter identification example from the field of radiation transport. The complexity of LNKS when automatic differentiation (AD) is employed in the Krylov matrix-vector operation is discussed in §3.5. Finally, in §3.6 we summarize our work and indicate some future directions.

3.2   Newton-Krylov-Schwarz

In this section, we describe the NKS framework from the inside outward, then illustrate it in a large-scale parallel context.

3.2.1   Schwarz

Schwarz [98, 166, 461] methods are solvers or preconditioners that create concurrency at a desired granularity algorithmically and explicitly through partitioning, without the necessity of any code dependence analysis or special compiler. Generically, in continuous or discrete settings, Schwarz partitions a solution space into n subspaces, possibly overlapping, whose union is the original space, and forms an approximate inverse of the operator in each subspace. Algebraically, to solve the discrete linear system, $Ax = f$, let Boolean rectangular matrix R_i extract the ith subset of the elements of x: $x_i = R_i x$, and let $A_i = R_i A R_i^T$. Then the Schwarz approximate inverse, B^{-1}, is defined as $\sum_i R_i^T A_i^{-1} R_i$. From the PDE perspective, subspace decomposition is domain decomposition. We form $B^{-1} \approx A^{-1}$ out of (approximate) local solves on (possibly overlapping) subdomains, as in Figure 3.1. This can be used to iterate in a stationary way, as a splitting matrix: $x^{k+1} = (I - B^{-1}A)x^k + B^{-1}f$. However, since $\rho(I - B^{-1}A)$ may be greater than unity in general, this additive splitting may not converge as a stationary iteration. "Multiplicative" Schwarz methods (Gauss-Seidel-like, relative to the Jacobi-like "additive" above) can be proved convergent when

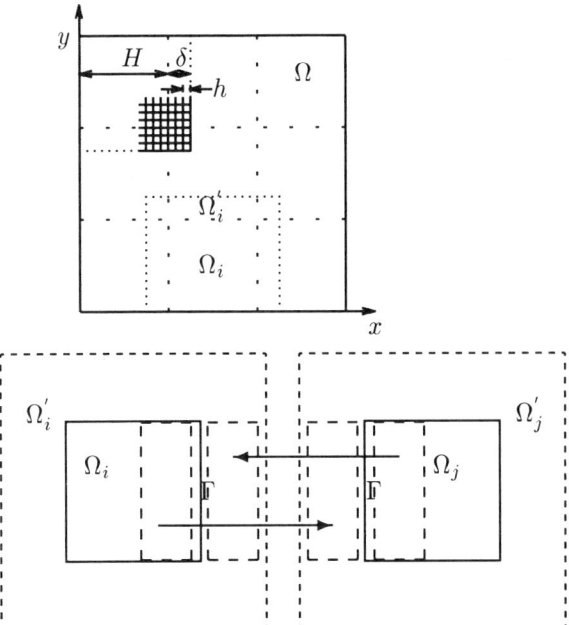

FIGURE 3.1. **Upper:** A domain Ω partitioned into nine overlapping subdomains, Ω_i, extended slightly by overlapping to subdomains Ω_i', showing the scales of the mesh spacing (h), the subdomain overlap (δ), and the subdomain diameter (H). **Lower:** Two adjacent subdomains with common edge Γ pulled apart to show overlap regions as separate buffers, which are implemented in the local data structures of each.

A derives from an elliptic PDE, under certain partitionings.

In the PDE context, Boolean operators R_i and R_i^T, $i = 1, \ldots, n$, represent gather and scatter (communication) operations, mapping between a global vector and its ith subdomain support. When A derives from an elliptic operator and R_i is the characteristic function of unknowns in a subdomain, optimal convergence (independent of $\dim(x)$ and the number of partitions) can be proved, with the addition of a coarse grid, which is denoted with subscript "0": $B^{-1} = R_0^T A_0^{-1} R_0 + \sum_{i>0} R_i^T A_i^{-1} R_i$. Here, R_0 is a conventional geometrically based multilevel interpolation operator. It is an important freedom in practical implementations that the coarse grid space need not be related to the fine grid space or to the subdomain partitioning.

The $A_i^{-1} (i > 0)$ in B^{-1} are often replaced with inexact solves in practice. The exact forward matrix-vector action of A in $B^{-1}A$ is still required, even if inexact solves are employed in the preconditioner.

Condition number estimates for $B^{-1}A$ are given in Table 3.2.2 for generous overlap $\delta = \mathcal{O}(H)$. Otherwise, if $\delta \ll H$, the two-level result is $\mathcal{O}(1 + H/\delta)$. The two-level Schwarz method with generous overlap has a condition number that is independent of the fineness of the discretization and the granularity of the decomposition, which implies perfect algorithmic scalability. However, there is an increasing implementation overhead in the coarse-grid solution required in the two-level method that offsets this perfect algorithmic scalability. In practice, a one-level method is often used, since it is amenable to a perfectly scalable implementation. These condition number results are extensible to nonself-adjointness, mild indefiniteness, and inexact subdomain solvers. The theory requires a "sufficiently fine" coarse mesh, H, for the first two of these extensions, but computational experience shows that the theory is often pessimistic.

3.2.2 Krylov-Schwarz

Although the spectral radius, $\rho(I - B^{-1}A)$, may exceed unity, the spectrum, $\sigma(B^{-1}A)$, is profoundly clustered, so Krylov acceleration methods should work well on the preconditioned solution of $B^{-1}Ax = B^{-1}f$. Krylov-Schwarz methods typically converge in a number of iterations that scales as the square-root of the condition number of the Schwarz-preconditioned system. For convergence scalability estimates, assume one subdomain per processor in a d-dimensional isotropic problem, where $N = h^{-d}$ and $P = H^{-d}$. Then iteration counts may be estimated as in the last two columns of Table 3.2.2.

TABLE 3.1. Theoretical condition number estimates $\kappa(B^{-1}A)$, for self-adjoint positive-definite elliptic problems [461] and corresponding iteration count estimates for Krylov-Schwarz based on an idealized isotropic partitioning of the domain in two or three dimensions.

Preconditioning	$\kappa(B^{-1}A)$	2D Iter.	3D Iter.
Point Jacobi	$\mathcal{O}(h^{-2})$	$\mathcal{O}(N^{1/2})$	$\mathcal{O}(N^{1/3})$
Domain Jacobi	$\mathcal{O}((hH)^{-1})$	$\mathcal{O}((NP)^{1/4})$	$\mathcal{O}((NP)^{1/6})$
1-level Additive Schwarz	$\mathcal{O}(H^{-2})$	$\mathcal{O}(P^{1/2})$	$\mathcal{O}(P^{1/3})$
2-level Additive Schwarz	$\mathcal{O}(1)$	$\mathcal{O}(1)$	$\mathcal{O}(1)$

3.2.3 Newton-Krylov-Schwarz

Let $F(x) = 0$ be a discrete system of nonlinear equations arising from an elliptically dominated system of PDEs. Let its Jacobian be denoted $J \equiv \partial F / \partial x$. Inexact Newton iteration on $F(x) = 0$, involves selecting an initial iterate $x^{(0)}$ and iterating for a correction to the current $x^{(k)}$: $x^{(k+1)} = x^{(k)} + \lambda_k \delta x$, where $||J(x^{(k)})\delta x + F(x^{(k)})|| < \eta_k$. A large body of literature exists on how to choose η_k and λ_k for robustness and efficiency. Any of these members of the inexact Newton family of algorithms may be implemented as a Newton-Krylov-Schwarz method, by iterating for δx with a linear Krylov-Schwarz method. Partitioning x induces block structure on the Jacobian matrix. As anticipated in the presentation of Schwarz above, we do not need any Jacobians explicitly; rather, matrix-vector action of the Jacobian at point $x^{(k)}$ may be performed with finite Fréchet differencing (FD) or automatic differentiation (AD) about the point, and preconditioning of the Jacobian is done with approximate local operators, approximately solved in accordance with overall performance trade-offs.

Newton-Krylov-Schwarz has been demonstrated to be an effective parallel implicit solver for large-scale nonlinear problems derived from PDEs (see, e.g., P. Brown and collaborators at LLNL [94, 95] and D. Knoll and collaborators at LANL [330, 377]). It has been applied to problems in aerodynamics, radiation transport, porous media, semiconductors, geophysics, astrophysical MHD, population dynamics, and other fields. It has been implemented in a parallel matrix-free object-oriented framework, including both FD and AD distributed matvecs, in PETSc software from Argonne [26].

We advocate using NKS in a split-discretization formulation, in which economizations are taken in the left-hand side preconditioner blocks of J relative to the more accurate, physical discretization-dictated right-hand operator for J. Examples of such economizations include sacrificed coupling for process concurrency, segregation of physics into successive phases with simple structure (operator-splitting), the Jacobian of a lower-order discretization for fewer nonzeros and fewer colors in a minimal coloring, the Jacobian of a related discretization allowing "fast" solves, a Jacobian with lagged values for any terms that are expensive to compute or small or both, and a Jacobian stored in half precision for superior (nearly doubled) memory bandwidth, as measured in words per second, in the bandwidth-limited linear algebra routines of a sparse, unstructured PDE solver.

3.2.4   Pseudo-Transient Newton-Krylov-Schwarz

NKS is commonly robustified with pseudo-transience (ΨNKS) [320, 457] or other continuation strategies. In ΨNKS one solves $F(x) = 0$ through a series of modified problems

$$H_\ell(x) \equiv \frac{x - x_{\ell-1}}{\delta t_\ell} + F(x) = 0, \quad \ell = 1, 2, \ldots,$$

each of which is solved (approximately) for x_ℓ. This sequence hugs a physical transient when δt_ℓ is small, for which the associated diagonally dominant Jacobians are well conditioned. δt_ℓ is advanced from $\delta t_0 \ll 1$ to $\delta t_\ell \to \infty$ as $\ell \to \infty$, so that x_ℓ approaches the root of $F(x) = 0$. Unlike many robustification techniques, ΨNKS does *not* require reduction in $\|F(x)\|$ at each step; its ability to climb hills in the residual norm is useful in problems with complex physics, such as combustion, in which a local minimum (e.g., extinction) may not be the physically desired one.

3.2.5   Example from computational aerodynamics

To illustrate the effectiveness of NKS in practice, we quote below some performance data for a computational aerodynamics problem, which won a 1999 Gordon Bell prize [14]. The Euler equations were solved on a tetrahedral unstructured grid for the flow over an ONERA M6 wing.

The finest-granularity decomposition consisted of 3072 subdomains on a grid of approximately 2.8M vertices. Each subdomain was computed on a pair of Intel Pentium Pro processors (6144 processors altogether) on the ASCI Red machine at Sandia, which executed in shared-memory OpenMP mode on the evaluation of $F(x)$, while the linear algebra portions of the computation were left single-threaded on each node. Up to 0.227 Tflop/s were achieved; see [14, 257] for details.

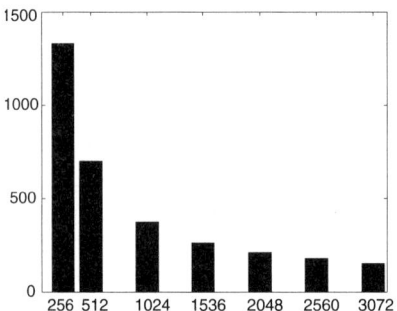 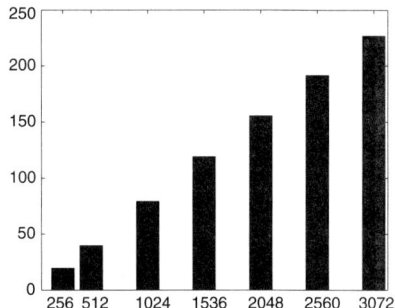

FIGURE 3.2. Execution time (left) and aggregate flop rate (right) for ΨNKS on incompressible Euler flow over an ONERA M6 wing, on a tetrahedral grid of 2,761,744 vertices, based on the KMeTiS-PETSc implementation of the NASA code FUN3D run on up to 3072 nodes of ASCI Red.

3.3   Implications of NKS for Optimization

Equality constrained optimization leads, through the Lagrangian formulation, to a multivariate nonlinear rootfinding problem for the gradient (the first-order necessary conditions), which is amenable to treatment by Newton's method. To establish notation, consider the following canonical framework, in which we enforce equality constraints on the state variables only. (Design variable constraints require additional notation, and inequality constraints require additional algorithmics, but these generalizations are well understood.) Choose m design variables u to minimize the objective function, $\phi(u, x)$, subject to n state constraints, $h(u, x) = 0$, where x is the vector of state variables. In the Lagrange framework, a stationary point of the Lagrangian function

$$\mathcal{L}(x, u, \lambda) \equiv \phi(x, u) + \lambda^T h(x, u)$$

is sought. When Newton's method is applied to the first-order optimality conditions, a linear system known as the Karush-Kuhn-Tucker (KKT) system arises at each step. There is a natural "outer" partitioning: the vector of parameters is often of lower dimension than the vectors of states and multipliers. This suggests a Schur complement-like block elimination process at the outer level, not for concurrency, but for numerical robustness and conceptual clarity. Within the state-variable subproblem, which must be solved repeatedly in the Schur complement reduction, Schwarz provides a natural "inner" partitioning for concurrency.

A major choice to be made in the Newton approach to constrained optimization is between exact elimination of the states and multipliers by satisfying constraint feasibility at every step (reduced system), and progress in all variables simultaneously, possibly violating constraints on intermediate iterates (full system). An advantage of the former is the existence of high-quality, robust black box software for this reduced sequential quadratic programming (RSQP) approach. The advantages of the latter are in reuse of high-quality parallel PDE software, the freedom to use inexact solves (since finely resolved PDE discretizations in 3D militate against exact elimination), and the ease of application of automatic differentiation software, without having to differentiate through the nonlinear subiterations that would be implied by repeated projection to the constraint manifold in RSQP.

We mention three classes of PDE-constrained optimization:

- **Design optimization** (especially shape optimization): u parametrizes the domain of the PDE (e.g., a lifting surface) and ϕ is a cost-to-benefit ratio of forces, energy expenditures, etc. Typically, m is small compared with n and does not scale directly with it. However, m may still be hundreds or thousands in industrial applications.

- **Optimal control**: u parametrizes a continuous control function acting in part of or on the boundary of the domain, and ϕ is the norm of the difference between desired and actual responses of the system. For boundary control, $m \propto n^{2/3}$.

- **Parameter identification/data assimilation**: u parametrizes an unknown continuous constitutive or forcing function defined throughout the domain, and ϕ is the norm of the difference between measurements and simulation results. Typically, $m \propto n$.

Written out in partial detail, the optimality conditions are

$$\frac{\partial \mathcal{L}}{\partial x} \equiv \frac{\partial \phi}{\partial x} + \lambda^T \frac{\partial h}{\partial x} = 0 \ , \tag{3.1}$$

$$\frac{\partial \mathcal{L}}{\partial u} \equiv \frac{\partial \phi}{\partial u} + \lambda^T \frac{\partial h}{\partial u} = 0 \ , \tag{3.2}$$

$$\frac{\partial \mathcal{L}}{\partial \lambda} \equiv h = 0 \ . \tag{3.3}$$

Newton's method iteratively seeks a correction,

$$\begin{pmatrix} \delta x \\ \delta u \\ \delta \lambda \end{pmatrix} \quad \text{to the iterate} \quad \begin{pmatrix} x \\ u \\ \lambda \end{pmatrix}.$$

With subscript notation for the partial derivatives, the Newton correction (KKT) equations are

$$\begin{bmatrix} (\phi_{,xx} + \lambda^T h_{,xx}) & (\phi_{,xu} + \lambda^T h_{,xu}) & h_{,x}^T \\ (\phi_{,ux} + \lambda^T h_{,ux}) & (\phi_{,uu} + \lambda^T h_{,uu}) & h_{,u}^T \\ h_{,x} & h_{,u} & 0 \end{bmatrix} \begin{pmatrix} \delta x \\ \delta u \\ \delta \lambda \end{pmatrix} = - \begin{pmatrix} \phi_{,x} + \lambda^T h_{,x} \\ \phi_{,u} + \lambda^T h_{,u} \\ h \end{pmatrix}$$

or

$$\begin{bmatrix} W_{xx} & W_{ux}^T & J_x^T \\ W_{ux} & W_{uu} & J_u^T \\ J_x & J_u & 0 \end{bmatrix} \begin{pmatrix} \delta x \\ \delta u \\ \lambda_+ \end{pmatrix} = - \begin{pmatrix} g_x \\ g_u \\ h \end{pmatrix}, \qquad (3.4)$$

where $W_{ab} \equiv \frac{\partial^2 \phi}{\partial a \partial b} + \lambda^T \frac{\partial^2 h}{\partial a \partial b}$, $J_a \equiv \frac{\partial h}{\partial a}$, and $g_a = \frac{\partial \phi}{\partial a}$, for $a, b \in \{x, u\}$, and where $\lambda_+ = \lambda + \delta \lambda$.

3.3.1 Newton Reduced SQP

The RSQP method [405] consists of a three-stage iteration. We follow the language and practice of [67, 68] in this and the next subsection.

- **Design Step** (Schur complement for middle blockrow):

$$H \, \delta u = f \, ,$$

where H and f are the reduced Hessian and gradient, respectively:

$$H \equiv W_{uu} - J_u^T J_x^{-T} W_{ux}^T + \left(J_u^T J_x^{-T} W_{xx} - W_{ux} \right) J_x^{-1} J_u$$
$$f \equiv -g_u + J_u^T J_x^{-T} g_x - \left(J_u^T J_x^{-T} W_{xx} - W_{ux} \right) J_x^{-1} h$$

- **State Step** (last blockrow):

$$J_x \, \delta x = -h - J_u \, \delta u$$

- **Adjoint Step** (first blockrow):

$$J_x^T \, \lambda_+ = -g_x - W_{xx} \, \delta x - W_{ux}^T \, \delta u$$

In each overall iteration, we must form and solve with the reduced Hessian matrix H, and we must solve separately with J_x and J_x^T. The latter two solves are almost negligible compared with the cost of forming H, which is dominated by the cost of forming the sensitivity matrix $J_x^{-1} J_u$. Because of the quadratic convergence of Newton, the number of overall iterations is few (asymptotically independent of m). However, the cost of forming H at each design iteration is m solutions with J_x. These are potentially concurrent over independent columns of J_u, but prohibitive.

In order to avoid computing any Hessian blocks, the design step may be approached in a quasi-Newton (e.g., BFGS) manner [405]. Hessian terms are dropped from the adjoint step RHS.

- **Design Step** (severe approximation to middle blockrow):

$$Q\, \delta u = -g_u + J_u^T J_x^{-T} g_x \ ,$$

  where Q is a quasi-Newton approximation to the reduced Hessian

- **State Step** (last blockrow):

$$J_x\, \delta x = -h - J_u\, \delta u$$

- **Adjoint Step** (approximate first blockrow):

$$J_x^T\, \lambda_+ = -g_x$$

In each overall iteration of quasi-Newton RSQP, we must perform a low-rank update on Q or its inverse, and we must solve with J_x and J_x^T. This strategy vastly reduces the cost of an iteration; however, it is no longer a Newton method. The number of overall iterations is many. Since BFGS is equivalent to unpreconditioned CG for quadratic objective functions, $\mathcal{O}(m^p)$ sequential cycles ($p > 0$, $p \approx \frac{1}{2}$) may be anticipated. Hence, quasi-Newton RSQP is not scalable in the number of design variables, and no ready form of parallelism can address this convergence-related defect.

To summarize, conventional RSQP methods apply a (quasi-)Newton method to the optimality conditions: solving an approximate $m \times m$ system to update u, updating x and λ consistently (to eliminate them), and iterating. The unpalatable expense arises from the exact linearized analyses for updates to x and λ that appear in the inner loop. We therefore consider replacing the exact elimination steps of RSQP with preconditioning steps in an outer loop, as described in the next subsection.

3.3.2   Full Space Lagrange-NKS Method

The new philosophy is to apply a Krylov-Schwarz method directly to the $(2n + m) \times (2n + m)$ KKT system (3.4). For this purpose, we require the action of the full matrix on the full-space vector and a good full-system preconditioner, for algorithmic scalability. One Newton SQP iteration is a perfect preconditioner—a block factored solver, based on forming the reduced Hessian of the Lagrangian H—but, of course, far too expensive. Backing off wherever storage or computational expense becomes impractical for large-scale PDEs generates a family of attractive methods.

To precondition the full system, we need approximate inverses to the three left-hand side matrices in the first algorithm of §3.3.1, namely, H, J, and J^T. If a preconditioner is available for H, and exact solves are available for J, and J^T, then it may be shown [321] that conjugate gradient Krylov iteration on the (assumed symmetrizable) reduced system and conjugate gradient iteration on the full system yield the same sequence of iterates. The iterates are identical in the sense that if one were to use the values of u arising from the iteration on the reduced system in the right-hand side of the block rows for x and λ, one would reconstruct the iterates of the full system, when the same preconditioner used for H in the reduced system is used for the W_{uu} block in the full system. Moreover, the spectrum of the full system is simply the spectrum of the reduced system supplemented with a large multiplicity of unit eigenvalues. If one retreats from exact solves with J and J^T, the equivalence no longer holds; however, if good preconditioners are used for these Jacobian blocks, then the cloud of eigenvalues around unity is still readily shepherded by a Krylov method, and convergence should be nearly as rapid as in the case of exact solves.

This Schur-complement-based preconditioning of the full system was proposed in this equality-constrained optimization context by Biros and Ghattas in 1998 [67] and earlier in a related context by Batterman and Heinkenschloss [34]. From a purely algebraic point of view, the same Schur-complement-based preconditioning was advocated by Keyes and Gropp in 1987 [321] in the context of domain decomposition. There, the reduced system was a set of unknowns on the interface between subdomains, and the savings from the approximate solves on the subdomain interiors more than paid for the modest degradation in convergence rate relative to interface iteration on the Schur complement. The main advantage of the full system problem is that the Schur complement never needs to be formed. Its exact action is felt on the design variable block through the operations carried out on the full system.

Biros and Ghattas have demonstrated the large-scale parallel effectiveness of the full system algorithm on a 3D Navier-Stokes flow boundary control problem, where the objective is dissipation minimization of flow over a cylinder using suction and blowing over the back portion of the cylinder as the control variables [68]. They performed this optimization with domain-decomposed parallelism on 128 processors of a T3E, using an original optimization toolkit add-on to the PETSc [26] toolkit. To quote one result from [68], for 6×10^5 state constraints and 9×10^3 controls, full-space LNKS with approximate subdomain solves beat quasi-Newton

RSQP by an order of magnitude (4.1 hours versus 53.1 hours).

Two names have evolved for the new algorithm: Lagrange-Newton-Krylov-Schwarz was proposed by Keyes in May 1999 at the SIAM Conference on Optimization, and Lagrange-Newton-Krylov-Schur by Biros and Ghattas in [68]. The former emphasizes the use of NKS to precondition the large Jacobian blocks, the latter the use of Schur complements to precondition the overall KKT matrix. Both preconditioner suffixes are appropriate in a nested fashion, so we propose Lagrange-Newton-Krylov-Schur-Schwarz (LNKSS) when both preconditioners are used (see Figure 3.3).

| **Lagrange** | **Newton** | **Krylov** |
| optimizer | nonlinear solver | accelerator |

| **Schur** | **Schwarz** |
| subspace precond. | subdomain precond. |

FIGURE 3.3. LNKSS: A Parallel Optimizer for BVP-constrained Problems

Automatic differentiation has two roles in the new algorithm: formation of the action on a Krylov vector of the full KKT matrix, including the full second-order Hessian blocks, and supply of approximations to the elements of J (and J^T) for the preconditioner. While the synergism of AD with LNKSS is in many ways obvious, advocacy of this novel combination is the primary thrust of this chapter.

3.4   Example of LNKS Parameter Identification

The effectiveness of Schwarz preconditioning is illustrated in the analysis context in §3.2.5. In this section, we illustrate Schur preconditioning and automatic differentiation in the parameter identification context. Schwarz and Schur techniques will be combined in a large-scale example from multi-dimensional radiation transport in the future. The governing constraint for our one-dimensional problem is the steady state version of the following radiation diffusion equation for material temperature:

$$\frac{\partial T}{\partial t} = \nabla \cdot (\beta(x)T^\alpha \ \nabla T). \tag{3.5}$$

Instead of solving the impulsive Marshak wave formulation of this problem, we ignore the time derivative and impose Dirichlet boundary conditions of 1.0 and 0.1, respectively, on $T(x)$ at the left- and right-hand endpoints of the unit interval. The resulting ODE boundary value problem is discretized with centered finite differences. The state variables are the discrete temperatures at the mesh nodes, and the design variables are the parameters α and $\beta(x)$. The cost function is temperature matching, $\phi(u, T) = \frac{1}{2}||T(x) - \bar{T}(x)||^2$, where $\bar{T}$ is based on a given α, $\beta(x)$ profile. These parameters are specified for the computation of $\bar{T}(x)$, and then "withheld," to be determined by the optimizer. More generally, $\bar{T}(x)$ would be a desired or experimentally measured profile, and the phenomenological law and material specification represented by α and $\beta(x)$ would be determined to fit. The Brisk-Spitzer form of the nonlinear dependence of the diffusivity on the temperature is $\alpha = 2.5$. For $\bar{T}$ we assume a jump in material properties at the midpoint of the interval: $\beta(x) = 1, 0 \leq x \leq \frac{1}{2}$ and $\beta(x) = 10, \frac{1}{2} < x \leq 1$.

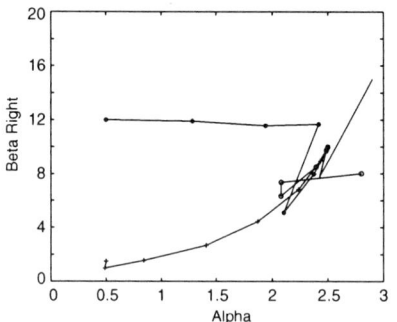 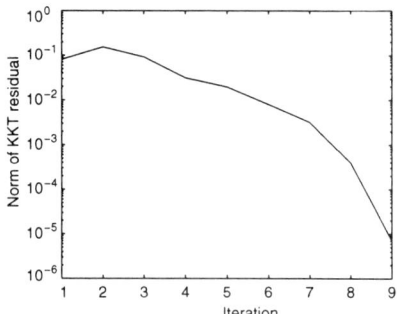

FIGURE 3.4. **Left:** Four convergence histories for (α, β_{right}) merged into one plot. **Right:** KKT norm convergence history for the parameter identification problem with initial iterate based on $\alpha = 0.5$, $\beta_{right} = 1.5$.

Our initial implementation of LNKS is in the software framework of MATLAB [476] and ADMAT [139]. ADMAT is an automatic differentiation framework for MATLAB, based on operator overloading. After an m-file is supplied for the cost function and constraint functions, all gradients, Jacobians, and Hessians (as well as their transposes and their contracted action on vectors) used anywhere in the LNKS algorithm are computed automatically without further user effort. There is one exception in the current code: our almost trivial cost function (with no parametric dependence and separable quadratic state dependence) is differentiated by hand,

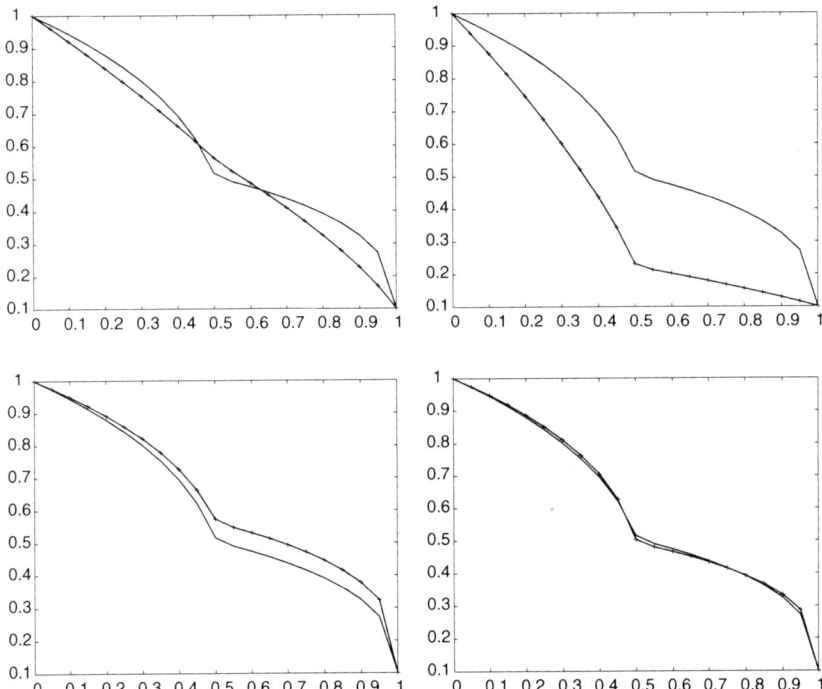

FIGURE 3.5. Initial and final temperature distributions for the radiation diffusion example with a starting point in each quadrant relative to the profile-matching parameters for $(\alpha, \beta_{right}) = (2.5, 10)$: upper left $(0.5, 1.5)$, upper right $(0.5, 12)$, lower left $(2.8, 8)$, lower right $(2.9, 15)$.

yielding an identity matrix for ϕ_{xx}. Our preconditioner is the RSQP block factorization, except that the reduced Hessian preconditioner is replaced with the identity. The reduced Hessian preconditioner block should be replaced with a quasi-Newton reduced Hessian in the future. In the present simple experiments, Newton's method is used without robustification of any kind. Figure 3.4 shows how the optimizer eventually finds the values of 2.5 for α and 10 for $\beta_{right} \equiv \beta(x)$ in the interval $\frac{1}{2} \leq x < 1$, from four different starting points of $(0.5, 1.5), (0.5, 12), (2.8, 8)$ and $(2.9, 15)$. A sample convergence history for the norm of the residual of (3.1)–(3.3) shows quadratic behavior.

Shown in Figure 3.5 are initial and final distributions of $T(x)$ for α and β_{right} displaced as in Figure 3.4 in all directions away from the "true" values of $(2.5, 10)$. The graphs show the "true" temperature profile to be matched at the final converged values of (α, β_{right}) (the curve common to all four plots) and the equilibrium temperature profile at the initial values of the parameters. Within each half-interval, the temperature gradient is sharper on the right (smaller values of $T(x)$), since the heat flux across every station is the same and the temperature-dependent diffusion coefficient factor inside the divergence operator of (3.5) is smaller in zones of smaller temperature. In regions with a larger β-factor, the overall average temperature drop is smaller by the same reasoning. Small α suppresses the nonlinear dependence of the diffusivity on temperature, so the initial temperature profiles are nearly linear within each constant-β region in the first two plots. Only in the first case is the approach to the true (α, β_{right}) monotonic, but plain Newton is robust enough to converge from all quadrants.

3.5   Complexity of AD-based LNKS

Although our demonstration example is low dimensional, LNKS will generally be applied to large problems of n state variables and m parameters. Upon surveying existing AD tools, we conclude that the preconditioned matrix-vector product can be formed in time linear in these two parameters. The shopping list of matrix actions in forming the preconditioned Jacobian-vector product of LNKS is $W_{xx}, W_{uu}, W_{ux}, W_{ux}^T, J_u, J_u^T, J_x^{-1}, J_x^{-T}$, and H^{-1}.

The first six are needed in the full-system matrix-vector multiplication. For this multiplication we require "working accuracy" comparable to the state of the art in numerical differentiation.

Accurate action of the last three is required in RSQP but not in the full system preconditioner. We recommend approximate factorizations of lower-quality approximations, including possibly just W_{uu} for H, or a traditional quasi-Newton rank-updated approximation to the inverse.

We estimate the complexity of applying each block of the KKT Jacobian, assuming only that $h(x, u)$ is available in subroutine call form and that all differentiated blocks are from AD tools, such as the ADIC [84] tool we are using in a parallel implementation of LNKSS. We assume that J_x is needed, element by element, in order to factor it; hence, J_x^T is also available. Since these are just preconditioner blocks, we generally derive these elements from a different (cheaper) function call for the gradient of the Lagrangian than that used for the matvec. Define C_h, the cost of evaluating h; p_x, 1 + the chromatic number of $J_x \equiv h_{,x}$; and p_u, 1 + the chromatic number of $J_u \equiv h_{,u}$. Then the costs of the Jacobian objects are shown in the first three rows of Table 3.5.

For the Hessian arithmetic complexity, we estimate the cost of applying each forward block to a vector. Assume that $h(x, u)$ and $\phi(x, u)$ are available and that all differentiated blocks are results of AD tools. Define C_ϕ, the cost of evaluating ϕ; q, 1 + number of nonzero rows in ϕ''; and r, an implementation-dependent "constant," typically ranging from 3 to 100. Then the cost of the Hessian-vector products can be estimated from the last two rows of Table 2.

For the inverse blocks, we need only low-quality approximations or limited-

TABLE 3.2. Complexity of formation of matrix objects or matrix-vector actions using forward or hybrid modes of modern automatic differentiation software. The asterisk signifies that the reverse mode consumes memory, in a carefully drawn time-space trade-off, so r is implementation-dependent.

Object	Cost: Forward Mode	Cost: Fastest (Hybrid) Mode
J_x, J_x^T	$p_x C_h$	$p_x C_h$
$J_u v$	$2C_h$	$2C_h$
$J_u^T v$	$p_u C_h$	$rC_h{}^*$
$W_{xx} v, W_{ux}^T v$	$p_x C_h + q C_\phi$	$r(C_h + C_\phi)^*$
$W_{uu} v, W_{ux} v$	$p_u C_h + q C_\phi$	$r(C_h + C_\phi)^*$

memory updates [96] of the square systems J_x^{-1}, J_x^{-T}, and H^{-1}.

The complexities for *all* operations required to apply the full-system matrix-vector product and its preconditioner are at worst linear in n or m, with coefficients that depend upon chromatic numbers (affected by stencil connectivity and intercomponent coupling of the PDE, and by separability structure of the objective function) and the implementation efficiency of AD tools.

3.6 Summary and Future Plans

As in domain decomposition algorithms for PDE analysis, partitioning in PDE-equality constrained optimization may be used to improve some combination of robustness, conditioning, and concurrency. Orders of magnitude of savings may be available by converging the state variables and the design variables within the same outer iterative process, rather than a conventional SQP process that exactly satisfies the auxiliary state constraints.

As with any Newton method, globalization strategies are important. These include parameter continuation (physical and algorithmic), mesh sequencing and multilevel iteration (for the PDE subsystem, at least; probably for controls, too), discretization order progression, and model fidelity progression. The KKT system appears to be a preconditioning challenge, but an exact factored preconditioner is known, and departures of preconditioned eigenvalues from unity can be quantified with comparisons of original blocks with blockwise substitutions in inexact models and solves. (For the full system, the preconditioned KKT matrix will be nonnormal, so its spectrum does not tell all.)

With the extra, but automatable, work of forming Jacobian transposes and Hessian blocks, but no extra work in Jacobian preconditioning, any parallel analysis code may be converted into a parallel optimization code—and automatic differentiation tools will shortly make this relatively painless.

The gamut of PDE solvers based on partitioning should be mined for application to the KKT necessary conditions of constrained optimization and for direct use in inverting the state Jacobian blocks inside the optimizer.

We expect shortly to migrate our ADMAT/MATLAB code into the parallel ADIC/PETSc framework, while increasing physical dimensionality and parameter dimensionality. We will also tune the numerous preconditioning parameters for optimal parallel execution time. In the multidimensional large-scale context, we will incorporate multilevel Schwarz linear preconditioning for the spatial Jacobian. Following the recent invention of the additive Schwarz preconditioned inexact Newton (ASPIN) [99], we will also experiment with full nonlinear preconditioning of the KKT system. This could include individual discipline optimizations as nonlinear preconditioner stages in a multidisciplinary computational optimization process— a key engineering (and software engineering) challenge of the coming years.

Acknowledgments: We acknowledge important discussions on algorithmics with George Biros (CMU), Xiao-Chuan Cai (UC-Boulder), Omar Ghattas (CMU), Xiaomin Jiao (UIUC), Tim Kelley (NCSU), Michael Wagner (ODU), and David Young (Boeing); on applications with Kyle Anderson (NASA Langley), Frank Graziani (LLNL), Dana Knoll (LANL), and Carol Woodward (LLNL); and on software with Satish Balay, Bill Gropp, Dinesh Kaushik, Barry Smith (all of Argonne) and Arun Verma (Cornell). This work was supported by the MICS subprogram of the Office of Advanced Scientific Computing Research, U.S. Department of Energy under Contract W-31-109-Eng-38; by Lawrence Livermore National Laboratory under ASCI Level-2 subcontract B347882 to Old Dominion University; by NASA under contract NAS1-19480 (while the second author was in residence at ICASE); and by the NSF under grant ECS-8957475. Computer facilities were provided by the DOE ASCI Alliance Program.

4

Performance Issues in Automatic Differentiation on Superscalar Processors

François Bodin and Antoine Monsifrot

ABSTRACT We overview code optimizations for superscalar processors (Pentium, Ultra Sparc, Alpha, etc.) in the context of automatic differentiation. Using an example we show the impact of program transformations used to increase code efficiency.

4.1 Introduction

Automatic differentiation (AD) is an efficient method to produce derivatives for programs. However, for most numerical codes, its usefulness depends on the resulting code performance. Computers used to run these codes are based on fast superscalar microprocessors (Pentium, Sparc, Alpha, etc.). Peak performance of these microprocessors is impressive, but exploiting a large part of it depends a lot on the structure of the source code and on the compiler [164, 362, 503]. Addressing performance issues in AD is not very different than for any other codes. However, because computing derivatives implies more computation and more memory, code optimization is a crucial step in the process of developing a derivative code.

In this chapter we overview code optimizations for superscalar processors. In §4.2, we introduce the main architectural features of superscalar microprocessors that impact performance. §4.3 gives an overview of source code optimization using an example, and §4.4 concludes by addressing some general issues for the optimization in the context of AD.

4.2 Superscalar Processors

Superscalar processor performance relies on a very high frequency (typically 1 gigahertz in 2001 corresponds to a processor cycle time of 1 nanosecond) and on executing multiple instructions in parallel (called *Instruction Level Parallelism,* ILP) [414, 462]. To achieve this, the internal architecture of superscalar microprocessors is based on the following features:

Multiple Pipelined Functional Units: The processor has multiple functional units that can run in parallel to execute several instructions per cycle

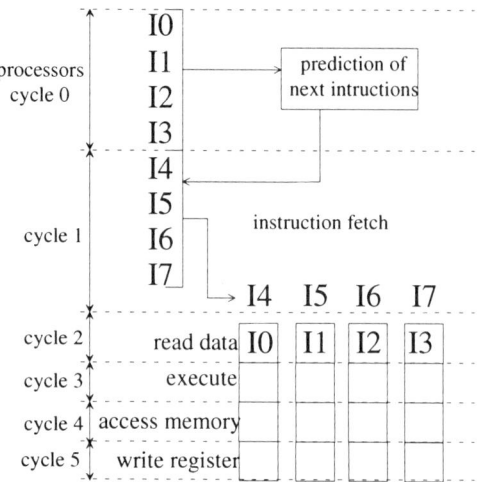

FIGURE 4.1. Multiple pipelined functional units

(typically an integer operation can be executed in parallel with a memory access and a floating point computation). Furthermore these functional units are pipelined. This divides the operation in a sequence of steps that can be performed in parallel. Scheduling instructions in the functional units is performed in an *out-of-order* or an *in-order* mode. Contrary to the *in-order*, in the *out-of-order* mode instructions are not always executed in the order specified by the program. The number of instructions executed per cycle depends on data hazards (i.e. waiting for operands to be available) and on the computation of the result of conditional branch instructions. This last computation is anticipated using a branch prediction mechanism.

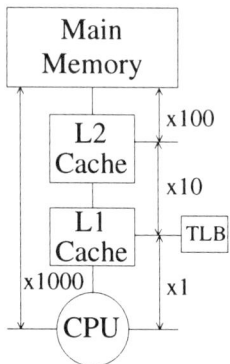

FIGURE 4.2. Memory hierarchy

Memory hierarchy: Main memory cycle time is typically hundreds of time slower than the CPU. To limit the slowdown due to memory accesses,

Original code	Derivative code

```
                              DO I = 1,PS
                                 DO J = 1,PS
                                    DO K = 1,3
DO I = 1,PS                                AD(J,K) = AD(J,K)
   DO J = 1,PS                                     +BD(I,J,K)+CD(J,I,K)
      A(J) = A(J)+B(I,J)+C(J,I)          ENDDO
   ENDDO                                 A(J) = A(J)+B(I,J)+C(J,I)
ENDDO                               ENDDO
                              ENDDO
```

FIGURE 4.3. Forward mode derivation with 3 derivatives.

a set of intermediate levels are added between the CPU unit and main memory; the closest to the CPU is the fastest, but also the smallest. The data or instructions are loaded from one level of the hierarchy to the next by blocks (sets of contiguous bytes in memory) to exploit the spatial locality of the memory accesses made by programs. In a classical configuration there are 2 levels, L1 and L2 of cache memories as shown on the figure. The penalty to load data from main memory tends to be equivalent to executing 1000 processor instructions. If the data is already in L2, it is one order of magnitude less. If the data is already in L1, the access can be made in a few CPU cycles.

The last component of the memory hierarchy is the *Translation Lookaside Buffer (TLB)* that is used to translate virtual memory page addresses into physical memory page addresses. This mechanism works similarly to cache memories. If a translation is not available in the TLB, it is reloaded from the main memory or the level 2 cache, resulting in hundreds or tens of cycles loss.

Efficiency of the memory hierarchy and ILP is directly related to the structure and behavior of the code. However memory hierarchy has to be considered first as its impact on performance is usually much more dramatic than instruction scheduling issues.

4.3 Code Optimizations

Code optimizations are code transformations that improve the execution time of a program for a given processor architecture while preserving its semantics [24]. They can be partitioned in the *classical*, or machine independent optimizations, and machine dependent optimizations. Most of them are performed by compilers.

Machine independent optimizations are performed during the process of translating the source code into machine code. They mainly consist in removing useless or redundant code introduced by the translation of the high level source code. These optimizations are not very relevant at the pro-

<table>
<tr><td>(a)</td><td>(b)</td></tr>
</table>

```
        (a)                          (b)
                            SUBROUTINE LOOP(PS)
SUBROUTINE LOOP(AD,BD,CD,N,PS)  INTEGER N,PS,I,J,K
REAL*8 AD(N,3),BD(N,N,3)     PARAMETER (N=1024)
REAL*8 CD(N,N,3)             COMMON/AR1/AD,BD
INTEGER N,PS,I,J,K           COMMON/AR2/CD,PAD
DO I = 1,PS                  REAL*8,TARGET :: AD(N,3),BD(N,N,3)
  DO J = 1,PS                REAL*8,TARGET :: CD(N,N,3)
    DO K = 1,3               REAL*8,POINTER :: PAD(:,:)
      AD(J,K) = AD(J,K)      DO I = 1,PS
        +BD(I,J,K)+CD(J,I,K)   DO J = 1,PS
    ENDDO                        DO K = 1,3
C     original statement           PAD(J,K) = PAD(J,K)
C     omitted for simplicity         +BD(I,J,K)+CD(J,I,K)
  ENDDO                          ENDDO
ENDDO                          ENDDO
END                          ENDDO
                             END
```

Execution Time for Loop (a) in seconds

Data Sizes (value of N)	256	1024	2500
Problem Sizes	160x200	10x800	1x2500
UltraSparc-200MHZ, f77 -O2	7.54	12.03	11.15
MIPS R10000-194MHZ, f77 -O2	29.72	33.97	16.95
UltraSparc-200MHZ, f77 -fast	1.13	3.19	2.24
MIPS R10000-194MHZ, f77 -Ofast=IP25	2.28	3.29	2.88

FIGURE 4.4. Loop example and corresponding performance

grammer level because they are usually efficiently performed by compilers.

Machine dependent optimizations are mainly performed at the source level and at the machine code level. They aim at exploiting possible re-ordering of the instructions to achieve a better throughput of the operations. Modifying the data layout is also to be considered. Typically such code transformations range from a few percent to **an order of magnitude in the execution time of a code**.

In the remainder we briefly present the main optimizations to be performed at the source code level using a simple example given in Figure 4.4. This example is extracted from the result of the derivation in forward[1] mode of the loop illustrated in Figure 4.3. Performance numbers are given for an UltraSparc at 200Mhz and a MIPS R10000 at 194Mhz. The compiler option -O2 was used, at which most of the target independent optimizations are performed. The execution times are given in seconds for three data sizes. To achieve a long enough execution time the routine, LOOP, is

---

[1] The reverse mode requires large I/O bandwidth and some prefetching techniques to ensure that the data are available when needed but does not introduce severe dependencies between statements. This problem is not addressed in this chapter.

called several times. This is given on the **Problem sizes** row where the first number is the number of repetition (for instance 160) and the second number the value of PS, the problem size (for instance 200). On this simple example for loop (a) in Figure 4.4, the compiler performs most of the optimizations needed (shown with the -fast option). If the compiler is unable to analyze the code, many optimizations are not performed by the compiler. This is the case for loop (b) in Figure 4.4[2].

4.3.1 Optimization examples

The first optimization that can be performed on the loop is to exchange the dimension of the arrays to get the dimension of size 3 to be the innermost one. As a consequence the stride of the data accesses becomes one with the benefit of improving the spatial locality and the TLB behavior. The result is shown below:

Array Dimension Interchange

	256	1024	2500
Data Sizes	256	1024	2500
Problem Sizes	160x200	10x800	1x2500
UltraSparc	4.52	5.72	5.40
MIPS	2.67	9.31	6.09

```
DO I = 1,PS
  DO J = 1,PS
    DO K = 1,3
      AD(K,J) = AD(K,J)
              + BD(K,I,J) + CD(K,J,I)
    ENDDO
  ENDDO
ENDDO
```

The second transformation that can be applied is to unroll the innermost loop. This improves both instruction parallelism and performance of the branch prediction mechanism. The impact on performance is limited as shown below:

Unrolling Loop K

	256	1024	2500
Data Sizes	256	1024	2500
Problem Sizes	160x200	10x800	1x2500
UltraSparc	4.15	5.30	4.82
MIPS	2.10	8.41	6.22

---

[2]The compilers must guarantee that the semantics of codes is not modified by program transformations. The pointer hides the effect of the assignment and so inhibits most optimizations. In the remainder we do not address the issue of the legality of the transformation used. The interested reader can refer to [24], where the -fast and -Ofast compiler options are ineffective.

```
DO I = 1, PS
 DO J = 1, PS
   AD(1,J) = AD(1,J) + ...
   AD(2,J) = AD(2,J) + ...
   AD(3,J) = AD(3,J) + ...
 ENDDO
ENDDO
```

The next optimization to be performed is called *array padding*. The array declarations are modified to avoid conflicts in the cache memories. When the leading dimension of an array is declared with a size that is a power of two, competition for the same cells in the cache memories is likely to occur. This is due to the structure of cache memories that uses the lower bits of the data addresses to compute where a data is stored.

Array Padding			
Data Sizes	**260**	**1030**	**2500**
Problem Sizes	**160x200**	**10x800**	**1x2500**
UltraSparc	3.63	4.29	4.82
MIPS	2.58	3.98	6.22

```
PARAMETER (N=1030)
REAL*8 AD(3,N)
REAL*8 BD(3,N,N)
REAL*8 CD(3,N,N)
```

Loop blocking can be used to limit performance degradation due to large data access stride or no exploitation of data reuses. In our example, the access BD(K,I,J) generates cache and TLB misses that directly impact performance. Blocking the loop improves performance by decomposing the iteration space in blocks. The size of the block must be chosen carefully so that the data used (in cache or TLB) can be cached between reuses. The transformation applied to the loop gives the following performance:

Loop Blocking (40x40)			
Data Sizes	**260**	**1030**	**2500**
Problem Sizes	**160x200**	**10x800**	**1x2500**
UltraSparc	2.44	2.59	2.58
MIPS	2.58	2.68	2.83

```
DO JJ = 1, PS, BS
 DO II = 1, PS, BS
   DO I = II, MIN(II+BS-1,PS)
     DO J = JJ, MIN(JJ+BS-1,PS)
       AD(1,J) = AD(1,J) + ...
       AD(2,J) = AD(2,J) + ...
       AD(3,J) = AD(3,J) + ...
     ENDDO
 ...
```

One last optimization that can be performed is unrolling the I loop inside the J loop. This transformation decreases the number of load and store operations (for array AD) executed by the loop. The resulting code is the following:

More Unrolling

Data Sizes	**260**	**1030**	**2500**
Problem Size	**160x200**	**10x800**	**1x2500**
UltraSparc	2.14	2.30	2.28
MIPS	2.25	2.44	2.48

```
DO I = II, MIN(II+BS-1,PS),4
  DO J = JJ, MIN(JJ+BS-1,PS)
    TMP = AD(1,J)
    TMP = TMP + BD(1,I,J) + CD(1,J,I)
    ...
    TMP = TMP + BD(1,I+3,J) + CD(1,J,I+3)
    AD(1,J) = TMP
  ...
  ENDDO
ENDDO
```

As it can be seen from the example, the effect of a source transformation depends on the target machine, the data size and the compiler used. This globally makes the optimization process difficult and time consuming. Furthermore, the performance may not be stable from one computer to another.

4.4   Conclusion

Most program optimizations complicate the structure of the code. This makes it even more difficult for the AD process to produce efficient and clean outputs. Optimization must be performed on the final code as adding data structures and computation might cancel the effect of most code transformations performed earlier. As a consequence, producing a derivative code using AD can be divided in the following steps: 1) code cleanup, 2) automatic differentiation, and 3) performance optimization.

Producing optimized derived code for the main libraries, such as BLAS, should be addressed to limit the issue of code optimization. Approaches such as Atlas [501] can potentially be applied. Finally, it is possible to semi-automate the optimization process of a given application. Flexible tools such as TSF [88] might be able to help with such a task.

5

Present and Future Scientific Computation Environments

Steve Hague and Uwe Naumann

ABSTRACT Problem Solving Environments and Environments for Scientific Computing play an increasingly important role in the modelling, simulation, and optimization of real-world applications. In this chapter we discuss some of the experiences of NAG Ltd. in several European projects aiming towards the development of such tools. It will also note some of the encouraging developments, particularly in the area of interface standards that might make PSE construction a more effective and realistic prospect in the future.

5.1   The Big Picture

Mathematical models of real-world applications have grown more complex over the past decades. The availability of high speed information technology has not only lead to a speedup in the solution of most problems, it has also caused a remarkable scale-up of the problems that scientists and engineers are willing and able to consider. Large numbers of often highly specialized software applications have been developed, each application seeking to address a major computational problem-solving task, or class of tasks. For most of these applications, the underlying philosophies, platforms and interfaces vary drastically. These variabtions make it very difficult to achieve desirable properties such as portability, extensibility, and ease-of-use. The

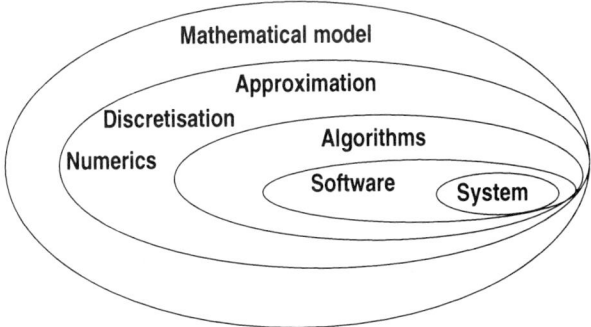

FIGURE 5.1. Applications

intention of this chapter is to present a (subjective) view on how software should be developed to support both application programmers and end-users from various fields in their attempts to understand, to simulate, and to optimize real-world processes.

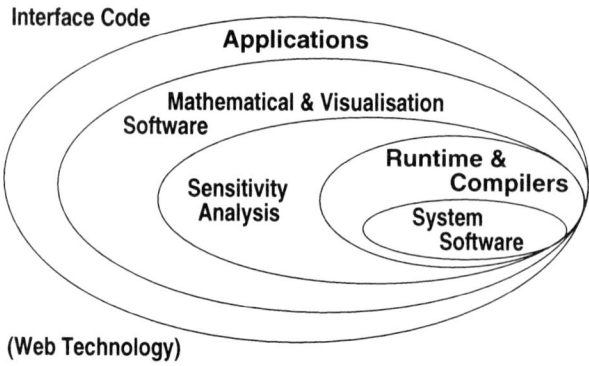

FIGURE 5.2. Software

The typical stages of the modern applications development process are illustrated by Figure 5.1 and Figure 5.2. Usually, this process requires contributions from experts from several fields. Due to their structure and complexity, most computational problems have to be approached using numerical methods which result in discretizations and approximations of the underlying continuous model. For example, algorithms are developed to generate meshes, for solving systems of linear or nonlinear equations, or for interpolating intermediate function values. These algorithms are usually implemented using some high-level programming language leading to highly complex software packages running on increasingly powerful computers. The development of mathematical and visualization software represents one of the major challenges faced by both industrial and academic experts. Furthermore, application code enhancement and transformation tools, such as compilers for scientific computing, automatic differentiation packages, and software for verified computation based on interval arithmetic, are gaining importance. Considerable effort has been put into the development of Problem Solving Environments / Scientific Computation Environments (PSE/SCE) with the aim to formalize and simplify both the modeling and the algorithmic aspects of the application development process. Simulation and optimization as well as visualization are addressed by various existing PSEs/SCEs. Different approaches lead to software packages typically based on some combination of symbolic and numeric computation, such as Maple, Mathematica, Matlab and its Toolboxes, or with a more statistical emphasis such as S-Plus and SPSS. These could be described as general or multi-purpose computational packages that can solve a variety of problems within their scope, sometimes with the aid of a supplementary package that has a particular theme (e.g., a toolbox for control engineers or one for financial analysts etc.). In addition to these wide-ranging packages, here are also a substantial number of systems that have a still broad but more field-specific emphasis. Examples include packages solving partial differential equations (e.g., Ellpack), for computational fluid dynamics (e.g., KIVA-3 and MIKE), or simulation packages (e.g., NASTRAN, PAM-CRASH). Yet another form of package available may not in itself provide the required computation tools directly but instead offer an application-building framework, in which such tools can be embedded. Such a framework would typically also provide strong support for visualisation, since that is now regarded as a "de rigeur" for any self-respecting modern computational application. NAG's IRIS Explorer [397] is one example of such visually-capable, application-building as a framework and has been used by NAG and a number of other organisations for PSE/SCE construction activities.

Considering the wide variety of computational problems now being solved (whether through the use of commercial packages such as those mentioned above or indeed of the "roll-your-own" type), the issue of model sensitivity is or should be of central importance. In the step from simulation to optimization, sensitivities of model outputs with respect to certain parameters have to be computed in most cases. Automatic Differentiation (AD) is able to deliver first and higher order derivatives with machine accuracy. Therefore AD has to deal with a large diversity of both current and potential applications as well as coping with a variety of data types and programming languages. This makes AD technology very complex. The underlying algorithms have to be both effective and efficient. Interfaces between the internal methods as well as intuitive user interfaces are essential. All these requirements suggest that, ideally, a PSE should be used to incorporate AD. This is the main reason why we regard AD as a challenging test case for the development of a PSE.

5.2   Looking at the Present

At the beginning of the 20th century, a typical inventor was a person who made an ingenious arrangement of wheels, levers and springs, and then claimed that the resulting mechanical contraption constituted an advance of civilization. In some cases, that claim would have some validity, and in other cases, it would not. Now, a hundred years later, it could be said that a person who is an application package builder is in fact a modern form of inventor, concocting an ingenious arrangement of computer code. The builder probably believes his or her construction to be intuitively obvious in its design and superior in its performance to any predecessor. There is no doubt that some inventions, both past and present, deserve all the accolades that they receive. However, as in the past, many present-day software "inventions" are at least partial re-inventions "of the wheel," in the sense that they unnecessarily replicate the function of existing code (probably of a more tried and therefore reliable pedigree), or they introduce new interface conventions of one form or another, when existing de facto or official standards for such interfaces would suffice just as well.

Against that background of a continuing "cottage industry" approach towards software package construction generally, and PSE/SCE construction in particular, NAG has increasingly invested effort in recent years in developing IRIS Explorer which, as described above, provides both an application building environment and powerful visualisation capabilities. It has been used as a framework for three European projects in which NAG has recently been involved, namely STABLE (EU 22832), DECISION (EU 25058), and JULIUS (EU 25050). All these projects were/are aiming towards the construction of PSEs. In other words, their primary objective is to build collaborative software development environments with the following main characteristics:

- use of the visual programming paradigm to provide enhanced ease-of-use and flexibility

- facilities for new approaches to visualizing data

- facilities for the development of end-user applications in 'canned' form

- open, extensible architecture for incorporating user-developed code

- gateways to existing external systems

- environment suitable for the research, implementation and promulgation of new algorithms

- components easily distributable over multiple processors in a heterogeneous environment

The main outcome of the STABLE project is a modern Statistical Application Building Environment. It provides statisticians with a 'next generation' statistical computing environment within which they can conduct all their statistical computations. The interface provides a standard to encourage cooperation between developers of statistical algorithms and methodology. STABLE also provides statisticians and statistical application builders with facilities for the prototyping and the building of customized applications for specific tasks or user groups thus making a wider range of statistical methodology available to non-specialist users of statistics in their problem domains.

The aim of the DECISION project was to develop an optimization platform which integrates new and existing optimization technology in a PSE suitable for tackling complex design optimization problems. In addition to

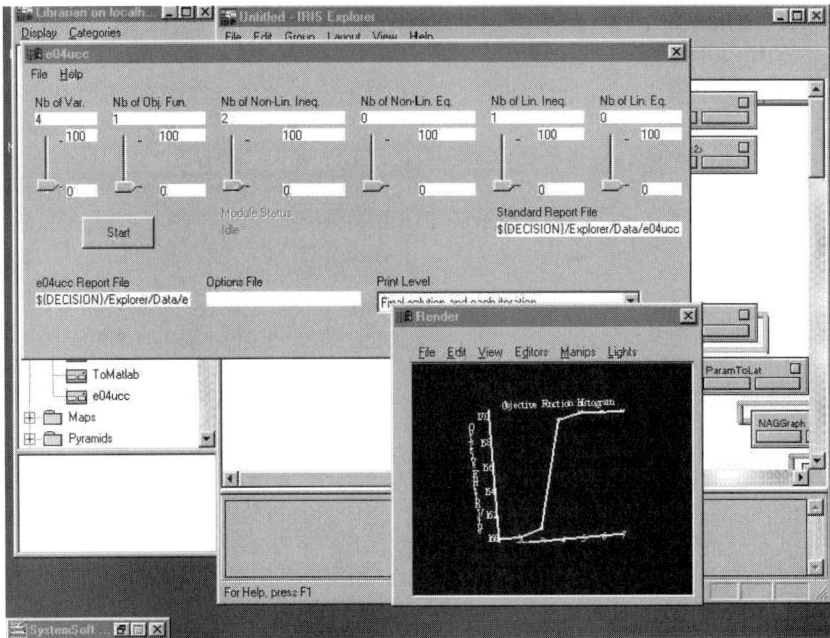

FIGURE 5.3. IRIS Explorer

this, it was to provide a database incorporating details of test problems from the end-user partners, specialized visualization software, and guidelines for optimization procedures. Complex design problems coming from three main industrial sectors - Aeronautics, Smart Materials and Forest Machinery - have been chosen to validate the efficiency of the design optimization system and to provide quality with background of a wide range of optimized results. The AD tool Odyssée [186] proved to be a reliable, robust and efficient way to generate the derivative code required by most of the optimization algorithms.

The third project, JULIUS, sought to create an integrated environment for multidisciplinary engineering simulation, with applications in the aerospace, automotive and manufacturing sectors. Its major outcomes were to include CAD input and repair, a range of 3D mesh generation capabilities, facilities for integrating numerical simulation packages, data extraction and visualization, data base integration, parallel tools, and resource scheduling. The complexity of this highly ambitious project led to extensive discussions between the partners regarding architectural issues, the definition of data types, and also the question concerning standard versus private communication protocols. We await with interest to see what commercially exploitable software tools and packages emerge from this project.

There are many interesting technical, commercial, and organisational issues that arise from any set of international collaborative projects lasting several years. Consideration of such issues lies beyond the scope of this chapter. However, let us restrict our attention in this context to the particular question: was IRIS Explorer as a generally available, supported package, an effective base for each of the three PSE-building efforts mentioned above? Or, would the project participants been better served by building their own frameworks, as has so often been the case with other PSE construction projects, both in Europe, North America and elsewhere?

Here, in brief, are answers to that question:

- for STABLE, the answer is generally yes; it proved possible within the framework provided to resolve the complex issue of interfacing statistical type data types and their visualisation counterparts in a sufficiently effective way.

- for DECISION, the answer is also positive; it took some time to decide how best to address the equivalent interface issue (between optimisation and visualisation objects in this case) but an effective solution was found with the framework provided, and a suitably designed user interface also constructed with the standard features of the package.

- In the case of JULIUS, IRIS Explorer's role at some stages in the evolving overall architecture was that of the visualisation engine.

However, There were some distinctive features of the size and type of simulation problems being solved that mitigated against the use of the type of flexible (but therefore generalised) framework provided by IRIS Explorer. For example, the particular data structures of interest (grid meshes) were very large and sparse, and so required specialised handling for the sake of high performance.

5.3   Looking to the Future

If one would like to give a definition of what is actually meant by a PSE, one might quote Ford and Iles [197] who said that a PSE is *an integrated, multitasking system that supports and assists the user in the solution of a given class of problems....* This leads to a variety of more practically oriented questions about how a PSE should actually be designed and implemented. Should it be a package of applications, an expert system, or some sort of operating system? Should it cover specific domains or rather be a more diffuse, multi-domain system? Do we require the user of such a system to be an expert, both in the underlying subject and in working with PSEs in general? Further questions to be answered would have to address issues such as loose coupling versus tight integration, extensible versus restricted use, the scalability of resource requirements, stand-alone use versus use via a network, blathering (that is, excessively verbose) or cryptic dialogue, etc. From an organizational point of view, the focus could be put on an operation-centered, a data-centered, or a display-centered approach. Probably, an answer that would be satisfactory in all situations can not be given. The most suitable approach will depend strongly on the given class of problems for which PSE is meant.

In the same way as STABLE addressed statistical application development or DECISION dealt with optimization problems, a PSE for AD would be likely to simplify the process of computing derivatives as well as the design and implementation of new, innovative algorithms. The use of different programming languages for the implementation of real-world applications, the large number of platforms available, and varying implementation philosophies would make a corresponding project highly ambitious. Similarly to optimization, AD comes as a large collection of different methods and strategies which can be combined to solve a given problem. Building on standard interfaces and communication protocols, the algorithms, once designed and implemented by some developer, could be used and extended by others. So, apart from potentially forming an important component of PSEs/SCEs, the further development of AD itself could derive considerable benefit from a specific application code enhancement and transformation PSE. First attempts to launch a corresponding European project [131] have shown a strong interest among both AD application programmers and end-users of AD. This should encourage further steps into this direction.

The last few years have led to an increased availability of components for common computational, presentational, data management and housekeeping tasks including modules for, e.g. numerical and symbolic computations, visualization, documentation, database application, and version control. Following the Plug & Play paradigm, the objective is to share these components among a variety of applications by simply plugging them into the PSE at hand. Naturally, this requires well-defined and standardized interfaces, protocols and data structures. Numerous attempts have been made to achieve this goal leading to "glue" technologies in the shape of distributed objects (e.g., CORBA, DCOM), scripting languages (e.g., Perl, Python), frameworks such as IRIS Explorer, Java, and a variety of GUI builders.

Math-friendly communication technologies are emerging to support the development, use and later re-use of PSEs/SCEs. Examples include XML, a mechanism for defining the structure of objects on the web, MathML, the first XML application with the *aim of facilitating use and re-use of mathematical content on the Web*, and OpenMath, an *extensible framework for semantically rich representation and exchange of mathematical objects on the Internet*.

5.4 Conclusions

Some encouraging progress has been made regarding the development of PSEs/SCEs, so far. This is also due to a more and more connected world.

Access to both remote resources and remote expertise has become easier. The general awareness of the various problems faced by scientists and engineers has grown. An increased emergence/availability of components, interfacing technologies, and frameworks has contributed to a wider implementation of the Plug & Play paradigm. Last but not least, the power/price ratio for computer technology has increased dramatically over the last decade leading to a new dimension of real-world problems that can be approached.

A variety of highly relevant questions arises from a deeper look into PSEs/SCEs. First, why not let *each flower bloom*? Undoubtedly, the development of PSEs/SCEs yields some scope of experimentation. However, too many parallel developments would not only lead to a mis-use of resources while trapping available expertise. It would also become very unlikely that effective standards will emerge. Looking at existing computational packages, one may well ask if they actually represent convincing evidence for the building of effective PSEs from re-usable software components. Considering the issues of extensibility and generalization, just how far can their current design be stretched?

Large-scale collaborative projects involving both developers and end-users are most likely to result in a PSE which satisfies the requirements outlined above. Funding is, of course, a major issue. The active involvement of end-users in the design and development process is another. Even more complicated appears the question of how to encourage the major software developers to adopt a component-based approach. The latter is surely crucial for the success of PSEs/SCEs.

Finally, the lessons learned from current/recent PSE projects will have to be applied to both future PSE construction in scientific research and the next generation of commercially supported computational environments. The question is: How and, indeed, how closely are the two related. This challenge is not easy but the price is great: Making the best algorithms and technology available in the most useful form to users.

Part II

Parameter Identification and Least Squares

6

A Case Study of Computational Differentiation Applied to Neutron Scattering

Christian H. Bischof, H. Martin Bücker and Dieter an Mey

ABSTRACT In a neutron scattering application, an unconstrained non-linear minimization problem is used for the fitting of model parameters to experimental data. Automatic differentiation enables, in a completely mechanical fashion, algorithmic changes by switching from a quasi-Newton method, where first order derivatives are approximated by finite differences, to a modified Gauss–Newton method using exact first order derivatives. Compared to the original code, the code generated by this black box approach produces reliable results rather than results of dubious quality. This approach also is faster in terms of execution time. Its performance is improved further by replacing the most time-consuming subroutine involved in the derivative evaluation by a surprisingly simple, hand-coded implementation of the corresponding analytic expression.

6.1 Introduction

Neutron scattering is an essential research tool used to understand precise details of the positions, motions and magnetic properties of atoms in solids and liquids. Neutron scattering plays a fundamental role in materials science, chemistry and condensed matter physics, and is also important in understanding aspects of large biological molecules and engineering materials. At the Institute of Physical Chemistry, Technical University of Aachen, neutron scattering experiments are carried out to investigate the molecular dynamics of isotropic liquids [416]. In these experiments, neutrons hit the atomic nuclei of the liquid molecules, and their resulting change of momentum and energy is measured.

A computer program is used to fit the data obtained from the experiments with a parameterized theoretical model used to predict certain properties of the liquid. Let e_j be the experimental scattering function representing the j-th data set obtained from the experiments. The experimental scattering function is used to count the number of neutrons of a particular type. Let Δe_j represent the associated error in measurement and let k

denote the total number of data sets obtained from the measurements. A typical value for the number of data sets is $k \approx 10,000$. Furthermore, let the theoretical scattering function $t_j(s)$ specify the corresponding number of neutrons predicted by the model that is characterized by a free real parameter s. When fitting the experimental data to the theoretical model, the underlying problem is

$$\text{Find} \quad s \in \mathbb{R} \quad \text{to minimize} \quad \sum_{j=1}^{k} \left(f_j(s) \right)^2, \tag{6.1}$$

where

$$f_j(s) := \frac{e_j - t_j(s)}{\Delta e_j}. \tag{6.2}$$

From a numerical point of view, (6.1) corresponds to finding an unconstrained minimum of a sum of squares of nonlinear functions $f_j(s)$. Effective minimization techniques for such problems typically make use of derivative information. In the neutron scattering application, however, the theoretical scattering function $t_j(s)$ appearing in (6.2) is given in the form of a computer program written in Fortran 77 with approximately 3,000 lines of code so that exact derivatives are not immediately available. In §6.2, automatic differentiation is used to obtain, in a completely mechanical fashion, exact first order derivatives of $t_j(s)$ enabling the use of a more sophisticated minimization technique. This black box approach is improved further in §6.3 where a time-consuming subroutine generated by automatic differentiation is replaced by a surprisingly simple, hand-coded implementation of the corresponding analytic expression.

6.2   Enabling Algorithmic Changes

In the original program, the nonlinear least-squares problem (6.1) is solved by calling the routine E04FDF of the NAG Fortran Library [396]. This routine implements a variable metric method with difference gradient, called quasi-Newton method hereafter, where the user has to supply a routine that evaluates $f_j(s)$ at a given point s. In the routine E04FDF, first order derivatives of f_j with respect to s are approximated by finite differences.

Automatic differentiation is used to generate a routine that computes accurate first order derivatives of $f_j(s)$ with respect to s. Throughout this note, the program variables associated with the result f_j and the input s are called dependent and independent, respectively. The availability of first order derivatives f_j' enables the use of a modified Gauss–Newton method rather than a quasi-Newton method. Hence, the solution of (6.1) is obtained by replacing the original call to E04FDF by a call to the corresponding routine E04GEF that takes as input a routine evaluating not only f_j but also f_j'. This routine is generated here by the forward mode of the ADI-FOR [77, 79] preprocessor. The two optimization algorithms implemented in E04FDF and E04GEF are described in [220].

For all numerical experiments in this chapter, we consider a run of the neutron scattering code where $k = 12,237$ sets of data are given. Here, switching from the quasi-Newton method to the Gauss–Newton method has the following significant effect. In the original code, the quasi-Newton method terminates with a warning indicating that the produced results are of dubious quality, whereas there is a normal termination using the Gauss–Newton method. However, the values of the sum in (6.1) computed by the two approaches differ only in the least significant digit when using double precision. The main advantage of automatic differentiation in this application is to avoid problems and errors as well as speeding up the computation. On a Hewlett Packard V-class server with 240 MHz PA8200 processors that is used throughout this study, the total execution time of the original code is 586 seconds using 13 evaluations of f_j, while the modified code takes 498 seconds involving 6 evaluations of the pair (f_j, f_j'). The execution time for evaluating f_j in the original code is 38.7 seconds. Evaluating (f_j, f_j') takes 69.2 seconds in the modified code.

6.3   Getting Improved Performance

Given the fact that absolutely no knowledge of the application code is necessary, the performance of the code generated by automatic differentiation is remarkably high. However, it is often possible to improve the performance further by exploiting some knowledge about the underlying application. Fairly simple changes may lead to significant performance improvements. Here, the combination of techniques of automatic differentiation and high-level knowledge is called *computational differentiation.* In this section, it is shown how the performance of the code using the Gauss–Newton method can be improved further exploiting mathematical knowledge.

A run time analysis shows that the total execution time of the Gauss–Newton code is dominated by a single routine of the ADIFOR-generated code whose name is g_w. An advantage of the source transformation approach implemented in ADIFOR or similar tools [72] is the simplicity of the generated code. Each subroutine is differentiated separately, thus enabling a human inspector to follow the action of the differentiated code. Hence, the subroutine of the original code from which g_w is deduced is easily determined as

subroutine w(u, v, x, y).

Its Fortran77-code consists of 53 lines without any subroutine calls and, though very short, is difficult to understand—not only because of the occurrence of six **goto**-statements. As the author of the original code pointed out to us, this routine takes as input a complex number

$$z = x + iy \tag{6.3}$$

and computes as output a complex number

$$w = u + iv, \tag{6.4}$$

where $i = \sqrt{-1}$ is the imaginary unit and $x, y, u, v \in \mathbb{R}$. In the parameter list of w, the variables u, v, x, and y are used to store $\mathrm{Re}(w)$, $\mathrm{Im}(w)$, $\mathrm{Re}(z)$, and $\mathrm{Im}(z)$, respectively. Furthermore, the subroutine w uses numerical quadrature to compute an approximation to the underlying function

$$w(z) = e^{-z^2} \cdot \mathrm{erfc}(-iz), \qquad \mathrm{Im}(z) > 0, \tag{6.5}$$

where

$$\mathrm{erfc}(z) = 1 - \frac{2}{\sqrt{\pi}} \int_0^z e^{-t^2} \mathrm{d}t.$$

Applying ADIFOR to the subroutine w as described in the previous section gives a subroutine g_w evaluating the function w together with its derivatives. In general, a human inspector does not know in advance with respect to which input variables derivatives of w are computed in g_w. The reason is that the user specifies $f_j(s)$ as the function being differentiated and the function w is used in a non-transparent fashion within the evaluation of $f_j(s)$. However, ADIFOR carries out a dependency analysis determining active and passive variables. A variable is called *active* if its value may depend on the value of an independent variable and, in addition, its value impacts the value of a dependent variable. Otherwise a variable is called *passive*. This dependency analysis is triggered by the specification of the independent and dependent variables when invoking ADIFOR.

Thanks to the transparent source transformation approach, it is possible to recognize active and passive variables from the parameter list of the ADIFOR-generated subroutine g_w and to exchange the subroutine g_w by a hand-coded subroutine implementing the corresponding analytic expressions. More precisely, the parameter list of the subroutine generated automatically by ADIFOR as the differentiated version of the original routine w is given by

subroutine g_w(u, g_u, v, x, y, g_y) .

Here, in addition to the variables given in the parameter list of w, two new variables, g_u and g_y appear, representing derivative objects associated with the output variable u and the input variable y, respectively. That is, u and y are active variables whereas v and x are passive. Since $u = \mathrm{Re}(w)$ and $y = \mathrm{Im}(z)$, the inspection of the parameter list shows that the rest of the differentiated code needs the information

$$\frac{\partial \, \mathrm{Re}(w)}{\partial \, \mathrm{Im}(z)}. \tag{6.6}$$

This information can be supplied by a hand-coded subroutine with the same functionality and, preferably, the same parameter list as g_w. To this end, suppose that z depends on a scalar parameter and let ∇ denote differentiation with respect to this parameter. Taking into account the relation

$$\frac{d \, \mathrm{erfc}(z)}{d \, z} = -\frac{2}{\sqrt{\pi}} \cdot e^{-z^2},$$

the differentiation of (6.5) results in

$$\nabla w = \mathrm{erfc}(-iz) \cdot \nabla e^{-z^2} + e^{-z^2} \cdot \nabla \, \mathrm{erfc}(-iz)$$

$$= -2z \, \mathrm{erfc}(-iz) \cdot e^{-z^2} \nabla z + e^{-z^2} \cdot \frac{2i}{\sqrt{\pi}} e^{z^2} \nabla z$$

$$= \left(-2wz + \frac{2i}{\sqrt{\pi}} \right) \nabla z. \tag{6.7}$$

Furthermore, from (6.3) and (6.4) the equations

$$\nabla z = \nabla x + i \nabla y \tag{6.8}$$

$$\nabla w = \nabla u + i \nabla v \tag{6.9}$$

follow. Inserting (6.3), (6.4), and (6.8) into (6.7) yields

$$\nabla w = -2(ux - vy)\nabla x + 2(uy + vx)\nabla y - \frac{2}{\sqrt{\pi}} \nabla y$$

$$+ i \left[-2(ux - vy)\nabla y - 2(uy + vx)\nabla x + \frac{2}{\sqrt{\pi}} \nabla x \right]. \tag{6.10}$$

Now, recall that the above mentioned dependency analysis shows that the variable x is a passive variable, and therefore $\nabla x = 0$ in (6.10). Hence, together with (6.9), we find the information (6.6) we are looking for, namely,

$$\nabla u = \left(2(uy + vx) - \frac{2}{\sqrt{\pi}} \right) \nabla y. \tag{6.11}$$

A hand-coded subroutine denoted by h_w with the same functionality and parameter list as the ADIFOR-generated subroutine g_w is now easily derived from (6.11). Its main ingredient is given by

```
g_u = ( 2.0d0*(u*y + v*x) - 1.12837916709551d0 )*g_y
```

where 1.128... approximates $2/\sqrt{\pi}$. This line of code needs as input correct values for u and v, the output of the function w. Therefore, the hand-coded subroutine h_w is obtained from appending the above line of code to the code of the routine w and adapting the parameter list correspondingly.

Comparing g_w and h_w in isolation from the rest of the code, the ratio of their execution times is $T_{g\_w}/T_{h\_w} = 6.9$, and the numerical difference of their output is $\|\nabla u_{g\_w} - \nabla u_{h\_w}\|_\infty = 6 \cdot 10^{-8}$ using double precision.

Summing up, we employed automatic differentiation to generate exact derivatives for roughly 3,000 lines of code and our mathematical insight to generate derivative code for the 53-line subroutine w that dominated the execution time. The resulting change of replacing g_w by h_w in the derivative code reduces the total execution time from 498 seconds to 333 seconds in a "global" view. The execution time to evaluate (f_j, f'_j) is reduced from 69.2 seconds to 41.5 seconds.

6.4 Concluding Remarks

In its original form, the investigated neutron scattering code produces numerically unreliable results throughout the course of the underlying minimization process. Exact first order derivatives, provided in a completely mechanical fashion by automatic differentiation, are used to generate reliable numerical results by switching from a quasi-Newton method to a modified Gauss–Newton method, which runs faster than the original code. In addition, the execution time of the modified code is significantly decreased by replacing the most time-consuming subroutine by a hand-coded implementation of the corresponding analytic expression.

Preliminary tests which involve a different theoretical scattering function $\tilde{f}_j(s, t)$ with two independent variables consisting of 5,000 lines of code indicate that a modified code obtained in a similar way by combining automatic differentiation and mathematical insight is also reliable and faster than the original code. This will be investigated further in an ongoing work.

Acknowledgments: Jürgen Ette of the Institute of Physical Chemistry, Technical University of Aachen, wrote the underlying computer code. This work was partially carried out while the second author was visiting the Mathematics and Computer Science Division, Argonne National Laboratory, supported by the Mathematical, Information, and Computational Sciences Division subprogram of the Office of Advanced Scientific Computing Research, U.S. Department of Energy, under Contract W-31-109-Eng-38.

7

Odyssée versus Hand Differentiation of a Terrain Modeling Application

Bernard Cappelaere, David Elizondo and Christèle Faure

ABSTRACT A comparison is made between differentiation alternatives for a terrain modeling problem, a sample application where strong nonlinearities are solved iteratively. Investigated methods include automatic differentiation (AD) with the Odyssée software (forward and reverse modes) and manual differentiation (MD) using the model's adjoint equations. The comparison mainly focuses on accuracy and computing efficiency, as well as on development effort. While AD ensures perfect consistency between the computer model and its derivative at a low development cost, MD shows significantly lesser computing costs. We discuss the perturbation method as well as hybrid strategies that combine advantages of AD and MD.

7.1 Introduction

Modeling natural systems means being able to fit complex models to large amounts of observed, diverse data. This leads to large inverse modeling problems calling for efficient numerical techniques such as quasi-Newton optimization methods [157]. Given a function expressing a diagnosis of the physical model's behavior, these methods require estimating the function's gradient with respect to all or part of the variables that control its behavior. We also need to differentiate sets of complex equations for the efficient solution of highly nonlinear direct problems, a frequent occurrence in many geoscience fields.

The intent of the work reported here is to make a comparison of possible alternative methods for differentiation of complex nonlinear models, whose nonlinearities require iterative solution procedures. An existing, time-independent terrain modeling application is used for this purpose. Methods of interest are code-based methods (differentiation of the original code, instruction by instruction, performed here by the Odyssée AD software) and equation-based methods (differentiation of the physical model's equations, with hand coding). Comparisons are also made with the perturbation method (finite-difference approximation through multiple runs of the direct model's code) as well as with intermediate approaches combining manual equation-based and automatic code-based features.

7.2   Terrain Modeling Application

The sample application is a terrain (ground surface) modeling method based on the numerical solution of a 2D nonlinear elliptic partial differential equation. This method is developed with the purpose of combining precise interpolation capabilities with a high degree of flexibility through spatially distributed control over the surface. The idea is to merge information of different natures, such as actual quantitative point observations of the variable being modeled and information on surrogate quantitative or qualitative variables that are related with the target variable. For instance, in geographical areas where topographic data are scarce, such as in developing countries, it should be possible to produce digital elevation models that are more reliable and more meaningful for various applications than those currently available, if these data were combined with geomorphologic information that reflect the characteristics of dominant landform patterns prevailing in the area. Mapping of geomorphologic features can be turned into parameter zonation for the surfacing model. Applying the model then becomes a question of data assimilation into the modeling structure, for identification of optimum parameters. Figure 7.1 is an example produced for a sahelian area of Niger, Africa.

The model is based on a nonlinear diffusion equation:

$$\text{div}\ (D(p_i, Z)\|\nabla Z\|^{n-1}\nabla Z) + S(p_i, Z) = 0 \tag{7.1}$$

where Z is the altitude, $n \geq 1$ is a real number, and D and S are two functions of m parameters $\{p_i, i=1, m\}$ which are space-dependent through the user-defined geomorphologic zonation. Inner and outer boundary conditions are applied, either as Dirichlet or Neumann conditions. The nonlinearity is due to D being a function of Z and to the exponentiation; it may be very strong since there is no upper bound on n.

Discretization of the PDE 7.1, using a 9-point finite-difference scheme, produces a nonlinear system of equations $E_M(Z, p)=0$, where p designates the parameter set. This nonlinear discrete system makes up our direct model M, implemented in a Fortran 77 program through a Newton-Raphson (NR) iterative solution procedure. The system is linearized by a first-order Taylor expansion around any given state Z_k: $R(Z_k+dZ) = R(Z_k) + J(Z_k) \bullet dZ$ where $R=E_M(Z, p)$ is the vector of model residuals at state Z, $J(Z_k)=\partial E_M/\partial Z(Z_k, p)$ is the system's Jacobian matrix with respect to the state variable at state Z_k, dZ is a vector of Z increments, and $\bullet$ represents the matrix-vector product. At each NR iteration the tangential linear model

$$J(Z_k) \bullet dZ = -R(Z_k) \tag{7.2}$$

FIGURE 7.1. Shaded view of a terrain modeling example (45×54 km^2) in Niger (Africa), produced with Equation 7.1 -based model.

is solved using the Lapack library's large sparse system solver [12]. The final solution is denoted by $Z_p = M(p)$. Because the Lapack routines raise some problems for AD, they were replaced by subroutines from the Numerical Recipes library [424] for the purpose of the present exercise. All results presented here are therefore obtained with the latter library.

The terrain modeling power of Equation 7.1 largely depends on the possibility of efficient data assimilation into the model, i.e. of identifying the model's control parameters (p_i) (the exponent parameter n and even some of the boundary conditions could also be included, although not considered here), given Z_{obs} a set of observations of the state variable. It is expressed through the introduction of a diagnostic criterion C that is an explicit function of the parameters p and of the associated discrete 2D distribution Z_p: $C = F(Z_p, p) = F_z(Z_p, Z_{obs}) + F_p(p)$, where the F_z component measures the agreement between modeled and observed values of Z (sum of squared differences), and F_p is an optional regularization term to constrain parameter values. Inverse modeling means finding a value of p that minimizes C. For this study, D and S are parameterized as spatially-uniform, piecewise-linear functions of Z; the case of a vector p of 8 parameters is used here. INRIA's m1qn3 library [216] provides the gradient-based quasi-Newton optimizer (a variable-storage BFGS algorithm) that iteratively calls for estimation of the value C and of its gradient vector $\nabla_p C \equiv {}^t dC/dp$ in the inverse problem, parameter space.

7.3   Differentiation Methods

Three distinct differentiated codes, MD, tAD and cAD, produced either by manual (MD) or automatic differentiation (AD), are compared for accuracy and computing efficiency. Additionally, we discuss the perturbation technique as well as hybrid approaches mixing AD and MD. All codes are run in double precision.

MD uses the equation-based, adjoint-state method [349]. Given the function $C = F(Z, p)$, where Z and p are linked by the simulation model M (discrete equations $E_M(Z, p) = 0$), the gradient $\nabla_p C$ is computed as $\nabla_p C = {}^t \partial F_p(p)/\partial p + {}^t \partial E_M/\partial p \bullet \gamma$, where the adjoint variable γ is the solution of the adjoint linear system: ${}^t \partial E_M/\partial Z \bullet \gamma = -{}^t \partial F_z(Z, Z_{obs})/\partial Z$. Hence, MD requires a number of differentiation operations. However the adjoint matrix is the transpose of the model Jacobian J, already coded to solve the original direct model (Equation 7.2). This approach is referred to as the *Two-phase method* in [238]. Whereas several NR iterations are needed for the direct model, only one linear system production/solution is subsequently needed to compute the gradient with the adjoint method (at the final state Z_p). This is a source of great efficiency for this manually-built algorithm, which, unlike a code-based method, does not merely follow the direct model's full computational path.

Automatic differentiation with respect to p, was performed by Odyssée 1.6 over the full collection of routines needed to evaluate C, producing tAD in forward mode and cAD in reverse mode, respectively. Applying Odyssée's forward mode to our pre-existing code was quite straightforward, with only minor modifications needed. Some more significant manual adaptation was required for the reverse mode, including the handling of extra memory allocation parameters that the "syntax reversal" differentiation algorithm introduces into the cAD code. Since grossly oversized parameter values may result in excessive memory requirements, some attention must be brought to these additional parameters. Although limited in volume, these tasks complicate somewhat the job of model maintenance and operation.

7.4   Accuracy

For most test cases the computed gradients are the same for the three codes. However slight differences have been observed for a few cases. This is illustrated in Table 7.1 showing three sample gradient components in the three differentiation modes, for a simple test case of $m=8$ input parameters on a 10×10 space grid with Dirichlet boundary conditions, and $n=1$.

TABLE 7.1. Gradient of C in 3 representative directions i.

i	MD	tAD	cAD
2	539.25487620235	540.31145861541	540.31145861799
6	-452330.15035587	-453111.00445059	-453111.00444789
7	8.1106014306548	8.1168627075904	8.1168627077914

It can be seen that the derivatives computed by tAD and cAD have between 10 and 11 common digits, against only between 2 and 4 when compared with MD. A natural question then arises: is the difference between MD and AD due to numerical errors (caused either by the numerical schemes used, or by computer rounding) or to some programming mistake in any of the codes. To check which of these codes computes the most exact derivatives, we compare the Taylor test curves obtained from C and its gradients computed by MD and AD. The Taylor test measures the discrepancy between an approximate gradient of F computed by finite differences for a perturbation ϵ denoted by $FD_F(\epsilon)$ and the exact gradient ∇_F : $\mathrm{taylor}(\epsilon) = \frac{FD_F(\epsilon)}{\nabla_F} - 1$. As the perturbation magnitude ϵ tends to zero, the finite difference gets more precise and the Taylor test value theoretically tends to zero if the gradient is exact. In practice when ϵ becomes too small the test degrades due to finite machine precision. The Taylor test curve is drawn by evaluating the Taylor test for decreasing perturbation magnitude. A factor $F/\nabla F$ is applied to ϵ to make it non-dimensional. We

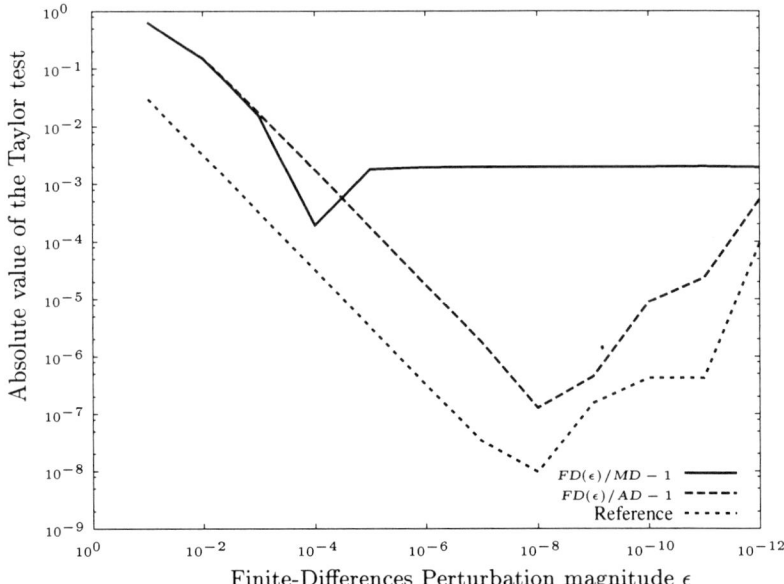

FIGURE 7.2. Taylor test curves for MD and AD compared to the reference.

use the curve obtained for the cosine $\cos(p)$ and its exact gradient $-\sin(p)$ around $p = 0$, as the reference Taylor test curve. The correctness of the MD and AD gradients is assessed by comparing the shape of their respective Taylor test curves to the reference one.

Figure 7.2 presents the reference Taylor test curve and the curves obtained for C and its gradients computed by MD and AD around the parameter set previously used. If the gradient is computed using MD, the test heads towards 0 for decreasing ϵ until ϵ reaches a value close to 10^{-4}, then it remains at a flat of around 10^{-3}. This is not the case for tAD and cAD: convergence continues until ϵ decreases to 10^{-8} (test value of 10^{-7}) before the test starts to degrade. The reference curve shows the same degradation indicating that the AD codes are as accurate as machine precision permits, whereas the MD gradient appears to be slightly biased. Similar results on the accuracy of AD are shown by [199] for the ADIFOR automatic differentiator, using *piggy-back* and *two-phase* differentiation.

In order to ensure that the MD code is not plagued by some programming mistake, Odyssée was used to verify the various steps of equation differentiation. The adjoint matrix can be obtained by AD, with respect to the state variable Z, of the original subroutine rhm that computes the residuals $R = E_M(Z, p)$ (right-hand member of Equation 7.2). Since it is a square matrix, the forward mode is preferable. MD- and AD- produced Jacobians were found to coincide perfectly. Similarly, Odyssée was used to perform the other differentiation steps involved in MD: differentiation of the same subroutine rhm with respect to the parameters p, and of the function F with respect to the state variables and parameters. All gradient and Taylor test results obtained with this equation-based, semi-automatic approach are identical to those of MD. The MD bias in Figure 7.2 and Table 7.1 may therefore be ascribed to the consistency shortcoming between the original program output and the gradient computed by the equation-based adjoint-state method: the latter assumes that the state equations are solved exactly, while only an approximate solution is produced by the former. Whether or not this bias is a real problem depends on how the gradient is used. It may alter the convergence of optimization procedures but such an alteration has not been demonstrated for our example. The equation-based, semi-automatic method used to reproduce the MD computations is a hybrid alternative to pure AD and MD: while outputs are those of MD, development and computer costs lie in between, closer to AD for the former and to MD for the latter.

7.5   CPU Time and Memory

To compare the three codes in terms of CPU time and memory-space, they were run for the 10×10 test-case previously used. MD and cAD need a separate run of the original direct code, while tAD computes the direct model and its derivatives concurrently. To enable comparison, the computational time of each code is defined as the total time required to obtain both the direct outputs and their derivatives (direct code time included when required). Wall-clock and CPU times are essentially equal for any of our runs, I/O being negligible. The first line of Table 7.2 shows the ratios of the measured run times to the run time of the original direct code, for the three differentiation modes. Only the forward mode's time depends on the number m of parameters, since tAD must be called once for each parameter whereas MD and cAD are only called once, irrespective of the number of parameters. The second line displays the corresponding theoretical bounds when available [215, 232].

TABLE 7.2. CPU-time ratios (direct code=1)

	MD	tAD	cAD
Measured	1.2	$1.8m$	12
Theoretical upper bound	-	$4m$	5

MD is by far the most computationally efficient method, the gradient computation representing only an additional 20% of the original direct code time. The reverse mode becomes more efficient than the forward mode for $m=7$ input parameters. The tAD time scales with m as expected, whereas the cAD time is above the theoretical bound. For the test case, the memory requirements are 1.09 Mbytes for the MD code, 1.97 Mbytes for tAD, and 79.1 Mbytes for cAD. The latter figure is a matter of concern for practical use of the method. Performances of the AD codes can be improved by restricting the differentiation computations to the last iterations of the NR algorithm as described in [238] as the *two-phase* differentiation. This requires proper hand modifications of the differentiated codes, leading to more possible hybrids between pure AD and pure MD codes. It appears that for the type of problems represented by our sample application, well thought-out "hand-driven" approaches to AD are to be preferred to its straight "black-box" use.

7.6   Conclusion

Based on the test of Taylor, the accuracy of the two AD modes proves to be excellent, as high as machine precision permits. A slight bias may appear for the equation-based adjoint code (MD), likely due to the fact that a theoretical model (discrete state equations) and a practical model (computer program) do not exactly coincide. This is not so much a matter of accuracy of the MD code as a question of consistency between the variable computed by the original code and the gradient produced by the MD code.

A major difference is noticed in computer resource requirements: the MD code largely outperforms all other methods for computing efficiency. The cost of this superior performance is the heavy, error-prone work necessary to produce the hand-coded adjoint. This cost may be considerably diminished without significant performance degradation by using AD in a combined equation-based / semi-automatic approach, to selectively produce the various derivatives specifically needed for the adjoint-state method. An additional, distinctive advantage of the equation-based adjoint method, be it manually or semi-automatically implemented, is that the construction and solution of the adjoint system is independent of the choice of the inverse problem unknowns.

Another possible hybrid strategy consists in hand-modifying *a posteriori* the fully AD-differentiated code for algorithmic performance improvement, such as loop count reduction in nonlinearity solving algorithms. All approaches but pure black-box AD require a precise knowledge of both the modeling methodology and the computer program structure.

Acknowledgments: The financial support brought to David Elizondo by the Ministère des Affaires étrangères (France), by the Pôle Universitaire Européen de Montpellier et du Languedoc-Roussillon (France) and by the Secretaría de Estado de Universidades, Investigación y Desarrollo of Ministerio de Educación y Cultura (Spain), is gratefully acknowledged.

8

Automatic Differentiation for Modern Nonlinear Regression

Mark J. Huiskes

ABSTRACT For modern nonlinear regression routines, the efficient computation of first and higher order derivatives is highly important. Automatic differentiation constitutes an opportunity to achieve both higher run-time efficiency and an increased feasibility of higher-order uncertainty analysis of complex models. In this article we present an overview of the derivative requirements of nonlinear regression routines. We further describe our experience in developing a C++ library for model analysis that uses the ADOL-C package for automatic differentiation. We show how the model analysis library, named MAP, has benefited from using automatic differentiation. Also a number of experiments are presented to show how more flexible and efficient execution trace management could further enhance the ease-of-use of ADOL-C.

8.1   Introduction

The ideas expressed in this chapter are based on experience obtained in developing a software library, named MAP, for nonlinear regression and subsequent uncertainty analysis. MAP is aimed at the analysis of models that are represented by smooth maps. The smooth maps that are to be analysed are provided as computer code. MAP consists of routines for model fitting, parametric uncertainty analysis, transformation of uncertainty information, model selection and various forms of sensitivity analysis. The library may be directly linked to C++ programs, or be used in combination with a graphic user interface, which can be used to control the various routines and view the resulting output. For the computation of first and higher-order derivatives MAP relies on the ADOL-C library [242]. This automatic differentiation tool is based on operator overloading (see [238] for an introduction). Further information about MAP can be found at www.cwi.nl/~markh.

In §8.2 we present a characterization of modern nonlinear regression. Particular emphasis is placed on the type of derivative information required by the various routines. In §8.3 we go deeper into our experience with the ADOL-C library and describe a number of experiments that indicate of the performance of the system. For these experiments we have linked a fishery stock assessment model to the library. This model is used in practice by the International Council for Exploration of the Sea (ICES) to reconstruct the population dynamics of North Sea herring [415].

8.2   Characterization of Modern Nonlinear Regression

8.2.1   Introduction

In nonlinear regression, model parameters are estimated by fitting a model to a set of observations. This is usually achieved by means of a likelihood function maximization. In the often used case that a normal error structure is assumed for the difference between observations and model estimates, the maximum likelihood method is equivalent to least squares minimization. See, for instance, [165]. In this case, after a suitable transformation to make the errors independent and identically distributed, we have

$$y = f(\theta) + \varepsilon, \qquad \varepsilon \sim N(0, \sigma^2 I_n), \tag{8.1}$$

where y is an n dimensional observation vector. The components of the function $f : \mathcal{P} \to \mathcal{S}$ from p dimensional parameter space $\mathcal{P}$ to n dimensional sample space $\mathcal{S}$ can be computed by means of a function that will be referred to as the *model map*. The model map, which will be described in more detail in §8.3, may also depend on quantities that are not estimated: the *regressor variables*. It is assumed that for a certain parameter vector θ^*, the n dimensional residual error vector ε is normally distributed with independent components of expectation zero and standard deviation σ. The least-squares and maximum likelihood estimator $\hat{\theta}$ of θ^* corresponds to the global minimum of the objective map $S(\theta)$:

$$S(\theta) = \|y - f(\theta)\|^2. \tag{8.2}$$

The analysis of a nonlinear regression problem may be characterized as a process consisting of three elements:

1. Computation of a parameter estimate by fitting the model to the data. Unlike for linear regression this estimate cannot be computed directly as a solution of the normal equations, but is obtained through an iterative optimization process.

2. Parametric uncertainty analysis, which is traditionally based on linearization of the model around the parameter estimate. Several computational procedures have been developed to take also higher-order derivatives into account in the computation of the parameter estimation bias, the variance-covariance matrix and the confidence intervals.

3. Model structure validation. The assumptions of the model must be checked, e.g., the error structure, corresponding to the assumed likelihood function. Further the sensitivity of the results with respect to the model structure and the observations must be investigated.

In the following sections we discuss the requirements of each of the nonlinear regression procedure elements with respect to derivative information.

8.2.2   Model fitting

The computational cost of the iterative optimization process depends on the complexity of the model to be fitted, i.e. the evaluation cost of the objective map depends mainly on the evaluation cost of the model map. For problems with a large number of parameters, derivative information is required for efficient optimization. With automatic differentiation using the reverse mode, the evaluation cost of the objective map derivative is only a small multiple of the evaluation cost of the objective map itself [238].

The nature of the derivative information required depends on the optimization algorithm chosen. For the conjugate gradient and quasi-Newton methods, the derivative of the model map or objective map with respect to the parameters is required. Newton type methods also require the second derivative with respect to the parameters .

In the case that the number of parameters is large, the problem generally must be scaled or preconditioned to obtain convergence of the optimization procedure. In this case derivatives with respect to the transformed parameters are required.

8.2.3   Higher-order parametric uncertainty analysis

The traditional nonlinear regression parametric uncertainty analysis, which is based on linearization, may be extended using higher-order model derivatives. These can be used both to obtain more accurate parameter uncertainty inferences and to obtain inference estimates for the case that a nonnormal error structure is assumed. In [448], an overview of higher order corrections to linear parameter inferences is presented. These are mainly based on the curvature measures described in [33]. In [267], second-order inference methods are developed. Computation methods for third-order accurate p-values, used to obtain confidence intervals for both normal and nonnormal error structures, are presented in [200].

In MAP, the corrections to the parametric uncertainty estimates are based on curvature arrays that quantify the nonlinearity of the model around the parameter estimate. The curvature arrays constitute normal and tangential projections of second-order model map derivatives.

8.2.4 Model structure validation

Model structure validation consists of more than checking the assumptions with respect to the assumed error structure. The sensitivity of the results with respect to model structure also should be investigated. Using first and higher-order derivatives not only the sensitivity of the model map responses with respect to the input variables can be quantified. It is also possible to evaluate the sensitivity of the estimated parameters with respect to all quantities that are not estimated by means of implicit sensitivity analysis (see [294]). The sensitivity of the results with respect to model structure can be further investigated by means of model modifications and their effect on parametric uncertainty measures and model selection criteria [355].

For these methods derivatives of the model map with respect to parameters and regressor variables are required.

8.3 Experience with the ADOL-C library

The procedures discussed in the previous section have been implemented in MAP using the ADOL-C library to compute the required derivatives. The most important function of which derivatives are computed is the model map which is used to compute the components of the function f in equation (8.1). These components may depend on both a vector of parameters that have to be estimated and a vector of measured quantities, the regressor variables. The model map is supplied by the user and has Function prototype 1.

```
ModelMap(adouble* Regs, adouble* Pars, adouble* Resp);
```

Function prototype 1: Model map that computes response variables for given parameters and regressor variables.

All input variables are declared active using ADOL-C datatype `adouble` so that derivatives with respect to both the parameter and the regressor variables may be obtained.

8.3.1 A North Sea herring case study

For the performance experiments that will be discussed in the next sections, we used a model map to generate estimates by means of an age-structured fishery model. The associated parameter estimation procedure, of type 8.1, constitutes a re-implementation of the Integrated Catch-Age (ICA) stock assessment procedure ([415]). This procedure is currently used by the International Council for Exploration of the Sea (ICES) to assess the North Sea herring stock. The goal of the fitting procedure is to reconstruct the history of the herring stock, to investigate its current state and to determine the influence of the commercial fishery. The results of the analysis are used to advise the European Commission on future fishing levels.

The data used to fit the model is obtained both from the commercial fishery and from several research vessel surveys. An example of the latter is an acoustic abundance index collected in cooperation by the various fishery institutes of the North Sea countries. Furthermore, data is used from surveys aimed at the measurement of larval stock size and spawning stock biomass.

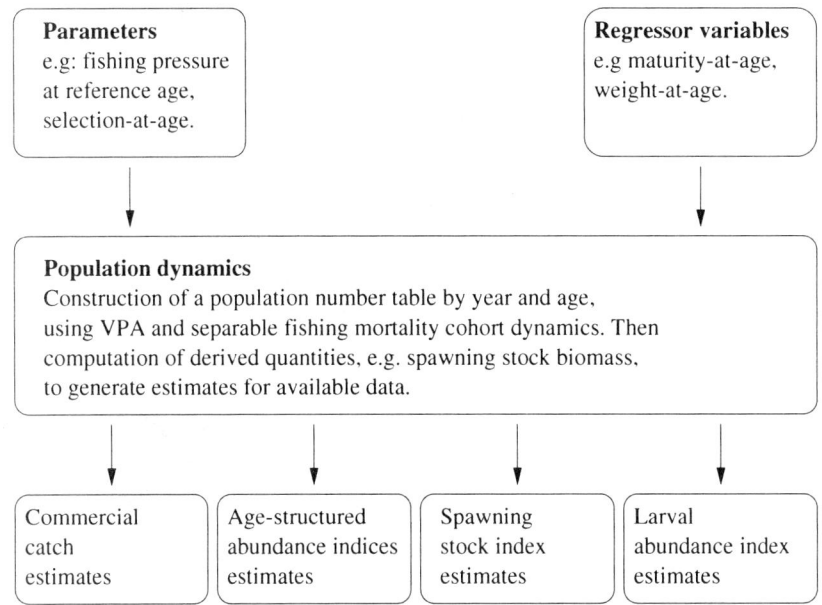

FIGURE 8.1. Schematic representation of the model map for the North Sea herring case study. The responses consist of estimates for stock observations from multiple information sources.

A schematic representation of the model map is shown in Figure 8.1. The parameters and regressor variables are used to generate a population numbers table over a certain time period and for a certain number of age classes. This table is generated using two methods: by means of virtual population analysis and separable fishing mortality cohort dynamics. Both methods are described in [415]. In each case the population is divided into cohorts, i.e. into groups of fish that were born in the same year. The cohort size dynamics are then determined by natural and fishing mortality. A number of parameters quantify the fishing pressure by parameterizing the fishing mortality. For a general overview of age-structured stock assessment methods of this type, see [378].

The model contains 45 parameters and computes 218 response variables. For a detailed description of the model and the results of the uncertainty analysis, see [296]. The maximum likelihood optimization was performed by a Polak-Ribiere conjugate gradient procedure (see for instance [222]) on a Pentium II machine (350 MHz, 64 Mb).

8.3.2 Computation of the objective map

For design reasons it is preferable to have only one model map function, which is represented by Function prototype 1. Since all variables for which derivatives are potentially required are active, all derivatives can be computed by defining those variables as independent. Furthermore, using this prototype, only one function has to be modified if the user wants to change the underlying model.

This means, however, that for evaluation of the model map, we must use a function that is less efficient than a function without overloaded operations, i.e. a function using variables of type double instead of adouble. ADOL-C offers the possibility to evaluate a function of which an execution trace has been recorded by means of a call to function(). Ideally, the overhead in evaluating a function in this manner is small.

The most critical function evaluation with respect to run time efficiency is that of the objective map, since it is required so often in the iterative optimization procedure. To investigate the overhead of using the adouble type, we compared the performance of a conjugate gradient procedure for two evaluation methods. First by means of the function() call and then by an objective map based on the original model map, i.e. a map that does not use any ADOL-C specific constructs.

For the fitting of the fisheries model the overhead turns out to be quite small, i.e. the conjugate gradient procedure using the function with variables of type double is faster only by a factor of 1.3.

8.3.3 Computation of the objective map derivative

One of the most important contributions of AD to regression analysis is the efficient computation of the objective map derivative. This derivative depends in turn on the model map derivative, according to

$$\frac{dS}{d\theta} = 2\left(f(\theta) - y\right)^T \frac{df}{d\theta}. \tag{8.3}$$

Then it is possible to compute the objective map derivative in two ways: (i) by tracing the entire objective map computation using the parameters as independent variables and the objective value as dependent variable, and by then using a gradient() call; (ii) by tracing the model map computation using the model map response variables as dependent variables; after a jacobian() call, the objective map derivative can then be computed using (8.3). We compared these methods and found that for the fisheries model map the first method turns out to be surprisingly more efficient, viz. approximately 8 times faster.

This can be understood by considering that for the first method the reverse mode is very efficient since the derivative of a scalar quantity is computed. For the second method one has to resort to the less efficient forward mode of differentiation since in this case the derivative with respect to a large number of components, 218 in fact, are required. The fact that the computation of the function of the first method is slightly longer because of the summing and squaring of the model map components, is in this example easily offset by the difference in efficiency of the derivative computation.

Next, we compared the performance of the automatically computed objective map derivative with one computed by means of central finite differences. We found that for the fishery model map the conjugate gradient procedure using the ADOL-C derivatives was approximately 7 times as fast as the procedure using the numerical approximation.

8.3.4   Computation of higher-order derivatives

An interface to facilitate the computation of higher-order derivatives was not yet available in the ADOL-C 1.6 version that we used (except for the computation of Hessians), but has now been implemented in newer versions.

In order to compute the curvature arrays mentioned in §8.2.3, first the Hessian matrix of each of the model map response vector components must be computed, i.e. a total of 218 Hessian matrices. Once again, we would like to make use of the function declared by Function prototype 1 to avoid the situation that if model modifications are desired this must occur in several functions. However, using this prototype as basis for the computation is not optimal for two reasons. The first is that we do not need active regressor variables, which constitutes an overhead comparable to that discussed in §8.3.2. The second reason is that the derivative is required with respect to only one of the response components.

We tested the performance of using prototype 1 with the regressor variables not defined as independent variables, and with only one of the components declared dependent. This was compared to the performance of a special function that only computes the required component and for which the regressor variables are not active at all. For the computation of the 218 Hessian matrices this special function was about twice as fast.

Finally, in many regression problems one is confronted with the situation that derivatives of the model map are needed with respect to the parameters, but each time for a different regressor vector. There are then two alternative evaluation methods: (i) Define both parameters and regressors as independent variables, then use only that part of the derivatives that corresponds to the parameters. In this case only one execution trace has to be recorded. (ii) Define only the parameters to be independent, and record a new trace for every regressor vector. Which alternative is more efficient depends on the number of regressor variables and the overhead associated with the recording of the trace. However, both options seem to be unnecessarily inefficient.

As a solution to this problem, a new type of variable can be introduced. Just as active variables can be labeled independent, it should then be possible to label active variables as of an auxiliary type. This type then indicates that no derivatives with respect to these variables are required, but that in the computation of the derivatives these variables are allowed to take a different input value. Taking this approach the execution trace would have to be computed only once.

8.4   Concluding Remarks

The MAP library has benefited substantially from the use of the ADOL-C library. ADOL-C has proved to be stable, and its interface is convenient to use. Automatic differentiation has made the parameter estimation more efficient and has made it feasible to perform higher order uncertainty analysis on complex models.

It is well-known that the current performance bottleneck for ADOL-C lies in the file access of the execution trace. Improvements using memory mapped I/O have been announced on the ADOL-C 2.0 web site.

One of the MAP design goals has been to use only a single model map function, i.e. for the actual overloaded model computation code, such that user modifications to the model are performed in only one place. This requires that the overhead in function evaluation from execution traces is small. In this respect we have also noted the desired feature to obtain derivatives with respect to only a subset of the active variables while keeping the possibility to change the values of the remaining active input variables without re-tracing the computation.

As an illustration, we investigated the North Sea herring stock assessment. As described in [296], we obtained more accurate parametric uncertainty information using higher-order model derivatives. Further by investigation of a number of model modifications, we showed that the precision of the estimates improves for models of smaller complexity. Also several types of analysis based on derivatives obtained by means of AD were used to investigate the sensitivity of quantities of interest to the fishery management with respect to various assumptions and input quantities.

9

Sensitivity Analysis and Parameter Tuning of a Sea-Ice Model

Jong G. Kim and Paul D. Hovland

ABSTRACT The values of many of the parameters in climate models are often not known with any great precision. We describe the use of automatic differentiation to examine the sensitivity of an uncoupled dynamic-thermodynamic sea-ice model to various parameters. We also illustrate the effectiveness of using these sensitivity derivatives with an optimization algorithm to tune the parameters to maximize the agreement between simulated results and observational data.

9.1 Introduction

The sea-ice cover in polar regions plays an important role in modeling the Earth's climate system. For example, sea-ice acts as a powerful insulating boundary layer, reducing the atmosphere-ocean heat exchange and reflecting incoming solar energy on the surface of sea-ice covers. Also sea-ice is a major factor in the thermohaline circulations [304, 395]. Thus, sea-ice is capable of profoundly amplifying and modulating the climate variations in regional and global-scale climate systems. For this reason, sea-ice models are being presently coupled to global atmospheric and ocean general circulation models in Earth climate studies.

The accuracy of a sea-ice model depends upon the accuracy of several model parameters. Yet tuning parameter values of the climate system has been largely an ad hoc procedure, because of the high computational cost of evaluating multivariate parametric sensitivities. Early sensitivity studies of sea-ice models include the works of Mykut and Untersteiner [394] and Semtner [450] with a one-dimensional thermodynamic sea-ice model. Full scale sensitivity studies of a dynamic-thermodynamic sea-ice model were carried out by Parkinson and Washington [412], Holland et al. [281], and Chapman et al. [120]. In all of these studies, the investigators calculate the sensitivities by the finite difference (FD) method, which obtains the derivatives by dividing the response perturbation of the simulation to input parameter variation. Automatic differentiation (AD) provides an attractive alternative, since sensitivity derivatives can be computed with much greater accuracy and typically at a lower cost relative to finite difference approximations [79, 232, 238].

In this chapter, we present AD as an alternative to the FD method for tuning a sea-ice model. Following a brief discussion of the model (§9.2) and the simulation conditions (§9.3), we discuss our experiments in evaluating the sensitivities of a dynamic-thermodynamic sea-ice model (§9.4). We also discuss the effectiveness of using AD-generated derivative codes coupled with minimization algorithms for tuning the model parameters to maximize the agreement between simulated result and observational data (§9.5). We conclude with a brief description of future work.

9.2   Model Description

Sea-ice models are comprised of two major components: thermodynamics and dynamics. Their formulations involve various diagnostic analyses of the interdependent physical and dynamical mechanisms to produce simulated seasonal changes of sea-ice. The thermodynamic routines of the code used in the study are essentially the same as those described by Hibler [275], while the dynamic component is the elastic-viscous-plastic (EVP) model presented by Hunke and Dukowicz [297]. We summarize the main elements of the formulation here in order to underline the parameters used in the numerical experiments of sensitivity studies and the tuning process. The Fortran code of the sea-ice model used in the study was written in rectangular coordinates [15]. Numerical experiments for Arctic sea-ice cover were performed with given atmospheric and oceanic forcing terms.

9.2.1   Thermodynamic component

The thermodynamic model for the calculations of ice thickness change and open-water formation are discussed in detail by Mykut and Untersteiner [394], Semtner [450], and Hibler [275], computing the time-dependent ice thickness and vertical ice temperature profile in vertical grid intervals. The vertical ice changes are modified according to the energy transfers between the ice layer boundaries: the ocean mixed layer and the atmosphere-ice and snow-ice interfaces. A basic surface heat balance equation is given by

$$k\frac{\partial T}{\partial h} = (1 - \alpha)F_{s\downarrow} + F_{L\downarrow} + D_1|U_G|(T_a - T_0)$$
$$+ D_2|U_G|[q_a(T_a) - q_s(T_0)] - D_3T_0^4, \qquad (9.1)$$

where T_0 is the surface temperature, α is the surface albedo, T_a is the air temperature, U_G is the geostrophic wind, q_a is the specific humidity of the air, q_s is the specific humidity at the ice surface, $F_{s\downarrow}$ and $F_{L\downarrow}$ are the incoming shortwave and longwave radiation terms, D_1 and D_2 are the bulk sensible and latent heat transfer coefficients, D_3 is the Stefan-Boltzmann constant times the surface emissivity, k is the ice conductivity, and h is the ice thickness. Key parameters for sensitivity studies are D_1, D_2, and α.

9.2.2 Dynamic component

Ice motion and deformation associated with vertical ice change modeled by the thermodynamic component are determined by the dynamics of a sea-ice model. Governing the dynamics component, the momentum balance equation is written for a mass of ice within a grid cell:

$$m \frac{\partial \mathbf{U}}{\partial t} = -mf\mathbf{K} \times \mathbf{U} + \tau_a + \tau_w - mg\nabla H + \nabla \cdot \sigma, \qquad (9.2)$$

where m is the ice mass per unit area, $\mathbf{U}$ is the ice velocity, f is the Coriolis parameter, $\mathbf{K}$ is a unit vector normal to the surface, H is the sea surface height, g is the gravity acceleration, and σ is the two-dimensional internal stress tensor. The forces due to air and water stresses, τ_a and τ_w, are given by

$$\tau_a = \rho_a C_a |U_G| (U_G \cos \phi + \mathbf{K} \times U_G \sin \phi),$$
$$\tau_w = \rho_w C_w |U_w - \mathbf{U}| [(U_w - \mathbf{U}) \cos \theta + \mathbf{K} \times (U_w - \mathbf{U}) \sin \theta],$$

where U_G is the geostrophic wind, U_w is the ocean current, C_a and C_w are the air and water drag coefficients, ρ_a and ρ_w are the air and water densities, and ϕ and θ are the turning angles at the ice-air and ice-ocean interfaces. The EVP rheology model that relates the internal ice stress σ and the rates of strain ϵ is given by

$$\frac{1}{E} \frac{\partial \sigma_{ji}}{\partial t_e} + \frac{1}{2\eta} \sigma_{ij} + \frac{\eta - \zeta}{4\eta\zeta} \sigma_{kk} \delta_{ij} + \frac{P}{4\zeta} \delta_{ij} = \epsilon_{ij},$$

where P is the internal (or hydrostatic) ice pressure, ζ and η are the bulk and shear viscosities, and E is Young's modulus. The stress tensor σ and the ice motion vector $\mathbf{U}$ are updated with the integration of Equation (9.2). In the integration, an effective model time scale of Δt is discretized by the subcyclic time step of $\Delta t_e = \Delta t / N$. In our study, the number of subcyclic iterations is set to $N = 72$. The pressure of the internal ice strength is determined by the ice thickness h and the fraction of ice-covered area A, given by

$$P = P^* h e^{[-C(1-A)]}.$$

The ice strength in this equation depends on both the ice thickness h and the ice area (or concentration) A. For this reason, a conservation law is derived for both quantities by applying a two-dimensional continuity equation. The continuity equations used in the study are the same as those described by Hibler [275]. The major parameters for the sensitivity study are C_a, C_w, and P^*.

9.3   Model Problem and Simulation Conditions

The domain data and the external forcing variables were provided by Arbetter in a personal communication. The model included the area from the Greenland-Iceland-Norwegian sea to portions of the Bering sea, an Arctic domain discretized by the resolution of an 80 km Cartesian grid (see Figure 9.1). The forcing variables of T_a, q_s, $F_{s\downarrow}$, $F_{L\downarrow}$, and U_G were derived from National Centers of Environmental Prediction reanalysis fields for 1992 at six hour intervals. A small nonzero constant was used for ocean currents. The domain boundaries were warped by Neumann boundary conditions. The constant ice thickness of 2 m and concentration of 0.5 were set for the initial conditions. The initial ice velocities were set to zero. For the details of the other input physical coefficients, see references [120, 281].

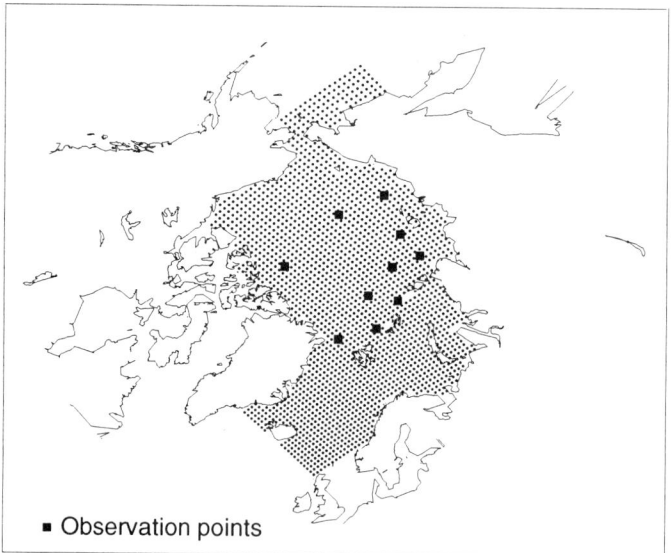

- Observation points

FIGURE 9.1. The model grid with resolution of 80 km and the locations of the observation data used in the study.

9.4   Sensitivity Experiments

Sensitivity experiments were carried out for the dynamic-thermodynamic ice model parameters. As a dependent variable, the average ice velocity was calculated over ten simulation points, which were randomly distributed in the domain (shown in Figure 9.1). A two-month period of simulation (January and February 1992) was integrated for each sensitivity experiment. The forward mode of the ADIFOR [79] tool was used to provide

TABLE 9.1. AD and FD derivatives of the average ice drift speed with respect to seven dynamic and thermodynamic parameters.

Parameters	ADIFOR	FD, $\triangle = 0.05$	FD, $\triangle = 0.001$
C_a	0.36078E-01	0.36247E-01	0.36042E-01
C_w	-0.16748E-01	-0.16185E-01	-0.16736E-01
P^*	-0.16192E-01	-0.15942E-01	-0.16215E-01
C	-0.11839E-21	0.69389E-15	0.41633E-13
D_1	-0.17610E-02	-0.16960E-02	-0.17580E-02
D_2	0.54740E-04	0.53690E-04	0.53620E-04
α	-0.58537E-11	-0.58539E-11	-0.58599E-11

code for the derivatives. The reverse mode or another automatic differentiation tool [212, 442] could have been used with similar results. AD and FD derivatives for seven control parameters are summarized in Table 9.1. As the table shows, AD-computed derivatives for the average ice drift speed are comparable with FD-computed derivatives. However, with larger increments of control parameters FD-derivatives are less accurate than AD-generated derivatives. Table 9.1 also indicates that the ice drift speed generally depends more on the dynamic parameters than on the thermodynamic parameters. The primary control parameters are the air and ocean drag coefficients determining the external stress terms in the dynamic equations. It is also indicated that the ice drift speed increases as the air drag coefficient increases. In contrast, the ice drift speed decreases with increasing ocean drag coefficient; this decrease occurs because the ocean currents impose a drag on the ice motion by reducing the motion force from wind stress. The effect of the ice strength constant is another substantial parameter for the sea-ice model. The ice motion tends to decrease with an increase of the ice strength constant. According to the ice rheology formulation, ice motion slows when the ice is compact with a larger ice strength constant. The thermodynamic parameters provide weaker impact on the sea-ice model, perhaps because the formulations of ice dynamics are more complicated than those of thermodynamics. This result requires that more general sensitivity analysis be addressed with comprehensive thermodynamic formulations.

9.5   Parameter Tuning

As demonstrated by the sensitivity results, the sea-ice model relies strongly on both dynamic and thermodynamic parameters. This interdependency of the model components requires accurately tuned parameters to simulate realistic average ice drift motions. Previous studies [290] have demonstrated the effectiveness of coupling optimization algorithms with automatic differentiation to tune algorithm parameters. This section describes the use of a similar approach to tune the physical parameters in the sea-ice model.

We used a bound constrained quasi-Newton minimization method, L-BFGS-B [508], which requires the user to provide the code computing the objective function (or cost function) and its gradients. The gradient information for the cost function was provided by the ADIFOR-generated derivative code. In this study, the cost function is defined by the norm of the difference between the observed and simulated average ice drift velocities. For the observational data, buoy ice drift velocities were provided by the International Arctic Buoy Programme (IABP) [142]. The five model parameters C_a, C_w, P^*, ϕ (or θ), and $D1$ were selected to control the simulated average ice velocities of ten separate points in the model domain (shown in Figure 9.1). The simulated average ice velocities were used to fit the observational data for a two-month period (January and February 1992). Five randomly selected parameter values (4.3450E-04, 0.0063, 2640, 0.2490, 1.65°) were used for the initial guess of the minimization. The convergence behavior of the cost function is shown in Figure 9.2. The norm of

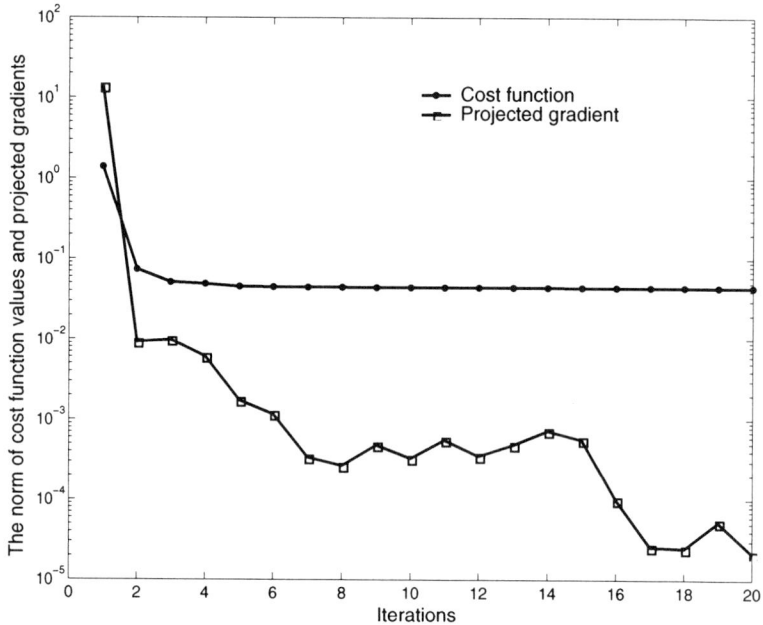

FIGURE 9.2. Convergence behavior of the cost function and its projected gradient with minimization iterations.

the cost function was reduced by an order of magnitude in about 20 minimization iterations. The tuned parameter values (0.0846, 0.0031, 50490, 3.5927, 53.1°) were obtained in the minimization process. The ratio between the two drag coefficients C_a/C_w, is 0.03654. This is a relatively low value compared with another parameter set (0.0055, 0.0012, 27500, 2.284,

25°), which has been used as default values in most sea-ice modeling studies. The ice motions simulated by these two different parameter sets are plotted in Figure 9.3. The ice motions of both are encouragingly comparable to the observed figures analyzed by IABP using buoy drift data (for the details of observed ice motions, see the web page of IABP [437]). Figure 9.3 also indicates that the drift speed is overestimated by the default parameters. This overestimation is clearly observed in the region of Fram strait.

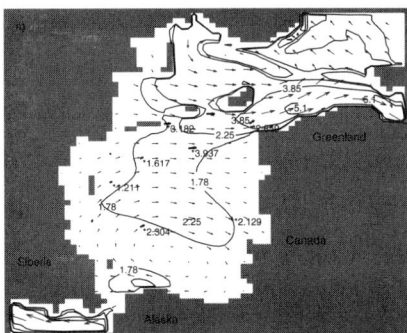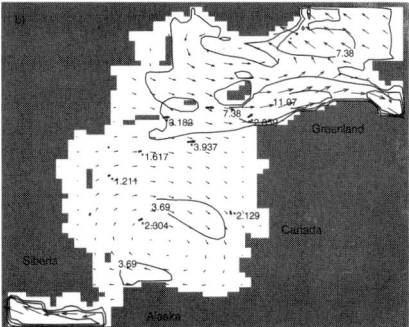

FIGURE 9.3. Average ice drift motion (cm/sec) plots for January and February 1992: (a) simulated with tuned parameters $(0.0846, 0.0031, 50490, 3.5927, 53.1^o)$ and (b) simulated with the default parameter set $(0.0055, 0.0012, 27500, 2.284, 25^o)$; * marks and thick vector lines are observational data.

9.6   Conclusion

Our experiments indicate that automatic differentiation can provide accurate sensitivity analysis of a multivariate sea-ice model. Our results are similar to those from sensitivity experiments described by Holland et al. [281] and Chapman et al. [120]. In general, we observed that the sea-ice model strongly depends on the dynamic parameters. In the parameter-tuning process with buoy drift data, we observed that the L-BFGS-B minimization algorithm coupled with the AD derivative code yields parameter values that produce realistic results. Thus, an optimization-based approach using the analytic derivatives computed by AD provides an effective basis for producing a model which better matches observational data. Future work includes tuning the parameters of multidecadal simulations of the sea-ice model, incorporating additional observational data, and applying the tuning process to other components of a GCM.

Acknowledgments: This work was supported by the Mathematical, Information, and Computational Sciences Division subprogram of the Office of Advanced Scientific Computing Research, U.S. Department of Energy, under Contract W-31-109-Eng-38.

We thank Todd Arbetter of the University of Colorado-Boulder for providing the Fortran program and the model problem sets. We are grateful to Michael Tobis for suggesting the idea of this sea-ice model parameter-tuning project. We are indebted to Ian Foster, Jay Larson, and John Taylor of Argonne National Laboratory for valuable discussions and assistance during the project. Finally, we thank Gail Pieper and Boyana Norris of Argonne National Laboratory for the careful review of the manuscript.

10

Electron Paramagnetic Resonance, Optimization and Automatic Differentiation

Edgar J. Soulié, Christèle Faure, Théo Berclaz and Michel Geoffroy

ABSTRACT This chapter describes an optimization problem applied to electron paramagnetic resonance spectroscopy. Levenberg-Marquardt fails to converge using a divided differences Jacobian approximation in single precision, while it succeeds using Odyssée-generated forward mode Jacobian values. In double precision, the optimizer returns a smaller "minimum" objective function with AD compared to DD in 46% more CPU time.

10.1 Introduction

Electron paramagnetic resonance (EPR) spectroscopy provides information about the interaction of electronic magnetic moments (spins) with the applied magnetic field, or Zeeman effect, and with the nuclear magnetic moments of neighbouring nuclei (hyperfine interactions). In the case of a single crystal, the EPR spectrum is recorded for several orientations. For each orientation, the spectroscopic parameters needed to define the "effective Hamiltonian" determine the values of the resonance fields. Parameter determination based on nonlinear least squares fitting only requires the knowledge of a variation law of the resonance fields with crystal orientation. For a powder, knowledge of the "effective Hamiltonian" does not suffice; for the variation laws of line intensities, line shapes and line widths are also required to simulate an EPR spectrum (under the assumption that the experimental powder spectrum is orientation invariant). The powder spectrum is calculated as the sum of many spectra corresponding to different orientations of a crystal in the applied field. As for the single crystal case, parameter fitting based on nonlinear least squares optimization may be applied to deduce the values of the spectroscopic parameters $\overrightarrow{p}$ using the following objective function

$$F(\overrightarrow{p}) = \sum_{j=1}^{m} (s_j^{obs} - s^{calc}(\overrightarrow{p}, B_j))^2$$

where m is the number of data points (B_j, s_j^{obs}) sampled to represent an experimental EPR spectrum, and $s^{calc}(\overrightarrow{p}, B_j)$ is the value of the EPR signal calculated according to a model specified by the simulator (as described below).

10.2   The Simulator

The simulator calculates a powder spectrum for a given set of parameters. In the course of optimization, it is evaluated repeatedly for several sets of parameters. We describe below the various elements of its calculation. Whereas three angles (for example Euler angles) are required to define the orientation of a solid (a single crystal) in the laboratory frame, only two angles are needed to define the orientation of the applied magnetic field with respect to a reference frame. Therefore, for each (tiny) single crystal, we select a different reference frame, bound to its crystal axes. Thus, only two integration variables are needed, instead of three. The intensity of the magnetic field at resonance can be calculated by an explicit analytic formula if the "effective Hamiltonian" only comprises the electronic Zeeman interaction. In general, the presence of hyperfine interaction terms requires that the value of the applied magnetic field B be determined by solving the *resonance equation* $h\nu = E_2(B) - E_1(B)$, where h stands for the Planck constant and ν for the microwave frequency; thus $h\nu$ represents the energy transferred to the spin system by the microwave oscillator. $E_1(B)$ and $E_2(B)$ stand for two eigenvalues of the effective Hamiltonian which depends on B. This equation may solved either by a numerical iterative method or in an approximate manner, by a perturbation calculation. The latter entails the major advantage of enabling analytical formulae to be used, thus avoiding the need to introduce another level of iterative calculation. Many formulae have been proposed for various particular cases. Machio Iwasaki [303] proposed formulae of fairly general validity. In the present case, the formulae are valid provided that the hyperfine interaction terms are small with respect to the Zeeman term. These formulae are vectorial and tensorial, thus concise. However, in Iwasaki's paper they were not given in the order corresponding to the writing of an algorithm. These formulae where given for the more general case where the effective Hamiltonian also entails fine structure terms, nuclear quadrupole interaction terms and nuclear Zeeman terms. The fine structure and nuclear quadrupole interaction terms are irrelevant in the examples described below. We omitted the nuclear Zeeman terms, which are much smaller than the hyperfine terms.

In the following, the four main components of the simulator are introduced: *a)* the intensity of the resonance, *b)* the resonance shape, *c)* the line width, and *d)* the calculated signal.

a) **The intensity of the resonance**

The electromagnetic wave induces a transition between the initial and final quantum states, the probability $P(\theta, \phi)$ of which is determined thanks to a general formula of quantum mechanics called the "Fermi golden rule." Application of this formula to the case of an effective Hamiltonian which only comprises the electronic Zeeman term leads to a formula which was determined by Bleaney [4] and Pilbrow [419]. Rigorously, the eigenstates of the complete effective Hamiltonian should be used.

b) **The resonance shape**

Resonance is not a phenomenon with an infinitely small width. The resonance signal intensity decreases as the applied magnetic field moves away from the resonance field. Let σ be the half-width at half-height, and x the reduced distance: $x = \frac{B - B_{res}}{\sigma}$ The most commonly found line shape functions are the Cauchy-Lorentz function: $f(B) = \frac{1}{1+x^2}$ and the Gauss-Laplace function; $f(B) = \exp(-Ln(2) \cdot x^2)$. Other functions are encountered, including the convolute between these two functions.

c) **The line-width**

Experiments conducted on single crystals indicated that the width σ_B expressed in magnetic field units varies much more rapidly than the width σ_ν expressed in frequency units. These two widths are related by the equation $h\sigma_\nu = \beta g_{eff}\sigma_B$, where β is the Bohr magneton, and g_{eff} is the current value of the factor g. Given the lack of an adequate theory, we assume that the width expressed in frequency units varies with the single crystal orientation according to the law:

$$\sigma_\nu(\theta, \phi) = \sqrt{(\sigma_x^2 \cos^2(\phi) + \sigma_y^2 \sin^2(\phi)) \sin^2(\theta) + \sigma_z^2 \cos^2(\theta)}.$$

d) **The calculated signal**

The resonance signal of the powder is the sum of the elementary signals due to the tiny single crystals and is calculated as a double integral over the angles. Experimentally, the derivative of the energy absorption with respect to the applied field B is recorded, so the calculated signal may be expressed as:

$$S(B) = \iint P(\theta, \phi) \cdot \frac{\partial f}{\partial B} \left[\frac{B - B_{res}(\theta, \phi)}{\sigma_B(\theta, \phi)} \right] \cos(\theta) \, d\theta \, d\phi.$$

The double integration over the sphere may not be achieved by formal calculus [466]. The numerical calculation of this double integral may be achieved by a number of methods. Numerical experiments have shown that substitution of the variable $z = \cos(\theta)$ and Gauss integration over 96 points [5] significantly improved the quality of the simulation. On the contrary, the introduction of this method for the angle ϕ gave erroneous results. In order to reduce the numerical noise in the calculated spectrum, we resorted to a random origin for the integration over angle ϕ. The spectrum so calculated must be brought to the scale of the observed spectrum and its baseline adjusted to that of the observed spectrum. Both operations are done by the resolution of a linear system. The simulator depends on several parameters: six parameters represent the independent components of the symmetrical tensor g. Three parameters represent the half line-widths for three orientations of the magnetic field. The next six parameters represent the independent components of the hyperfine interaction tensor with the first nucleus, expressed in Megahertz. The next six parameters do the same for the second nucleus, etc.

10.3   The Optimizer

The minimization of a sum of squares of twice differentiable nonlinear functions as encountered in the present case is approximated by the Levenberg-Marquardt algorithm [373, 388]. Since the domain of values which is accessible to a given parameter is limited in practice, it is important to assign lower and upper bounds to each adjustable parameter and to be able to let it vary or to have it blocked. The optimization software BSOLVE by Ball [27, 337] has these functionalities. This software enables us to calculate each line of the Jacobian matrix either by divided differences or by an auxiliary subroutine to be provided by the user. Upon each iteration, BSOLVE indicates the number of varying parameters at this step of optimization. When optimization ends normally, the value of this number is zero. However, given the large number of numerical operations involved in a spectrum calculation, rounding errors accumulate, and the optimization often stops with the diagnosis: "Impossible objective function improvement." We suspected that this diagnosis could result from the fact that divided differences provide a Jacobian matrix with an insufficient accuracy. Thus after each iteration, the optimizer would be poorly guided by the approximated Jacobian matrix towards the next iteration. This idea prompted us to resort to automatic differentiation to obtain from the code of the simulator a derivative code that computes the exact Jacobian matrix. We thus expected that the latter code would enable an improvement of the determination of the displacement of the vector of the adjustable parameters in the space $\mathbb{R}^n$.

   The simulator `simula` computes the signal `sigcal` from the vector x of parameters. The AD software Odyssée differentiates programs written in the Fortran 77 programming language. If a function is specified as a set of programming units, Odyssée differentiates it as a whole with respect to those input variables specified by the user. In this application, the forward mode of AD is used because the Jacobian matrix $\{\frac{\partial sigcal(i)}{\partial x(j)} i \in [1..N], j \in [1..n]\}$, where $n \ll N$ is required by the Levenberg-Marquardt algorithm. In forward mode, Odyssée produces a Fortran code that computes one directional derivative of the original simulator (subroutine `simula`). Applying

Odyssée to simula is straightforward. Neither the original code, nor the Odyssée-generated code requires manual modification. The automatically generated code simulatl thus computes the vector sigcalttl such that

$$sigcalttl(i) == \sum_j \frac{\partial sigcal(i)}{\partial x(j)} xttl(j).$$

The initial version of the optimizer computes an approximation of its Jacobian matrix column by column by divided differences. We have adapted the optimizer to compute each column k of the Jacobian matrix by a call to the tangent code of the simulator initialized with the canonical direction $xttl(j) = 0$ is $j \neq k$ and $xttl(k) = 1$.

10.4   Study of the Optimizer's Behavior

In this section, we compare the optimizer's behavior when using either AD or divided differences. The comparison is based on two optimizations starting from the same input data. In order to be able to draw conclusions, we have first treated three different test cases, and then we changed the precision from single to double on one test case. The tests presented in this chapter are performed on a SUN station with 512 Megabytes memory using OS5.

The three test cases studied are based on the same free radical, with at least two hyperfine interactions: one with a phosphor-31 ($I = 1/2$) and one with a carbon-31 ($I = 1/2$) nucleus. In Test case 1 only two interactions are taken into account; thus 21 parameters are involved among which 9 are adjustable. Test cases 2 and 3 incorporate one more hyperfine interaction with one proton (or hydrogen nucleus) ($I = 1/2$) which makes 27 parameters. Test case 2 considers 12 of these parameters to be adjustable whereas in Test case 3, 15 are adjusted.

For these three cases, in single precision, the original version of the optimizer produces the diagnosis "Impossible objective function improvement" for 11, 8 and 14 optimizations steps respectively. As expected, when AD is used, the optimization ends normally after 26, 37 and 61 optimization steps, respectively.

The final powder spectra obtained at the end of the optimization process look much alike whether divided differences or AD is used. We have chosen to compare some indicators computed at each iteration of the optimization to evaluate the consistency of the results. In the latter, we compare the optimizer's behavior for Test case 1 using AD or divided differences in single or double precision. The optimizer does not reach any solution for divided differences in single precision but reaches a solution in double precision. The number of truly modified adjustable parameters is computed at each step and its evolution during the optimization process (shown in Figure 10.1) indicates the speed of convergence. The curves are labelled DA-R4

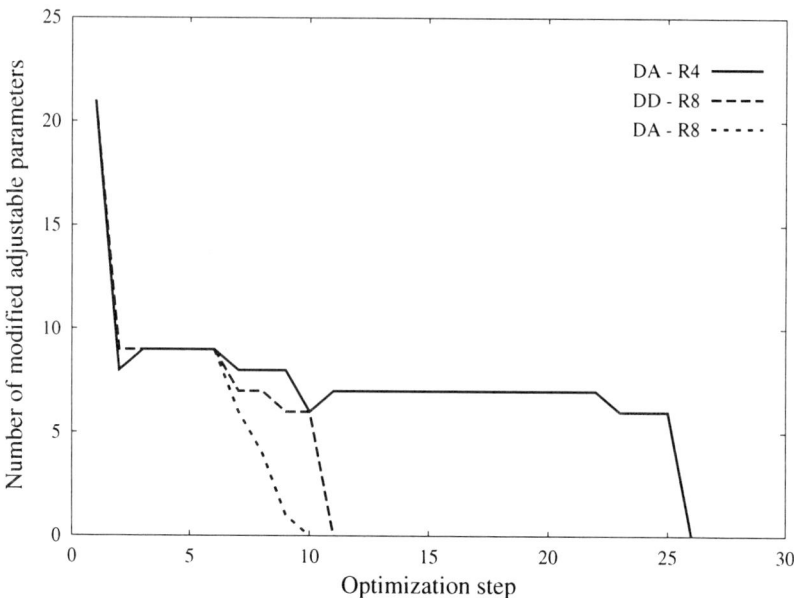

FIGURE 10.1. Number of truly modified adjustable parameters (Test case 1)

TABLE 10.1. Final value of the objective function

	Test case 1	Test case 2	Test case 3
DD-R4	0.11042409E+06	0.10719038E+06	0.10670412E+06
AD-R4	0.11039414E+06	0.10700169E+06	0.10662119E+06
DD-R8	0.11062249E+06	0.10705019E+06	–
AD-R8	0.11062246E+06	0.10698625E+06	–

for AD in single precision, DD-R8 for divided differences in double precision and DA-R8 for AD in double precision. The comparison of DD-R8 and DA-R8 shows that the number of truly modified adjustable parameters diminishes quicker when AD is used instead of divided differences. The comparison of DA-R4 and DA-R8 shows that using double precision increases the optimization speed measured by the number of steps by a factor of two.

We cannot show here the evolution of other parameters [187], but the number of simulator calls is smaller using AD, and the evolution of the angle between the modified Newton-Raphson direction and that of steepest descent is less chaotic.

Table 10.1 shows the final value of the objective function for the three test cases in single precision or double precision when available: this value is smaller using AD than using divided differences.

One can first conclude that changing divided differences to AD in the

TABLE 10.2. Execution time (Test case 1)

	One step		Total	
	Time (s)	DA/DD	Time (s)	DA/DD
DD-R4	159.7	–	1757.10	–
DA-R4	297.0	1.86	7723.57	–
DD-R8	283.6	–	3119.29	–
DA-R8	455.6	1.61	4555.98	1.46

optimizer studied here leads to correct results: the optimizer ends properly, and the quality of the final powder spectrum (evaluated by the optimizer itself) has improved.

We consider the cost of the change from divided differences to AD. Table 10.2 shows execution times in seconds and time ratios between AD and DD for Test case 1 in single and double precision. If one optimization step is considered (see column One step), the extra cost in execution time due to the use of AD is $0.86 * t$ in single precision and $0.61 * t$ in double precision. This is really small compared to the increase in quality of the solution. If the total optimization process (see column Total) is taken into account (only meaningful in double precision), the time overhead is even smaller $0.46 * T$ because the number of optimization steps is 10 instead of 11. We could use the vector forward mode of ADIFOR [78] or TAMC [211] to compute the Jacobian matrix in one shot. This would result in a runtime reduction when computing n directional derivatives. The theoretical cost of n directional derivatives is $(1 + C) * n$ using the forward mode, and $1 + n * C$ using the vector forward mode where $0 < C \le 2$ in practice. This first improvement will not lead to a great gain in execution time limited by the actual optimizer's capabilities.

In future work, we intend to change the optimizer to a gradient based one. Then the reverse mode AD will be applied on the fitting functional to generate its gradient (using Odyssée or TAMC). The calculated signal is a double integration over the sphere implemented as two nested loops. The fitting function is a sum (implemented as a loop) over the calculated signal: three loops are then necessary to compute this function.

A loop may be a bad case when using the reverse mode if all the iterations are recorded. This problem is being theoretically studied from several points of view. We summarize different possible optimizations that can be performed on a loop depending of its properties.

If the loop computes a contractive fixed point, it can be treated with contractive reverse [238, Ch. 11] with no extra storage. If the loop is truly iterative, the number of values to be recorded to compute backward the adjoint variables of all iterations is proportional to the number of iterations. The optimal checkpointing [235, 255] method can also be applied to replace at the lowest cost storage by recomputation.

If the loop is not iterative, the steps are independent from one another. As a result, the original and derivative loops can be merged to record the values modified by one step instead of all the steps [271]. All these optimizations are not fully automatic in AD tools and are left to the user.

In our case, the two external loops are not iterative (they are sums), whereas the inner one (Gauss) is iterative. We will demonstrate that this situation is very favourable for storage optimization.

We conclude that the use of automatic differentiation allows for an efficient solution of a problem for which divided differences did not lead to a reliable solution.

Part III

Applications in ODE'S and Optimal Control

11

Continuous Optimal Control Sensitivity Analysis with AD

Jean-Baptiste Caillau and Joseph Noailles

ABSTRACT In order to apply a parametric method to a minimum time control problem in celestial mechanics, a sensitivity analysis is performed. The analysis is continuous in the sense that it is done in the infinite dimensional control setting. The resulting sufficient second order condition is evaluated by means of automatic differentiation, while the associated sensitivity derivative is computed by continuous reverse differentiation. The numerical results are given for several examples of orbit transfer, also illustrating the advantages of automatic differentiation over finite differences for the computation of gradients on the discretized problem.

11.1 Introduction

This chapter is concerned with the use of automatic differentiation (AD) in the context of sensitivity analysis of optimal control problems (here, the minimum time transfer of a satellite to a geostationary orbit [107, 404]). Whereas AD is commonly employed on approximated optimal control problems [117], it is seldom used before discretization, in the infinite dimensional setting typical of control. The originality of this article is a use of AD not only to compute gradients of the discretized problem, but also to perform a continuous sensitivity analysis (see also [90] in the case of PDEs). AD then turns to be an efficient way to deal with the cumbersome computations involved in real-life control problems.

The minimum time orbit transfer problem is briefly stated in §11.2. Then, an outline of the specific parametric technique developed to solve it is presented in §11.3; its use requires the sensitivity analysis of interest here, which essentially amounts to integrating a Riccati equation evaluated by AD. The associated sensitivity derivative is computed by reverse differentiation. Some numerical results for the orbit transfer are given in §11.4, especially for very low thrust transfers. Besides, they demonstrate the relevance of AD to evaluate the gradients of the discrete algorithm.

11.2   Low Thrust Orbit Transfer

The problem motivating this study is the minimum time transfer of a satellite towards a geostationary orbit. The dynamics is expressed using the orbital parameters that define the ellipse osculating to the trajectory (since these coordinates are first integrals of the unperturbed motion, they are slowly varying. On the other hand, the expression of the dynamics becomes intricate). We use a more realistic model than in [107], taking into account the variation of the mass m, so that, on a suitable open submanifold of $\mathbb{R}^n$ (n is the dimension of the system; $n = 4$ for the 2D model, $n = 6$ for the 3D one), the dynamics can be written as:

$$\dot{x} = f_0(x) + B(x)u/m$$
$$\dot{m} = -\delta|u| \ ,$$

where the control u is the thrust of the engine, and $|.|$ is the Euclidean norm (see [107, 404] for more details). There are also boundary conditions defining the initial and the final orbit,

$$x(0) = x^0, \ m(0) = m^0, \ h(x(t_f)) = 0 \ ,$$

together with a constraint on the maximum modulus of the thrust:

$$|u| \leq F_{max}$$

with F_{max} small (low thrust transfer). The problem of finding an absolutely continuous state (x, m), and an essentially bounded control u that minimize the transfer time t_f will be referred to as $(SP)_{F_{max}}$. Among other results, it is proven in [104] that any optimal control has finitely many switchings so that $|u| = F_{max}$ almost everywhere. As a consequence, $m(t) = m^0 - \delta F_{max}t$ and $(SP)_{F_{max}}$ is reduced to a non-autonomous problem. The technique used to solve it is described in the next section.

11.3   Continuous Sensitivity Analysis

Rather than using direct methods (e.g., direct transcription) that lead to nonlinear programming, we emphasize indirect approaches. They are faster and more accurate for our problem (see [103, 105, 347] for comparisons). Their main drawback is the loss of robustness: the sensitivity of single shooting to the initialization of the adjoint state is well-known [16]. In an attempt to deal with these difficulties, a new parametric technique is introduced in [107] for minimum time control problems. We here give an outline of the method for the problem of §11.2. If we denote by $\phi(\beta)$ the value function of the optimal control problem $(SP)^{\beta}_{F_{max}}$ with fixed final time β (reformulated for convenience on $[0, 1]$ with an obvious change of variables)

$$1/2|h(x(1))|^2 \to \min$$
$$\dot{x} = \beta f(\beta t, x, u)$$
$$x(0) = x^0$$
$$|u| \leq F_{max}$$

(with $f(t, x, u) = f_0(x) + B(x)u/m(t)$) then the original problem is clearly equivalent to finding the first zero $\bar{\beta}$ of ϕ (which gives a measurement of the non-controllability of the system with respect to the end-point constraint for a prescribed final time β). The advantages of this approach are of three kinds [107]: first, thanks to the separate management of the criterion (that would be treated like any other variable by single shooting), the sensitivity to the initialization of t_f is reduced. Moreover, the ordered search provided by a Newton-like search on ϕ prevents the algorithm from finding too coarse local minima. Finally, since shooting (that will be used on the auxiliary problems, see §11.4) is embedded in this Newton process, the sensitivity to the adjoint state is attenuated too.

Here, though we can prove that ϕ is Lipschitz–and hence almost everywhere differentiable–we need C^1-regularity in order to apply Newton's method to the equation $\phi(\beta) = 0$. To this end, we use the recent sensitivity results for optimal control of [368, 369]. Of course, one could apply a direct method to $(SP)^\beta_{F_{max}}$, approximate ϕ by the resulting value function, and use AD to compute the gradients involved in the verification of the usual finite-dimensional sufficient conditions for sensitivity analysis. Again, in order to preserve the continuous information, we prefer to postpone the discretization process and perform the analysis on the continuous form. As in finite dimensions, the idea is to construct an extremal family and to ensure local optimality by sufficient second order conditions that will also be checked by means of AD. For β in $]0, \bar{\beta}[$, if $(x(., \beta), u(., \beta))$ is a solution to $(SP)^\beta_{F_{max}}$, the Pontryagin maximum principle holds, and there is an absolutely continuous adjoint state $p(., \beta)$ such that $y = (x, p)$ is a solution of the boundary value problem $(BVP)_\beta$

$$\dot{x} = \partial_p H(t, x, p, u(x, p), \beta) \tag{11.1}$$
$$\dot{p} = -\partial_x H(t, x, p, u(x, p), \beta) \tag{11.2}$$
$$x(0) = x^0, \ p(1) = {}^t h'(x(1))h(x(1)) \tag{11.3}$$

with $H(t, x, p, u, \beta) = \beta(p|f(\beta t, x, u))$ the Hamiltonian and

$$u(., \beta) = u(x(., \beta), p(., \beta))$$
$$= -F_{max} {}^t B(x(., \beta))p(., \beta)/|{}^t B(x(., \beta))p(., \beta)|$$

whenever ${}^t B(x(., \beta))p(., \beta)$ does not vanish. Actually, we assume that

(I1) $u(.,\beta)$ is continuous

Then, if $Z(t, y, \beta)$ denotes the second member of (11.1-11.2), $Z = (Z_1, Z_2) = (\partial_p H(t, x, p, u(x, p), \beta), -\partial_x H(t, x, p, u(x, p), \beta))$, if $\varphi(t, y, \beta)$ is the smooth maximal flow of $\dot{y} = Z(t, y, \beta)$, $(BVP)_\beta$ is equivalent to the shooting equation: find p^0 in $\mathbb{R}^n$ such that

$$S(p^0, \beta) = 0$$

with $S(p^0, \beta) = b(\varphi(1, x^0, p^0, \beta))$ (where $b(y) = p - {}^t h'(x)h(x)$ is the boundary condition of (11.3)). Finally, we need a second regularity condition

(I2) $\partial_p S(p(0, \beta), \beta)$ belongs to $\mathrm{GL}_n(\mathbb{R})$

together with a coercivity condition

(I3) the symmetric Riccati equation below has a bounded solution on $[0, 1]$:

$$\dot{Q} = -Q\mathcal{A}(t, \beta) - {}^t\mathcal{A}(t, \beta)Q + Q\mathcal{B}(t, \beta)Q - \mathcal{C}(t, \beta) \tag{11.4}$$
$$((R^f - Q(1))v|v) \geq 0, \ v \in \mathbb{R}^n \tag{11.5}$$
$$\mathcal{A}(t, \beta) = \partial_x Z_1(t, y(t, \beta), \beta), \ \mathcal{B}(t, \beta) = \partial_p Z_1(t, y(t, \beta), \beta)$$
$$\mathcal{C}(t, \beta) = \partial_x Z_2(t, y(t, \beta), \beta) ,$$

where R^f is a fixed n by n matrix. Then, we are able to prove that ϕ is $\mathbb{C}^1$ and to give a very simple closed form of its derivative. Indeed, taking advantage of the fact that the constraint on the control is active everywhere (assumption (I1)), the parametric problem $(SP)^\beta_{F_{max}}$ can be rewritten as an abstract optimization problem with equality constraints

$$J(z, \beta) \to \min$$
$$F(z, \beta) = 0$$

with $z = (x, u)$ and obvious expressions for J and F. Then, if we define the Lagrangian $L(z, \lambda, \beta) = J(z, \beta) + \langle \lambda, F(z, \beta) \rangle$ (where $\langle ., . \rangle$ is the duality pairing on the codomain of F), since the dependence $\beta \mapsto (z(\beta), \lambda(\beta))$ is $\mathbb{C}^1$ under the previous assumptions, and since $\partial_z L(z(\beta), \lambda(\beta), \beta) = 0$ (KKT condition), we can compute $\phi'(\beta) = d/d\beta \, J(z(\beta), \beta)$ by reverse differentiation [176] (here on the continuous problem), and get

$$\phi'(\beta) = \partial_\beta L(z(\beta), \lambda(\beta), \beta) . \tag{11.6}$$

As a result, we have

Proposition 1 *Under assumptions (I1)-(I3), ϕ is $\mathbb{C}^1$ on $]0, \bar{\beta}[$ and*

$$\phi'(\beta) = H(1, \beta)/\beta . \tag{11.7}$$

Proof. We just need to check the assumptions of the sensitivity analysis result of [369]. For a given β_0 in $]0, \bar{\beta}[$, $(SP)^{\beta_0}_{F_{max}}$ has a solution (x_0, u_0) and an adjoint state p_0. The control is smooth by virtue of (I1) and, if $\tilde{H}(t, x, p, u, \mu, \beta) = \beta(p|f(\beta t, x, u)) + 1/2\mu(|u|^2 - F^2_{max})$ is the augmented Hamiltonian (μ scalar multiplier associated with the inequality constraint $1/2(|u|^2 - F^2_{max}) \leq 0$), one has $\nabla_u \tilde{H}(t, x_0, p_0, u_0, \mu_0, \beta_0) = 0$ by taking $\mu_0 = \beta_0|^tB(x_0)p_0|/(m(t)F_{max}) \geq 0$. Accordingly, μ_0 is smooth and (I1) implies that strict complementarity holds ($\mu_0 > 0$ on $[0, 1]$). Moreover, $\nabla^2_{uu}\tilde{H}(t, x_0, p_0, u_0, \mu_0, \beta_0) = \mu_0\mathbf{I}$ (**I** identity matrix) in order that the strict Legendre-Clebsch condition is fulfilled. Then, with (I2) and (I3), all the assumptions of [369] are valid so that, for any β in an open neighbourhood of β_0,

$$\phi'(\beta) = \int_0^1 \partial_\beta H(t, \beta)dt$$

by virtue of (11.6) (lemma 1 of [107]). Now, along the optimal trajectory,

$$d/dt(tH) = H + t\dot{H} = H + t\partial_t H = \beta\partial_\beta H \ ,$$

so $\phi'(\beta) = H(1, \beta)/\beta$, which concludes the proof.   □

Both conditions (I2) and (I3) are only verifiable numerically: (I2) is simply the regularity of the Jacobian of the shooting function (checked when solving the auxiliary problem $(SP)^{\beta}_{F_{max}}$ by shooting). The Ricatti equation (11.4-11.5) requires the computation of $3n^2$ partial derivatives and is assembled using AD as explained in the next section.

11.4 Numerical Results

The numerical computation is done in two steps. First, for a given thrust, $(SP)_{F_{max}}$ is solved by the parametric approach of §11.3; $\phi(\beta)$ is evaluated by shooting on the auxiliary problem $(SP)^{\beta}_{F_{max}}$ (which allows the numerical verification of (I2)); and ϕ' is evaluated using either (11.7), which is extremely easy to compute, or a finite differences approximation, depending on the precision of the shooting resolution. Then, the Riccati equation (I3) is integrated to check coercivity, backwards since it is straightforward to find a matrix $Q(1)$ matching the boundary condition (11.5), and since $y(1, \beta)$ is provided by shooting. AD is used to generate the gradients. Indeed, we deal with a large system, so there are many derivatives to determine ($3n^2$, that is 48 in 2D, and 108 in the 3D case). Moreover, computing the required second order derivatives of the dynamics is a cumbersome process because of the choice of coordinates (one has to deal with trigonometric rational fractions). In the 3D case for instance, even the first order derivatives of the dynamics for the adjoint equation are evaluated by AD. As the Fortran code for the boundary value problem was available, ADIFOR2.0 [79] was a natural choice. AD is also used in a more classical but yet efficient fashion to find the exact Jacobian of the shooting function, $\partial_p S(p^0, \beta)$, by differentiating the numerical integrator (Runge-Kutta of order 4). A comparison with finite differences (FD) is provided Table 11.1 where the transfer times for various thrusts are detailed in the 2D case (compare [347, 208, 404, 103]). The results we obtained using AD are systematically more accurate. Two examples of optimal trajectories and verification of the coercivity condition are given in Figures 11.1 and 11.2.

TABLE 11.1. Minimum transfer times

F_{max}	t_f FD	AD	$\phi(t_f)$ FD	AD	Execution FD	AD
60	14.732	14.732	5e-28	7e-29	12	14
24	34.133	34.133	2e-22	3e-27	25	25
12	69.294	69.294	2e-25	2e-21	60	40
9	93.187	91.930	3e-19	1e-26	54	70
6	141.64	139.37	3e-13	2e-17	122	86
3	278.98	278.98	1e-24	1e-27	285	217
2	420.10	420.10	1e-17	1e-26	257	485
1.4	597.92	598.12	4e-18	5e-13	485	648
1	839.97	836.86	5e-12	3e-13	496	504
0.7	1195.7	1195.7	2e-12	9e-15	1084	1106
0.5	1685.2	1674.8	3e-12	2e-12	1978	1391
0.3	2838.4	2797.7	7e-10	4e-13	2128	1938

Thrusts are in Newtons, transfer times in hours, and execution times (on an HP PA-C160) in seconds.

11.5   Conclusion

Automatic differentiation has been used in two ways: on the discretized problem to evaluate the Jacobian of the shooting function, but also on the original control problem to check a continuous sensitivity condition (and then to find a continuous exact gradient by reverse differentiation). In the first case, AD provides a much more accurate computation than finite differences. In the second, it is the most practical way to assemble the Riccati equation connected to the coercivity condition. For the real-life control problem considered, the large dimension of the dynamics together with its complexity make any hand-made computation cumbersome, if not impracticable. A similar analysis with respect to the parameter F_{max} is currently worked out [102].

Acknowledgments: This work was supported in part by the French Space Agency (CNES-ENSEEIHT contract 871/94/CNES/1454) and the French Ministry for Education, Higher Education and Research (grant 20INP96). The authors are also indebted to Professor Y. Evtushenko for enlightening discussions during the Conference.

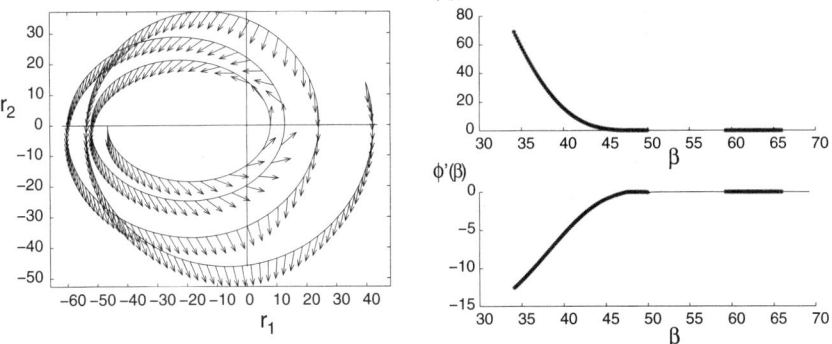

FIGURE 11.1. Thrust of 12 Newtons (3 day transfer). Left, the optimal trajectory (the arrows represent the control). Right, the evaluation of ϕ and ϕ'. Points where the Riccati equation has been successfully integrated are marked with a $*$.

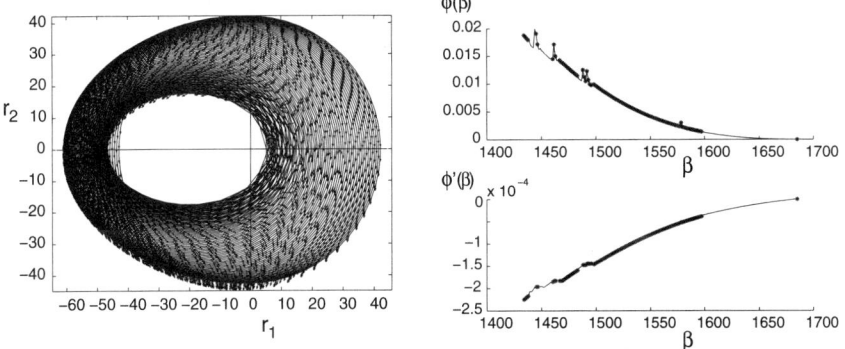

FIGURE 11.2. Thrust of 0.5 Newton (2 month transfer). The result is typical of the low thrust case: the coercivity condition is more difficult to check in the neighbourhood of the solution, and jumps are observed on the function, due to local minima.

12

Application of Automatic Differentiation to Race Car Performance Optimisation

Daniele Casanova, Robin S. Sharp, Mark Final, Bruce Christianson and Pat Symonds

ABSTRACT A formal method for the evaluation of the minimum time vehicle manoeuvre is described. The problem is treated as one of optimal control and is solved using a direct transcription method. The resulting nonlinear programming problem is solved using the sequential quadratic programming algorithm SNOPT for constrained optimisation. The automatic differentiation software tool **AD**opt is used for the evaluation of the first-order derivatives of objective and constraint functions with respect to the control variables. The implementation of automatic differentiation is more robust and ten times as fast compared to the use of a finite difference determination of the Jacobian.

12.1 Introduction

Computer simulation is now widely used in circuit racing to predict the maximum performance of a racing car as a function of vehicle parameter changes. The methods which are currently implemented assume that the racing line is known by some means and is divided into arbitrarily short segments with constant curvature where the vehicle operates at its steady-state performance limits [252]. However, the assumption that the race car maneuvring may be treated as a sequence of steady-state conditions is a strong limitation, since many vehicle design parameters have a significant effect on the vehicle transient behaviour, while they are not influential in steady-state conditions. Furthermore, the assumption that the racing line is known is seen as a constraint, as it is known that the optimal driving strategy is different for different vehicle set-ups.

An alternative approach is to define the minimum race car lap time problem as one of optimal control. The task is to find the optimal vehicle controls, i.e. steer angle and driving/braking torque, which allow the vehicle to traverse a lap of the circuit in as short a time as possible without violating the road boundary constraints. This problem was firstly solved by Hendrikx et al. [274] for a single lane change manoeuvre using the Pontryagin's Minimum Principle [324]. More recently Allen [9] developed a solution equivalent to that of Hendrikx using a direct shooting method [333] and mathematical programming techniques. This avoided the need to derive the adjoint equations, which may become very difficult or even impossible as the complexity of the mathematical model of the mechanical system grows, and offered the possibility of being computationally more efficient than Hendrikx's approach.

In this chapter the work of Allen is extended by applying automatic differentiation (AD) to find the sensitivity of the objective function and the constraints with respect to the controls. The vehicle representation includes five degrees of freedom, giving a system of ten first-order differential equations, and various nonlinear functions representing realistic aerodynamic and tyre force systems. The results for a simple, one-turn vehicle manoeuvre are presented and are compared with the solutions obtained by a finite difference determination of the derivatives.

12.2   Problem Definition

Let the differential equations for a general vehicle model, representing the rate of change of its n state variables $\mathbf{x}$ with respect to the travelled distance s, measured on the road centre line, be defined as follows

$$\dot{\mathbf{x}} = \mathbf{a}\left(\mathbf{x}\left(s\right), \mathbf{u}\left(s\right), s\right) \qquad \mathbf{x}\left(0\right) = \mathbf{x}_0 \qquad s \in [0, S] \;, \qquad (12.1)$$

where $\mathbf{u}\left(s\right)$ is a two element vector function returning the vehicle lateral (i.e. steer angle) and longitudinal (i.e. driving/braking torque) controls.

The objective function for the optimal control problem is the manoeuvre time. Let us adjoin one more state variable x_{n+1} which represents the time elapsed from the beginning of the manoeuvre, and which satisfies

$$\dot{x}_{n+1} = \frac{dt}{ds} = \frac{1}{\mathbf{V}} \;,$$

where the components of the velocity $\mathbf{V}$ are already part of the state vector. Then, the objective function is

$$F = x_{n+1}\left(S\right) \;. \qquad (12.2)$$

The road is described by the co-ordinates of its centre line and by its width, expressed as functions of the path co-ordinate s, i.e. the travelled distance. Since the vehicle states are functions of the same independent variable, the position of the car relative to the road centre line may be evaluated directly during the integration of Equation 12.1. The condition for the car to stay within the road boundaries is imposed by identifying m segments along the vehicle trajectory where the racing line is expected to lie on the road edges in order to maximise the cornering speed. Within each segment the distance d of the car's centre of gravity from the road centre line is monitored, and the maximum value is used to evaluate the constraints

$$C_i = \left(\frac{d_{\mathrm{MAX}}}{w}\right)^2 - 1 \qquad 0 \le s_i \le S \qquad i = 1, \ldots, m \; ,$$

where w is half the road width. Hence, the road boundary constraints assume negative values when the vehicle is within the road boundaries and positive values otherwise.

A *direct shooting method* [333] has been implemented for the solution of the minimum time vehicle manoeuvring. The original optimal control problem is converted into a nonlinear programming problem by replacing the continuous control histories with discrete approximations. It is assumed that the control action can only be adjusted at a number of fixed positions along the trajectory, while the intermediate values are estimated by means of interpolation techniques. For simplicity, a linear interpolation method has been used for the present work. Formally, if $\mathbf{u}_n$ is the vector of discrete control parameters and $\mathbf{s}_n$ is the vector of the corresponding instances within the interval $[0, S]$, one may write

$$\mathbf{u}(s) = \mathrm{lin\_interp}(\mathbf{s}_n, \mathbf{u}_n, s) \; .$$

The control parameters uniquely determine the vehicle state trajectories which, in turn, determine the objective and constraint functions. Therefore, the objective and the constraints may be expressed directly as functions of these control parameters. Hence, the resulting nonlinear programming problem may be stated as follows

$$\begin{aligned}
\min_{\mathbf{u}_n} \quad & F(\mathbf{u}_n) \\
\text{such that} \quad & C_i(\mathbf{u}_n) \le 0 \qquad i = 1, \ldots, m \\
\text{and} \quad & \mathbf{u}_n^L \le \mathbf{u}_n \le \mathbf{u}_n^U \; ,
\end{aligned} \qquad (12.3)$$

where $\mathbf{u}_n^U$ and $\mathbf{u}_n^L$ represent the upper and lower control bounds, respectively.

12.3   Application of Automatic Differentiation

Problem 12.3 is solved using the sequential quadratic programming (SQP) algorithm SNOPT [221]. This requires the evaluation of the gradient of the objective function and the Jacobian of the constraints with respect to all the control parameters $\mathbf{u}_n$. Although finite difference determination of the derivatives may be used, they are time consuming and noisy. A further complication lies in the nature of the minimum time vehicle manoeuvring problem, which implies driving the vehicle to the very limit of the tyre contact forces. In this condition the vehicle may become unstable, e.g., if the rear tyres saturate before the front ones. For the numerical simulation this makes the objective and the constraint functions very sensitive with respect to a tiny variation of the control input. Approximating the derivatives by applying a small variation to the controls becomes very difficult.

The derivatives required for the optimisation problem presented in this chapter are obtained using an (AD) research tool called $\mathbf{AD}opt$, developed at the Numerical Optimisation Centre, Hatfield. $\mathbf{AD}opt$ is an object oriented, interpretative AD tool written in the C++ programming language. Both forward and reverse schemes are implemented to differentiate scalar or vector-valued functions. They use an arithmetic type called scalar Doublets in order to calculate derivatives.

$\mathbf{AD}opt$ uses a Wengert-type list [498] to represent the target function. Consider such a representation of a scalar function f of dimension n. Then the scalar Doublet corresponding to an intermediate result, t, in that representation of f is defined as

$$\mathcal{T} = (t, \dot{t}) = (t, \mathbf{p} \cdot \nabla t) \ ,$$

where $\mathbf{p}$ is an arbitrary vector of order n, and ∇ is the gradient operator. Hence, at the conclusion of a forward pass of the representation of f, the scalar Doublet corresponding to the dependent variable is $\mathcal{F} = (f, \mathbf{p} \cdot \nabla f)$, i.e. the function value and a directional gradient.

Scalar Doublets may also be used in reverse passes, and the adjoint scalar Doublet of the same variable t as above, is defined as

$$\bar{\mathcal{T}} = (\bar{t}, \mathbf{p} \cdot \nabla \bar{t}) \ .$$

Griewank [232] defines the adjoint of the variable t as $\bar{t} = \frac{\partial f}{\partial t}$. Hence the 'dotted' half of the scalar Doublet adjoint of the ith independent variable corresponds to the dot product of $\mathbf{p}$ with the ith row (or column, due to symmetry) of the second derivative matrix of f. Thus a reverse pass of the function's representation yields a gradient and a directional Hessian vector.

Intermediate results in the forward and reverse passes of the function representation are calculated on a run-time stack using the arithmetic of scalar Doublets—directional first-order calculus. For example, the product of two scalar Doublets $\mathcal{X} = (x, \dot{x})$ and $\mathcal{Y} = (y, \dot{y})$ is given by a version of the product rule

$$\mathcal{X}\mathcal{Y} = (xy, y\dot{x} + x\dot{y})$$

The advantage of using scalar Doublet arithmetic is that both components of the data type are scalar values and so the evaluation of all expressions involves only calculations with scalars.

The derivatives available from $\mathbf{AD}opt$ are sufficient and necessary for unconstrained optimisation methods, such as the Truncated Newton as well as constrained optimisation methods such as the SQP algorithm, which is used to solve the problem described in this chapter.

Consider the initial value problem described in Equation 12.1. This problem is solved using a fixed step, second order Runge-Kutta integration formula (midpoint method). Assuming Δs as the length of the integration step, one may write

for $0 \leq s_i \leq S$

$$\mathbf{x}\left(s_i + \frac{\Delta s}{2}\right) = \mathbf{x}\left(s_i\right) + \frac{\Delta s}{2} \cdot \mathbf{a}\left(\mathbf{x}\left(s_i\right), \mathbf{u}\left(s_i\right), s\right)$$

$$\mathbf{x}\left(s_i + \Delta s\right) = \mathbf{x}\left(s_i\right) + \Delta s \cdot \mathbf{a}\left(\mathbf{x}\left(s_i + \frac{\Delta s}{2}\right), \mathbf{u}\left(s_i + \frac{\Delta s}{2}\right), s_i + \frac{\Delta s}{2}\right)$$

(12.4)

The above computation is implemented using the overloaded C++ operators defined in **ADopt**. This involves declaring all the variables of a type defined in **ADopt** identifying them as active in the AD sense, and defining which are the dependent and which are the independent variables. Specifically for the present problem:

- The control parameters $\mathbf{u}_n$ are the independent variables, i.e. the variables with respect to which derivatives are required;

- The state variables $\mathbf{x}$ and the actual control input $\mathbf{u}$ are intermediate variables, i.e. their derivatives are computed but not shown;

- The objective F and the constraints $\mathbf{C}$ are the dependent variables, i.e. the variables of which derivatives are required.

As a side-effect of implementing the method described in Equation 12.4 using AD arithmetic, the evaluation of the state differential equations provides the Jacobian of the state variables at any intermediate integration step

$$\frac{\partial \mathbf{x}\left(s_i\right)}{\partial \mathbf{u}_n} \qquad 0 \leq s_i \leq S .$$

Therefore, the evaluation of the road boundary constraints automatically yields their Jacobian

$$C_i = c\left(\mathbf{x}\left(s_i\right)\right) \qquad \frac{\partial C_i}{\partial \mathbf{u}_n} = \frac{\partial c}{\partial \mathbf{x}} \frac{\partial \mathbf{x}\left(s_i\right)}{\partial \mathbf{u}_n} \qquad 0 \leq s_i \leq S \qquad i = 1, \ldots, m .$$

Analogously, evaluation of the manoeuvre time at the end of the integration yields objective and gradient

$$F = x_{n+1}\left(S\right) \qquad \frac{\partial F}{\partial \mathbf{u}_n} = \frac{\partial x_{n+1}\left(S\right)}{\partial \mathbf{u}_n} .$$

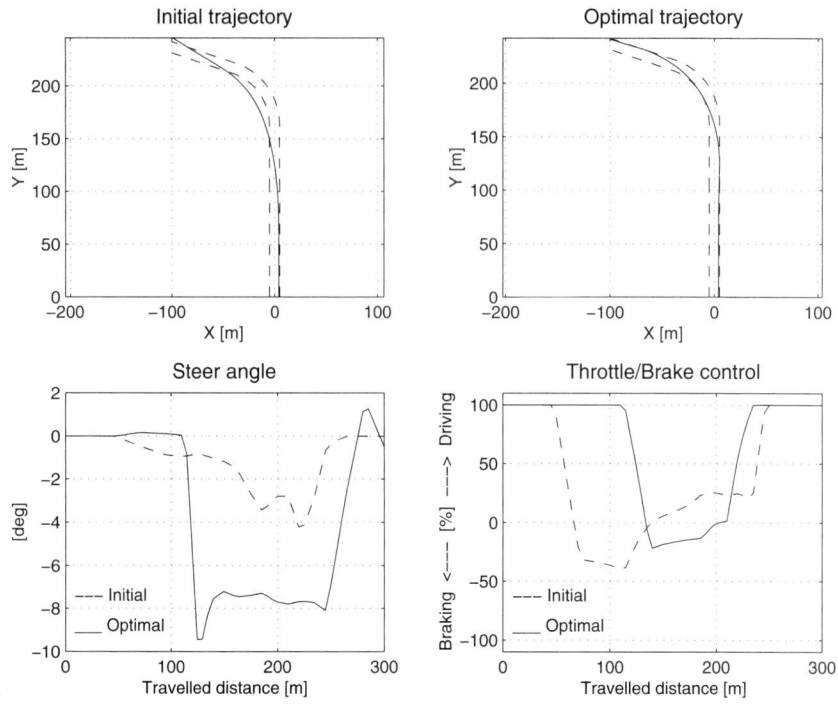

FIGURE 12.1. Optimisation results

12.4 Numerical Results

As an example for the application of the above framework, consider a manoeuvre involving a single corner. Figure 12.1 shows the initial and optimal
vehicle trajectories and controls for the manoeuvre. The road section is 300
[m] long. The control interval has been chosen to be equal to 12 [m] for
both steer angle and throttle/brake controls. Thus, the problem has 50 independent variables in total (two control variables for 25 control nodes).
The objective function is the time it takes for the vehicle to traverse the
road section and is evaluated using Equation 12.2. Finally, there are three
road boundary constraints which are located where the optimal trajectory
is tangent to the road boundaries.

Initially, a convergence test is performed to identify the integration step
ensuring sufficient accuracy for the objective and constraint values as well as
for their derivatives. Then, the same optimisation problem is solved several
times while steadily decreasing the optimality tolerance in order to find
the maximum precision achievable. The same procedure is finally repeated,
computing the derivatives by means of finite differences for comparison. All
the optimisation cases are solved starting from the same initial guess for

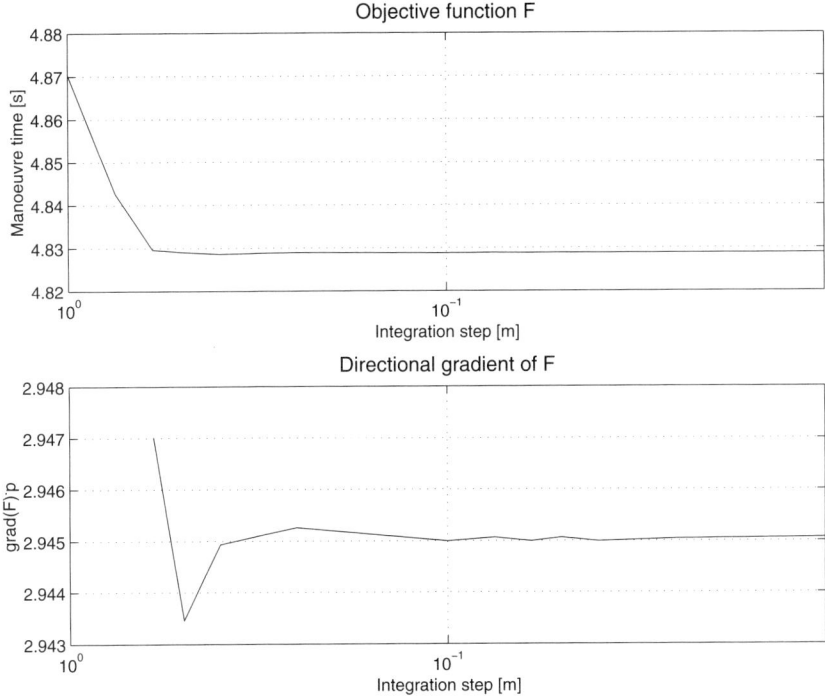

FIGURE 12.2. Convergence test for the identification of the integration step

the solution. The initial value for the objective function was 4.829 [s].

Figure 12.2 shows the initial objective value and its corresponding directional gradient as functions of the integration step. A value equal to 0.05 [m] was chosen for Δs, which ensured an absolute precision for the objective function of the order of 10^{-4} [s]. Table 12.1 reports the numerical results,

TABLE 12.1. Summary of the results (* did not converge)

$OptTol$	Automatic Differentiation			Finite Differences		
	n_ite	$CPU\ time$	obj	n_ite	$CPU\ time$	obj
2.0e-3	42	263 s	4.213 s	42	3137 s	4.213 s
1.0e-3	62	378 s	4.201 s	52	4206 s	4.207 s
5.0e-4	93	563 s	4.194 s	133*	12616 s	4.194 s
2.0e-4	115	746 s	4.193 s	-	-	-

i.e. the number of iterations, the CPU time (referring to a single processor workstation Alpha Digital EV6 at 700 MHz with 512 Mb of RAM, Unix operating system and "gcc" compiler version 2.95.2) and the final objective function values, obtained for four values of optimality tolerance. The program using finite differences failed to converge for tolerances smaller than 10^{-3}. The optimisation algorithm stopped as it was unable to improve the objective function further to meet the required tolerance. Furthermore the second run stopped after only 52 iterations, returning a larger value for the objective function compared to the analogous case using AD, indicating that the corresponding tolerance was too large.

12.5   Conclusions

The minimum time vehicle manoeuvring problem has been set up and solved as an optimal control problem using a direct method. The AD tool **AD**$_{opt}$ has been successfully integrated into the optimisation software. For the particular case of the vehicle simulation, the application of a simple second order Runge-Kutta method allows us to achieve sufficient accuracy for the solution of the equations of motion as well as for the evaluation of the derivatives. The results indicate that considerable time saving and greater accuracy are obtained compared to the same problem solved using more traditional numerical differentiation techniques.

13

Globalization of Pantoja's Optimal Control Algorithm

Bruce Christianson and Michael Bartholomew-Biggs

ABSTRACT In 1983 Pantoja described a stagewise construction of the Newton direction for a general class of discrete time optimal control problems. His algorithm incurs amazingly low overheads: the cost (measured in target function evaluations) is independent of the number of discrete time-steps. The algorithm can be modified to verify that the Hessian contains no eigenvalues less than a postulated quantity, and to produce an appropriate descent direction in the case where the Hessian fails to be positive definite and global convergence becomes an issue. Coleman and Liao have proposed a specific damping strategy in this context. Here we describe how automatic differentiation can be used to implement Pantoja's algorithm, and we briefly consider some alternative globalization strategies, within which AD techniques can be further deployed.

13.1 Introduction

Consider the following optimal control problem: choose $u_i \in R^p$ for $0 \leq i < N$ to minimize

$$z = F(x_N) \, ,$$

where x_0 is some fixed constant, and $x_{i+1} = f_i(x_i, u_i)$ for $0 \leq i < N$. Here each f_i is a smooth map from $R^q \times R^p \to R^q$, F is a smooth map from R^q to R, and N is the number of discrete time-steps. The dimensions p_i and q_i may depend upon i, but for notational convenience we omit these subscripts.

The more usual formulation of a discrete time optimal control problem, where z has the form $z = \sum_{i=0}^{N-1} F_i(x_i, u_i) + F_N(x_N)$, can be reduced to the form $z = F(x_N)$ by adjoining to each state x_i a new component $v_i \in R$ defined by $v_0 = 0, v_{i+1} = v_i + F_i(x_i, u_i)$ and then defining $F(x_N, v_N) = v_N + F_N(x_N)$. Consequently, we lose nothing by restricting attention to minimizing target functions of the form $z = F(x_N)$.

An obvious approach to solving this problem is to apply Newton's method. Dealing explicitly with quadratic approximations to the f_i is extremely inconvenient when performing actual calculations, because of interactions between the very large number of cross terms. It would be much more convenient if we could calculate the Newton direction in a way which required only linear approximations to the problem dynamics. Pantoja [409, 410] has shown that this can be done, provided we also consider linear approximations to the adjoint problem dynamics.

13.2   The Generalized Pantoja Algorithm

Let g be the block vector with i-th block given by $g_i = \bar{u}_i = [\partial z/\partial u_i]$. Let H be the block matrix with (i, j)-th block given by $H_{ij} = [\partial \bar{u}_i/\partial u_j] = [\partial^2 z/\partial u_i \partial u_j]$. Let Λ be a symmetric block diagonal matrix. For block vectors t and b we write $Ht = b$ to signify $\sum_{j=0}^{N-1} H_{ij} t_j = b_i$ for $0 \leq i < N$.

Given a starting position u_i and arbitrary values b_i for $0 \leq i < N$, to obtain values for t_i such that $(H + \Lambda)t = b$ proceed as follows.

Step 1. For i from 1 up to N calculate $x_{i+1} = f_i(x_i, u_i)$ where x_0 is a fixed constant. Define $\bar{x}_N = F'(x_N), a_N = 0 \in R^q; D_N = F''(x_N) \in R^{q \times q}$.

Step 2. For i from $N-1$ down to 0 calculate $\bar{x}_i, a_i \in R^q; \bar{u}_i, c_i \in R^p; A_i, D_i \in R^{q \times q}; B_i \in R^{p \times q}; C_i \in R^{p \times p}$ by

$$\bar{x}_i = \bar{x}_{i+1} \left[f'_{x,i} \right], \qquad \bar{u}_i = \bar{x}_{i+1} \left[f'_{u,i} \right]$$

$$A_i = \left[f'_{x,i} \right]^T D_{i+1} \left[f'_{x,i} \right] + (\bar{x}_{i+1}) \left[f''_{xx,i} \right]$$

$$B_i = \left[f'_{u,i} \right]^T D_{i+1} \left[f'_{x,i} \right] + (\bar{x}_{i+1}) \left[f''_{ux,i} \right]$$

$$C_i = \left[f'_{u,i} \right]^T D_{i+1} \left[f'_{u,i} \right] + (\bar{x}_{i+1}) \left[f''_{uu,i} \right] + \Lambda_i$$

where $[\cdot]$ denotes evaluation at (x_i, u_i), and we write (for example)

$$\left(\left[f'_{u,i} \right]^T D_{i+1} \left[f'_{x,i} \right] \right)_{j,k} \quad \text{for} \quad \sum_{\ell=1}^{q} \sum_{m=1}^{q} \left[\frac{\partial (x_{i+1})_\ell}{\partial (u_i)_j} \right] (D_{i+1})_{\ell,m} \left[\frac{\partial (x_{i+1})_m}{\partial (x_i)_k} \right] \quad \text{etc.}$$

If C_i is singular, the algorithm fails. Otherwise set

$$D_i = A_i - B_i^T C_i^{-1} B_i, \qquad E_i = C_i^{-1} B_i$$

$$c_i = a_{i+1} \left[f'_{u,i} \right] - b_i, \qquad a_i = a_{i+1} \left[f'_{x,i} \right] - c_i E_i.$$

Step 3. For i from 0 up to $N - 1$ calculate $t_i \in R^p, s_{i+1} \in R^q$ by

$$t_i = -E_i s_i - C_i^{-1} c_i^T, \qquad s_{i+1} = \left[f'_{x,i} \right] s_i + \left[f'_{u,i} \right] t_i$$

where $s_0 = 0 \in R^q$.   STOP.

Either this algorithm fails because some C_i is singular, or else at the end the t_i satisfy $\widehat{H}t = b$, where $\widehat{H} = H + \Lambda$. Consequently if all the C_i defined in Step 2 of the algorithm are invertible, then so is $\widehat{H}$.

If all the C_i are positive definite, then so is $\widehat{H}$. Conversely, if $\widehat{H}$ is positive definite then all the C_i are positive definite (and hence are invertible). In this case all eigenvalues of every C_i are bounded below by the smallest eigenvalue λ_0 of $\widehat{H}$, and furthermore $(t, b) = \sum_i t_i \cdot b_i > 0$ provided $b \neq 0$.

In the particular case where $b = -g$ and $\Lambda = O$, the algorithm is equivalent to the following modified form: in Step 1, replace the definition $a_N = 0$ by $a_N = \bar{x}_N$; in Step 2, simplify the calculation for c_i to $c_i = a_{i+1} \left[f'_{u,i} \right]$. This is the original algorithm, given by Pantoja in [410].

13.3 Implementation using AD

Automatic differentiation (AD) combined with checkpointing [235, 250] can be used to provide efficient implementations of the algorithm introduced in §13.2 [31, 128]. Accurate (truncation-free) second derivative values are required, because they occur in the recurrence relations which generate the linear equations to be solved at each time stage. Using AD, existing code to calculate the target function z does not require extensive (and error-prone) manual re-writing to evaluate these derivatives.

The forward accumulation technique of AD associates with each program variable v a vector $\dot{v}$, which contains numerical values for the partial derivatives of v with respect to each of the r independent variables. The combined structure $V = (v, \dot{v})$ is called a *doublet*. When the doublets U_k corresponding to the independent variables u_k are initialized by setting $\dot{u}_k$ to be the k−th Cartesian unit vector, we write this (rather loosely) as $[\dot{u}] = [I_r]$.

In the reverse accumulation technique of AD, a floating point *adjoint* variable $\bar{v}$ (initially zero) is associated with each program variable v. The adjoint variables $\bar{v}$ are updated, in the reverse order to the forward computation, so that at each stage they contain the numerical value of the partial derivative of the dependent variable y with respect to v at the corresponding point in the forward computation.

The forward and reverse techniques can be combined to calculate Hessian matrices. We embed doublet arithmetic into an implementation of reverse AD: each program variable value is a doublet rather than a real, and so is each corresponding adjoint variable value. After initializing $[\dot{u}]$ we calculate $Y = f(U)$ giving $\dot{y} = f'(u)\dot{u}$ as before. We then initialize $\bar{Y}$ and perform the reverse pass in doublet arithmetic, following which we have $[\bar{u}] = \bar{y}[f'(u)]$ as before, and

$$[\dot{\bar{u}}] = \bar{y}[f''(u)]\dot{u} + \dot{\bar{y}}[f'(u)]$$

The values required by Pantoja's algorithm can be evaluated by choosing suitable initial values for $\dot{u}$ and $\bar{Y}$.

For example, in Step 1 of the algorithm, we define doublets X_N with scalar parts x_N respectively, and vector parts (of length q) given by $\dot{x}_N = [I_q]$. Evaluate $Z = F(X_N)$ in doublets, recording a trajectory. We now have $\dot{z} = [F'(x_N)]$. Define $\bar{Z}$ by setting $\bar{z} = 1.0, \dot{\bar{z}} = 0_q$. Reverse through the computation for Z to obtain the doublets $\bar{X}_N$. We have $\bar{x}_N = [F'(x_N)]^T, \dot{\bar{x}}_N =$

$[F''(x_N)]$. Set $D_N = \dot{\bar{x}}_N$.

Similarly, consider Step 2 of the algorithm. We wish to calculate $\bar{x}_i, a_i$ and D_i assuming that the corresponding quantities are available for $i+1$. Define doublets X_i and U_i with scalar parts x_i and u_i respectively, and vector parts (of length $q+p$) given by

$$\begin{bmatrix} \dot{x}_i \\ \dot{u}_i \end{bmatrix} = \begin{bmatrix} I_q & O \\ O & I_p \end{bmatrix}.$$

Evaluate $X_{i+1} = f_i(X_i, U_i)$ in doublets. Now we have $[\dot{x}_{i+1}] = [f'_{x,i}\ f'_{u,i}]$. Calculate the vectors $a_{i+1}\left[f'_{x,i}\right]$ and $c_i = a_{i+1}\left[f'_{u,i}\right] - b_i$. Define $\bar{X}_{i+1}$ by setting $\bar{x}_{i+1}$ to the supplied value and setting

$$\begin{bmatrix} \dot{\bar{x}}_{i+1} \end{bmatrix} = \begin{bmatrix} D_{i+1}f'_{x,i} & D_{i+1}f'_{u,i} \end{bmatrix}.$$

Reverse through the trace for X_{i+1} to obtain the doublets $\bar{X}_i$ and $\bar{U}_i$. Then

$$\begin{bmatrix} \dot{\bar{x}}_i \\ \dot{\bar{u}}_i \end{bmatrix} = \begin{bmatrix} A_i & B_i^T \\ B_i & C_i \end{bmatrix}.$$

Row reduction gives

$$\begin{bmatrix} A_i - B_i^T C_i^{-1} B_i & O \\ C_i^{-1} B_i & I \end{bmatrix} = \begin{bmatrix} D_i & O \\ E_i & I \end{bmatrix}.$$

Now $\bar{x}_i, a_i = a_{i+1}\left[f'_{x,i}\right] - c_i E_i$ and D_i are available for the next iteration.

An implementation of Pantoja's algorithm using AD typically requires on the order of $p+q$ times the floating point cost of an evaluation of z, independent of N, together with order $(p+q)^3$ multiply-and-add operations per time step (less if there is structural sparsity). Checkpointing can be used to reduce the total storage requirement to the order of $4p$ floating point stores per time step. Detailed time and space bounds are reported in [128].

13.4   Globalization Strategies

If H is positive definite, and we take $\Lambda = 0$, then all the C_i are invertible. Hence, near a second order minimum, the algorithm will successfully produce the Newton direction t. We can verify that we are at such a minimum by checking for each i that $\bar{u}_i = 0$ and C_i is positive definite, even if we did not use Newton to find the optimal point. The algorithm can also be used to inform us efficiently if H fails to be positive definite at an arbitrary point of interest, or has an eigenvalue falling below a specified threshold.

However, if good initial estimates are not available for the control variables, then H may be indefinite. In this case the algorithm may fail, or may produce a t which is not a descent direction. An alternative strategy to obtain a search direction is required in order to ensure global convergence of the algorithm. A straightforward way of doing this is to choose Λ to ensure that all the C_i are positive definite.

Coleman and Liao [135] have proposed modifications to Pantoja's original algorithm to achieve this, based upon the Nocedal-Yuan trust-region method. This approach sets $\Lambda = \lambda I$ and adjusts λ so that the resulting step falls within the trust region r. The adjustment requires the solution for t and d of the pair of equations $(H + \Lambda)t = -g, (H + \Lambda)d = t$, (since to first order $\delta t = \delta \lambda d$, whence $\delta \lambda = (r^2 - t^2)/2t \cdot d$, see [135, §2.2]).

Essentially Coleman and Liao construct a solution of $(H + \lambda I)t = b$ for given λ and b by applying Pantoja's algorithm to a modified target function. In contrast, the approach formulated in §13.2 constructs such solutions by applying the generalized algorithm, with $\Lambda = \lambda I$, to the original target function. As well as producing a modest reduction in the operation count [128], the generalized algorithm additionally allows the components of b and Λ to be specified dynamically rather than in advance.

Thus rather than restrict Λ to a multiple of the identity, we can choose Λ as the algorithm proceeds to ensure that $\widehat{H} = H + \Lambda$ has no eigenvalues smaller than the positive value $\lambda_0 = \|g\|/r$. In this case the equation $Ht = -g$ is guaranteed to have a solution with $\|t\| \leq \|g\|/\lambda_0 = r$. To ensure this, we conjointly evaluate the sequence $\widehat{C}_i$ corresponding to $\widehat{\Lambda}_i = \Lambda_i - \lambda_0 I_p$ and choose Λ_i to ensure that the $\widehat{C}_i$ are positive definite at each stage.

A trust region approach is not the only established mechanism for ensuring global convergence. The requirement that Λ_i be chosen to make C_i positive definite in the algorithm amounts just to saying that if, in the course of inverting C_i, we hit a pivot which is not significantly positive, then we can replace that pivot by a positive value (say 10^{-3} scaled by the norm of the corresponding row) and carry on. This process builds up a diagonal matrix Λ with the amounts added to the pivots at stage i forming the corresponding diagonal Λ_i.

The eigenvalues of C_i are bounded below by the smallest eigenvalue of $\widehat{H}$, and the resulting value for t is guaranteed to be a descent direction, with the property that $g + Ht$ is zero everywhere except for blocks corresponding to those time steps at which the pivots required adjustment. Furthermore the algorithm can readily be extended to produce, corresponding to each such time step i, a direction $n^{(i)}$ of negative curvature for H, with $n_j^{(i)} = 0, j < i; (Hn^{(i)})_j = 0, j > i$. These concave directions $n^{(i)}$ can be combined with the descent direction t to establish a search direction in a number of ways. This type of combining is examined in a different context in [30].

In one alternative, suggested by Conforti and Mancini [143], a curvilinear search can be performed, selecting among steps of the form $\alpha t + \alpha^2 t'/2$, where $\widehat{H}t' + g' = 0$ and g' is the gradient $\nabla_u (t^T Ht)$.

Another possibility is to use AD to establish a quartic model on the subspace of directions, or to carry out a multi-dimensional search in the spirit of Miele and Cantrell [381].

These types of nonlinear search, combining directions using accurate higher-order derivative information obtained by AD, offer a promising avenue for future research, particularly where state or control constraints have been incorporated into z by penalty or barrier terms.

13.5   Concluding Remarks

Pantoja originally introduced his algorithm as a variant of Differential Dynamic Programming (DDP) [305, 392]. Traditional DDP algorithms and other state-control feedback regimes can also be implemented using AD: for example traditional DDP just puts $\bar{x}_{i+1}$ in place of a_{i+1} in the initialization of $\bar{X}_{i+1}$. One advantage of Pantoja's algorithm relative to other state-control feedback methods is that, because it provides the exact Newton step even far from the optimal point, we can use all the theory and tools (e.g., trust regions, line searches, termination criteria etc.) already available from the extensive literature on Newton's method.

Pantoja's algorithm operates on a discrete time optimal control problem. An optimal control problem may also be formulated as a continuous problem, see [169, §3]. Continuous variations of Pantoja's algorithm exist in other forms: for example the matrix Riccati method can be viewed in this light. However, continuous problems must be discretized one way or another before being solved on digital computers.

In this chapter we take the view that the problem for which the exact Newton step is appropriate is the actual discretization being used. This discretization may change during the course of problem solution (for example by inserting finer time steps in regions where the control variables are changing rapidly), but verification of descent criteria such as Wolfe conditions, and the introduction of devices to enforce global convergence of Newton's method, should be applied to the numerical values actually being calculated.

Pantoja's Algorithm is extremely useful even when a Newton optimization technique is not employed: it allows inexpensive determination of whether a Hessian is positive definite at a given point. As well as helping enforce global convergence, this feature also allows post-hoc verification of the inclusions required for interval fixed point constructions.

14

Analytical Aspects and Practical Pitfalls in Technical Applications of AD

Florian Dignath, Peter Eberhard and Axel Fritz

ABSTRACT We apply AD to three technical problems from multibody dynamics, vehicle dynamics and finite element material modeling. The multibody dynamics application investigates threshold and maximum-value optimization criteria in direct representation using AD for the sensitivity analysis. Secondly, an adaptive cruise controller of a vehicle convoy is optimized with respect to control error and control effort. Two experimental cars equipped with state-of-the-art sensors and actuators are used to verify the designed controllers. Since the engine voltage of the actuator is restricted to ±10 V, there are non-differentiable points in the time trajectory, but AD techniques can still be used. Thirdly, AD is used for the evaluation of constitutive relations for hyperelastic materials. Some important continuum mechanical quantities correspond to the first and second derivatives of the scalar stored energy function with respect to the strain tensor. These quantities can be derived analytically using complicated analysis but it is fascinating that with minimal preparation and only few mathematical insight exactly the same values can also be computed using AD methods.

14.1 Optimization of Multibody Systems

The dynamics of a multibody system with generalized coordinates y and generalized velocities z may be described by the equations of motion

$$\dot{y} = v(t, y, z), \quad y(t_0) = y_0, \qquad (14.1)$$

$$M(t, y)\dot{z} + k(t, y, z) = q(t, y, z), \quad z(t_0) = z_0, \qquad (14.2)$$

where M is the symmetric, positive definite mass matrix, k is the vector of generalized centrifugal and Coriolis forces and gyroscopic moments, and q is the vector of generalized applied forces. For actively controlled multibody systems these equations may also include the controller dynamics and actuator forces. The vector of generalized coordinates may be extended by additional components for the controller dynamics.

The aim of optimization is often to improve the damping of vibrations or the controller behavior. In the nonlinear case, the optimization criteria have to be computed from a time integration of the equations above. Typically, this is done by integral criteria of the form

$$\psi^I(x) = G(t_{end}, y_{end}, z_{end}, x) + \int_{t_0}^{t_{end}} F(t, y, z, \dot{z}, x) dt \qquad (14.3)$$

with design variables x, using continuously differentiable vector functions G and F. Criteria of this type can be handled by standard methods of optimization using the adjoint variable method for the computation of sensitivities [171], which are necessary for efficient optimization algorithms. Sometimes design goals cannot be formulated using continuously differentiable functions only, e.g., if point- and piecewise defined control laws or criteria (e.g., for the evaluation of the maximum overshoot) or threshold criteria (e.g., for the definition of a desired window for the dynamic response) are considered. In this case, the computation of gradients can be accomplished either by numerical approximations or the application of AD.

14.2   Application of Point- and Piecewise Defined Optimization Criteria

As an example for the application of point- and piecewise defined maximum-value and threshold criteria, we consider a space station truss, modelled as a double pendulum with an additional rotation of the space station. Between the space station and the truss a torque actuator is used to control the active damping of the structure. A simple PID-controller is used to determine the appropriate torque of this actuator M_{ACT}, depending on the measured and the desired truss angles β and β_S. The three gains of this controller are adapted by an optimization in order to reduce (a) the maximum overshoot and (b) the offset from a given design window. For the evaluation of the dynamic behavior the reduction of an initial positive offset in the truss angle β is analyzed. The maximum overshoot then is $\psi^U = \max\{\beta_S - \beta(t)\}$, $t_0 \leq t \leq t_{end}$, where the desired truss angle β_S is zero for the considered situation. In order to apply efficient optimization algorithms, a sensitivity analysis of this criterion is needed. Haftka and Gürdal [264] show that criteria of this type are continuously differentiable at each local maximum, since

$$\frac{d\psi^U}{dx} = \frac{dy_j}{dx}\bigg|_{t=t_{mi}} = \left(\frac{\partial y_j}{\partial x}\bigg|_{t=t_{mi}} + \frac{\partial y_j}{\partial t}\bigg|_{t=t_{mi}} \frac{dt_{mi}}{dx} \right),$$

and the last term of this expression always vanishes since either $\partial y_j/\partial t = 0$ (necessary condition for an extremal point) or $dt_{mi}/dx = \mathbf{0}$ for a maximum at the border of the allowed time range. While this would allow the computation of sensitivities by standard methods, e.g., the adjoint variable method, a non-differentiable point may occur where one local maximum is overtaken by another, as it can be seen in the optimization results for the maximum-value criterion in Figure 14.3. In principle, the adjoint variable method could also be applied to such criteria, but would have to be adapted to each individual case. Alternatives are the application of finite difference approximations or AD as shown in [315] and [273]. A comparison of these two methods shows that the finite difference approximations are relatively simple to implement, but lead to numerical problems with maximum value functions in combination with automatic step size control of the integration algorithms, while AD on the other hand needs some care during the implementation but computes exact gradients, see Figure 14.1.

The problem of finite differences is due to the fact that the change in the time discretization for the integration contributes significantly to the change of the criteria function. This problem can be reduced by the use of large parameter variations, but this has the drawback of a loss of accuracy of the results [238]. For example, around the non-differentiable point at $K \approx 1\,000\,000$, finite differences compute an "average" gradient instead of the mathematical exact gradient of AD, see Figure 14.2, which may lead to wrong evaluations of the Kuhn-Tucker optimality conditions. To use

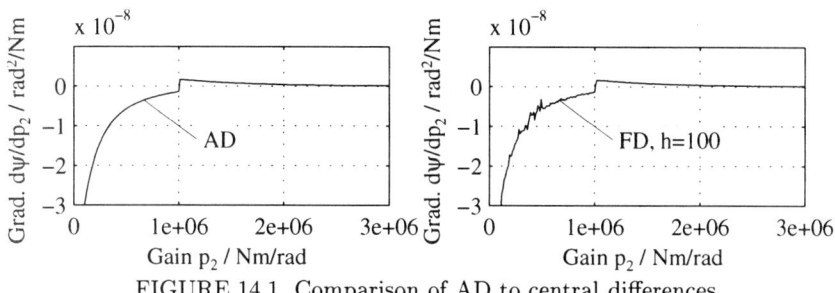

FIGURE 14.1. Comparison of AD to central differences

AD, a Runge-Kutta integrator was differentiated using ADIFOR [77]. After actions had been taken to prevent differentiation of the automatic step size control, which would introduce the additional term dh/dx for the step

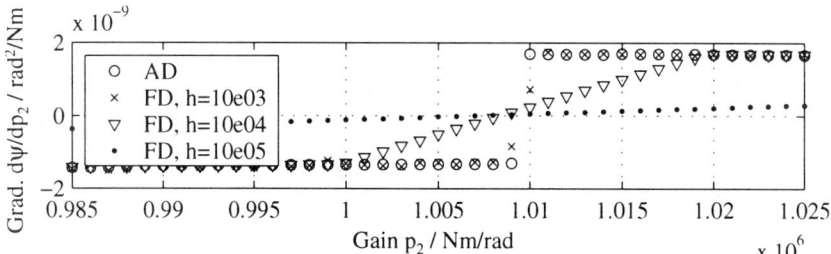

FIGURE 14.2. Discontinuous gradient at a non-differentiable criterion point

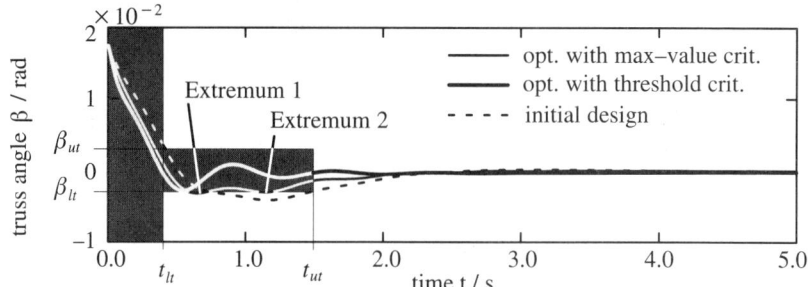

FIGURE 14.3. Optimization results and switch between local minima

width h in the result [173], AD computed the exact sensitivities, which was shown by comparisons to the adjoint variable method. This shows that AD cannot always be applied as a black box. It would be interesting to get tools by the AD developers to detect and manually to cut undesired dependencies. Some optimization results are shown in Figure 14.3.

14.3   Adaptive Cruise Control of a Vehicle Convoy

The second example deals with the parameter optimization of an adaptive cruise controller used for the longitudinal control of a vehicle convoy, see Figure 14.4. The convoy consists of two vehicles, where the following car is equipped with a velocity sensor determining the longitudinal velocity v_{x2}, an acceleration sensor and a laser scanner measuring the range R_x between the two cars. Communication between the vehicles is not available. The simulation model for the longitudinal dynamics of each car is based on the method of multibody systems. The model of the following car is extended by a nonlinear model of a controlled servo-motor used as actuator. For the controller design the concept of flat outputs in connection with

FIGURE 14.4. Vehicle convoy consisting of two BMW vehicles

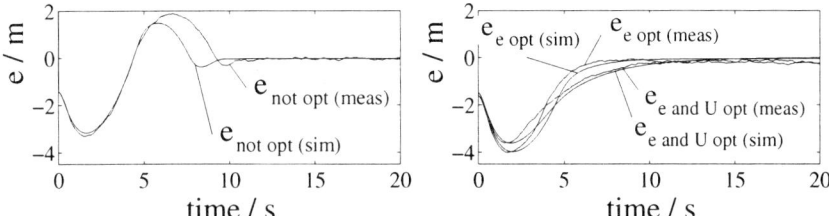

FIGURE 14.5. Control error of a vehicle convoy (two cars) using the initial parameter set (left) and the optimized parameter sets (right)

exact state linearization is applied, see [201]. Using this compensation, a linear system in Brunovsky canonical form, see [302], can be obtained as an alternative mechanical system for the parameter optimization problem, see equations (14.1) and (14.2). The time trajectories of this model contain non-differentiable points because of a restriction of the servo motor to $\pm 10V$ and the restriction of the throttle angle to $0 - 90°$. Two performance criteria are formulated in the form of equation (14.3): The first one is to minimize the control error e and the derivative of the control error which is given as the difference between the desired range $R_{xdes} = d_0 + T_h v_{x2}$ and the measured range R_x. The second one serves to reduce the consumed energy U of the servo component. Additionally some inequalities, see [201], guaranteeing individual vehicle and string stability for a convoy consisting of more than two vehicles are formulated as explicit criteria in the optimization problem. Because of the piecewise defined functions in the model, the performance criteria contain non-differentiable points, and the standard methods could not be applied without adaptations to the special situation. However, using AD for the computation of sensitivities as in §14.2 the optimization method can be applied even to such complex systems without any changes, provided the optimization algorithm can handle the non-differentiable criteria.

For the following simulations and measurements a step disturbance in the throttle position of the leading vehicle from $0°$ to $10°$, $d_0 = 10m$ and $T_h = 1s$ is used. Since there is an initial headway error of $1.5 \div 2m$ the regarded maneuver can be characterized as closing in to the vehicle in front and then keeping the desired range. In Figure 14.5 real measurements and computed simulation results of the control error are shown. The left graph shows the results with an parameter set $(\lambda_{1/2/3} = -7)$ which was successfully used in measurements without initial errors, while the right graph shows the errors using the controller parameters from the optimization with one (e opt.: $\lambda_1 = -1, \lambda_{2/3} = -13 \pm i11$) and with both performance criteria (e and U opt.: $\lambda_1 = -0.4, \lambda_{2/3} = -17 \pm i17$). The measurements and the simulation results agree sufficiently well. A comparison of both plots shows that there is a big overshoot for $\lambda_{1/2/3} = -7$, while both optimized controllers reduce the headway error very smoothly. Therefore, no braking is necessary to achieve good performance.

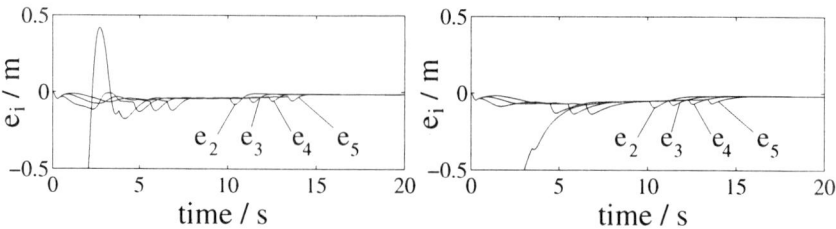

FIGURE 14.6. Control error of a vehicle convoy (five cars) using the initial parameter set (left) and the optimized parameter sets (right)

Simulation results for a convoy consisting of five vehicles using the controller parameters $(\lambda_{1/2/3} = -7)$ and $(\lambda_1 = -1, \lambda_{2/3} = -13 \pm i11)$ are shown in Figure 14.6. The error e_2 between the first two vehicles behaves similar to the plots shown in Figure 14.5. Further, string stability is given, since the error of the i−th vehicle e_i is smaller than that of the $(i-1)$−th vehicle. Therefore the inequalities are still satisfied after the optimization.

14.4 Hyperelastic Material Behavior

For many engineering problems the finite element approach can be used. The material behavior is determined by the constitutive relations. Hyperelastic materials which are determined by a scalar function of the stored strain energy $W(E)$ are appropriate for many applications and situations. To use the nonlinear FEM, the first and second derivative of W with respect to the Green-Lagrange strain tensor E have to be determined in order to compute the 2nd Piola Kirchhoff stress tensor $S(E) = \partial W(E)/\partial E$ and the material tensor $\partial S(E)/\partial E = \partial^2 W(E)/\partial E^2$.

Isotropic hyperelastic material may depend only on the invariants of the correct Cauchy-Green strain tensor $C = I + 2E$, i.e., $W(I_C, II_C, III_C)$ with $I_C = \mathrm{tr}(C)$, $II_C = 0.5(\mathrm{tr}(C)^2 - \mathrm{tr}(C \cdot C))$, $III_C = \det(C)$. An important example is the compressible neo-Hooke material with

$$W(I_C, III_C) = \frac{\mu}{2}(I_C - 3) - \mu \ln(\sqrt{III_C}) + \frac{\lambda}{2}(\sqrt{III_C} - 1)^2.$$

For a manual computation of the derivatives formulae such as

$$\frac{\partial I_C}{\partial C} = I, \quad \frac{\partial III_C}{\partial C} = III_C C^{-1}, \quad \frac{\partial \left[C^{-1}\right]_{ij}}{\partial C_{kl}} = -\left[C^{-1}\right]_{ik}\left[C^{-1}\right]_{lj}$$

have to be derived analytically, which can only be done by mathematical experts since it requires a lot of mathematical insight and experience [172].

An alternative which yielded correct results and small computation times is the use of AD programs such as ADIC [84]. Here the computation sequence is very simple: $E \rightarrow C \rightarrow I_C, III_C \rightarrow W$ and it is easy to get code to compute the first and second derivatives of $W(E)$ with respect to E.

15

Nonlinear Observer Design Using Automatic Differentiation

Klaus Röbenack and Kurt J. Reinschke

ABSTRACT State feedback controllers are widely used in real-world applications. These controllers require the values of the state of the plant to be controlled. An observer can provide these values. For nonlinear systems there are some observer design methods which are based on differential geometric or differential algebraic concepts. The application to non-trivial systems is limited due to a burden of symbolic computations involved. The authors propose an observer design method using automatic differentiation.

15.1 Introduction

During the last few years, researchers have developed many algorithms for controller and observer design for nonlinear state-space systems

$$\dot{\mathbf{x}}(t) = \mathbf{f}(\mathbf{x}(t)) + \mathbf{g}(\mathbf{x}(t)) \cdot u(t), \quad y(t) = h(\mathbf{x}(t)). \tag{15.1}$$

The vector-valued state is denoted by $\mathbf{x}$, u denotes the scalar input signal, and y denotes the scalar output signal. Some design methods are based on differential geometric or differential algebraic concepts [195, 302]. These approaches are based on time-consuming symbolic computations. We show how the use of automatic differentiation (AD) avoids cumbersome computations.

State feedback control requires the value of the state $\mathbf{x}$. An observer is a dynamical system, whose state $\hat{\mathbf{x}}$ asymptotically tracks the state $\mathbf{x}$ of the original system (15.1). As for general nonlinear systems, observer design is still an active area of research [447]. To simplify presentation, this chapter considers the unforced system case. In this case, several observer design methods are known [65]. Their application to non-trivial systems is limited due to the symbolic computations involved [154, 443]. The authors propose an observer design method based on automatic differentiation which is applicable to single output systems.

In §15.2, we review observer design for linear time-varying systems. The use of AD for the design of nonlinear observers is described in §15.3. The method derived there will be applied in §15.4 to an example.

15.2 Theory

In this section, we summarize some results of observer design. Consider a linear time-varying system

$$\dot{\mathbf{x}}(t) = A(t)\,\mathbf{x}(t), \quad y(t) = C(t)\,\mathbf{x}(t) \tag{15.2}$$

with a scalar-valued output y. For an observer

$$\dot{\widehat{\mathbf{x}}}(t) = A(t)\,\widehat{\mathbf{x}}(t) + \mathbf{k}(t)\,(y(t) - C(t)\,\widehat{\mathbf{x}}(t)),$$

we want to compute the gain vector $\mathbf{k}(t)$ in such a way that the eigenvalues of the observer are assigned to desired places. This task is straightforward if the system matrices were given in observer canonical form

$$\tilde{A}(t) = \begin{pmatrix} 0 & \cdots & 0 & -a_0(t) \\ 1 & \ddots & \vdots & -a_1(t) \\ & \ddots & 0 & \vdots \\ 0 & \cdots & 1 & -a_{n-1}(t) \end{pmatrix}, \quad \tilde{C}(t) = \begin{pmatrix} 0 & \cdots & 0 & 1 \end{pmatrix}. \tag{15.3}$$

Then, the eigenvalues to be assigned are the roots of the characteristic polynomial

$$\det(sI - \tilde{A}(t) + \tilde{\mathbf{k}}(t)\tilde{C}(t)) = s^n + p_{n-1}s^{n-1} + \cdots + p_1 s + p_0,$$

where the coefficients $p_0, \ldots, p_{n-1}$ are constant, and the components of the vector $\tilde{\mathbf{k}}(t) = (\tilde{k}_1(t), \ldots, \tilde{k}_n(t))^\top$ are $\tilde{k}_i(t) = p_{i-1} - a_{i-1}(t)$ for $i = 1, \ldots, n$.

To get $\mathbf{k}(t)$ for the general case, a well-known generalization of Ackermann's formula for linear time-varying systems is used here [65, 196, 447]. The observer canonical form (15.3) can be obtained from (15.2) by a change of coordinates $\tilde{\mathbf{x}}(t) = T(t)\,\mathbf{x}(t)$ if the observability matrix

$$Q(t) := \begin{pmatrix} C(t) \\ \mathcal{L}C(t) \\ \vdots \\ \mathcal{L}^{n-1}C(t) \end{pmatrix} \tag{15.4}$$

has full rank. The differential operator $\mathcal{L}$ in (15.4) is defined by

$$\mathcal{L}C(t) = \dot{C}(t) + C(t)\,A(t). \tag{15.5}$$

The inverse transformation matrix T^{-1} can be computed as $T^{-1}(t) = (\mathbf{q}(t), \tilde{\mathcal{L}}\,\mathbf{q}(t), \ldots, \tilde{\mathcal{L}}^{n-1}\mathbf{q}(t))$, where $\mathbf{q}$ denotes the last column of the inverse observability matrix, i.e., $\mathbf{q}(t) = Q^{-1}(t)\,(0, \ldots, 0, 1)^\top$, and the differential operator $\tilde{\mathcal{L}}$ is defined by

$$\tilde{\mathcal{L}}\mathbf{q}(t) = -\dot{\mathbf{q}}(t) + A(t)\,\mathbf{q}(t). \tag{15.6}$$

Finally, the desired gain vector $\mathbf{k}(t)$ can be calculated as follows:

$$\mathbf{k}(t) = \left(p_0 I_n + p_1\tilde{\mathcal{L}} + \cdots + p_{n-1}\tilde{\mathcal{L}}^{n-1} + \tilde{\mathcal{L}}^n\right)\mathbf{q}(t). \tag{15.7}$$

15.3   Observer Design by Automatic Differentiation

Consider an autonomous nonlinear state-space system

$$\dot{\mathbf{x}} = \mathbf{f}(\mathbf{x}), \quad y = h(\mathbf{x}), \quad \mathbf{x}(0) = \mathbf{x}_0, \tag{15.8}$$

where the maps $\mathbf{f} : \mathbb{R}^n \to \mathbb{R}^n$ and $h : \mathbb{R}^n \to \mathbb{R}$ are assumed to be sufficiently smooth. We consider the associated variational equation (15.2) with $A(t) = f'(\mathbf{x}(t))$ and $C(t) = h'(\mathbf{x}(t))$. The Taylor coefficients A_i and C_i of

$$\begin{aligned} A(t) &= A_0 + A_1 t + A_2 t^2 + \cdots \\ C(t) &= C_0 + C_1 t + C_2 t^2 + \cdots \end{aligned} \tag{15.9}$$

can be computed by means of a Taylor expansion of the curves $\mathbf{x}(t) = \mathbf{x}_0 + \mathbf{x}_1 t + \mathbf{x}_2 t^2 + \cdots$, $y(t) = y_0 + y_1 t + y_2 t^2 + \cdots$, and $\dot{\mathbf{x}}(t) \equiv \mathbf{z}(t) = \mathbf{z}_0 + \mathbf{z}_1 t + \mathbf{z}_2 t^2 + \cdots$ using standard AD tools such as ADOL-C [242]. More precisely, the coefficient matrices occurring in (15.9) are partial derivatives of Taylor coefficients [125], i.e., $A_{j-i} = \frac{\partial \mathbf{z}_j}{\partial \mathbf{x}_i}$ and $C_{j-i} = \frac{\partial y_j}{\partial \mathbf{x}_i}$. We can get the total derivatives $B_k = \frac{d\mathbf{x}_{k+1}}{d\mathbf{x}_0}$ and $D_k = \frac{dy_k}{d\mathbf{x}_0}$ as follows [238]:

$$\begin{aligned} B_k &= \frac{1}{k+1} \frac{d\mathbf{z}_k}{d\mathbf{x}_0} \\ &= \frac{1}{k+1} \sum_{j=0}^{k} \frac{\partial \mathbf{z}_k}{\partial \mathbf{x}_j} \frac{d\mathbf{x}_j}{d\mathbf{x}_0} \\ &= \frac{1}{k+1} \left(A_k + \sum_{j=1}^{k} A_{k-j} B_{j-1} \right), \end{aligned} \qquad \begin{aligned} D_k &= \frac{dy_k}{d\mathbf{x}_0} \\ &= \sum_{j=0}^{k} \frac{\partial y_k}{\partial \mathbf{x}_j} \frac{d\mathbf{x}_j}{d\mathbf{x}_0} \\ &= C_k + \sum_{j=1}^{k} C_{k-j} B_{j-1}. \end{aligned}$$

The observability matrix Q of the linear time-varying system (15.2) was defined in terms of the differential operator $\mathcal{L}$, see Equations (15.4) and (15.5). We can express the function values $\mathcal{L}^k C(0)$ in terms of Taylor coefficients generated by AD tools: $\mathcal{L}^k C(t)\big|_{t=0} = \frac{dy^{(k)}}{d\mathbf{x}(t)}\big|_{t=0} = k! \frac{dy_k}{d\mathbf{x}_0} = k! D_k$.

For the nonlinear system (15.8), the observability matrix is defined by

$$\begin{pmatrix} dL_{\mathbf{f}}^0 h(\mathbf{x}) \\ \vdots \\ dL_{\mathbf{f}}^{n-1} h(\mathbf{x}) \end{pmatrix}, \tag{15.10}$$

where $L_{\mathbf{f}} h(\mathbf{x})$ denotes the Lie derivative of h along the vector field $\mathbf{f}$ with $L_{\mathbf{f}}^k h(\mathbf{x}) := \frac{\partial L_{\mathbf{f}}^{k-1} h(\mathbf{x})}{\partial \mathbf{x}} \mathbf{f}(\mathbf{x})$ and $L_{\mathbf{f}}^0 h(\mathbf{x}) := h(\mathbf{x})$. Because of the identity (see [302, p. 140])

$$y(t) = \sum_{k=0}^{\infty} L_{\mathbf{f}}^k h(\mathbf{x}_0) \frac{t^k}{k!},$$

we can express the function values of the Lie derivatives $L_f^k h(\mathbf{x}_0)$ and the corresponding gradients $dL_f^k h(\mathbf{x}_0)$ in terms of the Taylor coefficients calculated above [438]:

$$L_f^k h(\mathbf{x}_0) = k! \, y_k \quad \text{and} \quad dL_f^k h(\mathbf{x}_0) = k! \, D_k.$$

This shows that the observability matrix (15.10) of the nonlinear system (15.8) coincides with the observability matrix (15.4) of the associated variational equation (15.2).

Next, we compute the columns of $T^{-1}(t)$. For the k-th order total derivative (along the flow of (15.2)) of $Q(t) \, \mathbf{q}(t) = (0, \ldots, 0, 1)^\top$ we obtain

$$0 = \frac{d^k}{dt^k}[Q(t) \, \mathbf{q}(t)] = \sum_{i=0}^{k} \binom{k}{i} \left[\frac{d^i}{dt^i} Q(t)\right] \left[\frac{d^{k-i}}{dt^{k-i}} \mathbf{q}(t)\right], \quad k > 0. \quad (15.11)$$

The i-th row of the observability matrix Q is the Jacobian of the $(i-1)$-th output derivative, see (15.5). Let Q_i denote the matrix consisting of the rows $1+i, \ldots, n+i$ of the (extended) observability matrix (15.4) for $t = 0$. Then the total derivatives of Q can be computed as follows:

$$\frac{d^i}{dt^i} Q(t) \bigg|_{t=0} = Q_i = \begin{pmatrix} i! \cdot D_i \\ \vdots \\ (n+i-1)! \cdot D_{n+i-1} \end{pmatrix}.$$

Let q_j denote the j-th derivative of $\mathbf{q}$ along the flow of (15.2) for $t = 0$, i.e., $\mathbf{q}_j = \frac{d^j}{dt^j} \mathbf{q}(0)$. We have $\mathbf{q}_0 = Q_0^{-1} e_n$ by definition. From (15.11) one obtains

$$\mathbf{q}_j = -Q_0^{-1} \cdot \sum_{i=1}^{j} \binom{j}{i} Q_i \, \mathbf{q}_{j-i} \quad \text{for} \quad j > 0.$$

In particular, we have

$$\begin{aligned} \mathbf{q}_1 &= -Q_0^{-1} \cdot Q_1 \mathbf{q}_0 \\ \mathbf{q}_2 &= -Q_0^{-1} \cdot (Q_2 \mathbf{q}_0 + 2Q_1 \mathbf{q}_1) \\ \mathbf{q}_3 &= -Q_0^{-1} \cdot (Q_3 \mathbf{q}_0 + 3Q_2 \mathbf{q}_1 + 3Q_1 \mathbf{q}_2). \end{aligned}$$

The differential operator $\tilde{\mathcal{L}}$ defined by (15.6) represents the total derivative of $\mathbf{q}$ along the flow of (15.2) for *reverse* time. This means that we have to change the sign of odd Taylor coefficients, i.e., $\tilde{\mathcal{L}}^j \mathbf{q}(t)|_{t=0} = (-1)^j \, \mathbf{q}_j$. According to (15.7) we finally obtain the gain vector

$$\mathbf{k}(0) = p_0 \, \mathbf{q}_0 - p_1 \, \mathbf{q}_1 + \cdots + (-1)^{n-1} p_{n-1} \mathbf{q}_{n-1} + (-1)^n \mathbf{q}_n.$$

Since the original ODE (15.8) is autonomous, we can apply the algorithm to compute $\mathbf{k}$ at an arbitrary time t_i. In fact, the gain vector has to be computed for every time step t_i during the simulation of the nonlinear observer

$$\dot{\hat{\mathbf{x}}}(t) = \mathbf{f}(\hat{\mathbf{x}}(t)) - \mathbf{k}(t) \cdot (y(t) - h(\hat{\mathbf{x}}(t))). \quad (15.12)$$

15.4   Example

Consider the 2-dimensional state-space system

$$
\left.\begin{array}{rcl}
\dot{x}_1 & = & x_1 - x_1\,x_2 \\
\dot{x}_2 & = & x_1\,x_2 - x_2
\end{array}\right\} \quad \dot{\mathbf{x}} = \mathbf{f}(\mathbf{x}) \quad \text{with} \quad y = h(\mathbf{x}) = \sinh(x_2). \quad (15.13)
$$

The simulation result started at an initial value $\mathbf{x}_0 = (3,1)^\top$ is shown in Figure 15.1(a). The matrices A and C of the variational equation (15.2) are

$$
A(t) = \begin{pmatrix} 1 - x_2(t) & -x_1(t) \\ x_2(t) & x_1(t) - 1 \end{pmatrix} \quad \text{and} \quad C(t) = \begin{pmatrix} 0 & \cosh(x_2(t)) \end{pmatrix}.
$$

The solution is periodic, and the eigenvalues of $A(t)$ vary along the trajectory. Figure 15.1(b) shows their real and imaginary part as a function of t. Due to this large variation of the spectrum of A, the dynamics of an observer with an constant gain vector $\mathbf{k}$ would be difficult to analyze, i.e., it would be difficult to guarantee certain (prescribed) dynamical properties such as the convergence rate of the observer.

As an AD tool we used ADOL-C 1.8 with the GNU-Compiler gcc 2.95.2. At first we generated the tapes for the functions $\mathbf{f}$ and h with C++. The low-level C interface of ADOL-C was used for the differentiation process itself. The Taylor coefficients x_i and y_i are computed in the forward mode (e.g., with forodec), whereas the coefficient matrices A_i and C_i are computed in the reverse mode (e.g., with the driver hov_reverse). Moreover, ADOL-C provides functions for the accumulation of the total derivatives (e.g., accodec, see [242]).

For the system (15.13), we designed an observer with eigenvalues $s_{1,2} = -5$. The observer was simulated with the ODE solver of Scilab 2.5 (function ode). To do this, the object code generated by ADOL-C and gcc was dynamically linked with Scilab. Then, the C functions which compute the required Taylor coefficients were called from Scilab by the built-in function call.

The simulation results are shown in Figure 15.1, (c) and (d). For the observer we used the initial value $\hat{\mathbf{x}}_0 = (0.1, 0.1)^\top$, which differs significantly from the initial value $\mathbf{x}_0 = (3,1)^\top$ of the system (15.13). The numerical results illustrate that the state $\hat{\mathbf{x}}$ of the observer converges to the state $\mathbf{x}$ of the original system (15.13), i.e., the observer serves his purpose.

Note that Scilab was not compiled with the C++ compiler g++. In order to use the high-level C++ interface of ADOL-C, it is necessary to link additional libraries to Scilab. The libraries required there can be found during the compilation process using "g++ -v".

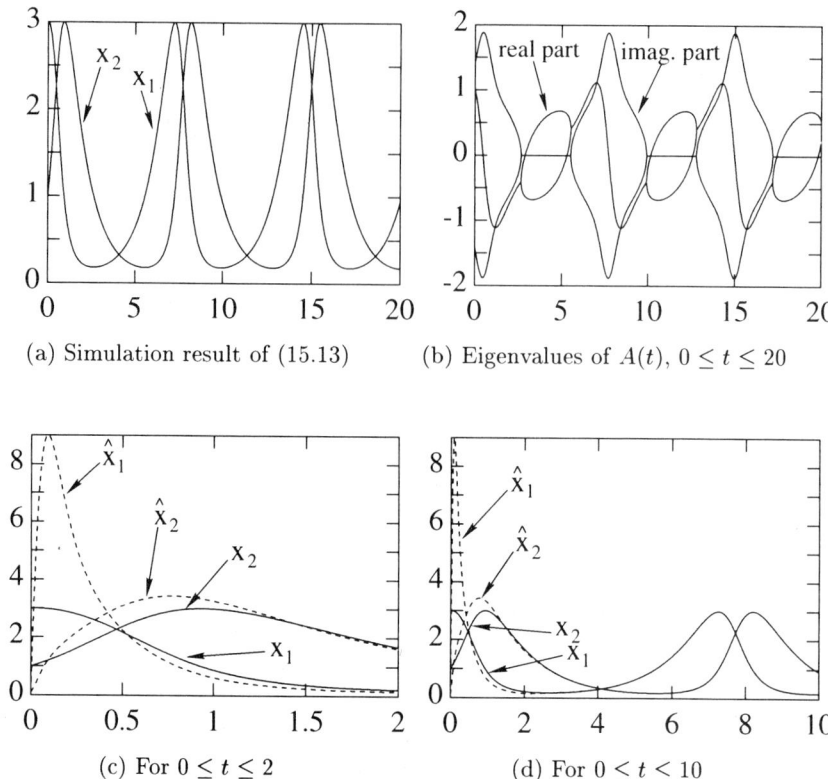

(a) Simulation result of (15.13)

(b) Eigenvalues of $A(t)$, $0 \le t \le 20$

(c) For $0 \le t \le 2$

(d) For $0 \le t \le 10$

FIGURE 15.1. Trajectories of (15.13) with $\mathbf{x}_0 = (3, 1)^\top$ and the associated observer (15.12) with $\widehat{\mathbf{x}}_0 = (0.1, 0.1)^\top$.

15.5   Conclusions

A method for observer design by means of automatic differentiation was presented. In contrast to symbolic approaches that are based on a direct computation of Lie derivatives and Lie brackets, the AD-based algorithm presented here is applicable to large-scale systems.

Part IV

Applications in PDE'S

16

On the Iterative Solution of Adjoint Equations

Michael B. Giles

ABSTRACT This chapter considers the iterative solution of the adjoint equations which arise in the context of design optimisation. It is shown that naive adjoining of the iterative solution of the original linearised equations results in an adjoint code which cannot be interpreted as an iterative solution of the adjoint equations. However, this can be achieved through appropriate algebraic manipulations. This is important in design optimisation because one can reduce the computational cost by starting the adjoint iteration from the adjoint solution obtained in the previous design step.

16.1 Introduction

In computational fluid dynamics (CFD), one is interested in solving a set of nonlinear discrete flow equations of the form

$$N(U, X) = 0.$$

Here X is a vector representing the coordinates of a set of computational grid points, U is the vector of flow variables at those grid points, and $N(U, X)$ is a differentiable vector function of the same dimension as U.

To solve these equations, many CFD algorithms use iterative methods which can be written as

$$U^{n+1} = U^n - R(U^n, X) N(U^n, X),$$

where R is a non-singular square matrix which is a differentiable function of its arguments. If R were defined to be L^{-1}, where $L = \partial N / \partial U$ is the non-singular Jacobian matrix, this would be the Newton-Raphson method which converges quadratically in the final stage of convergence. However, in the iterative methods used in CFD, R is a poor approximation to L^{-1} and therefore they exhibit linear convergence asymptotically, with the final rate of convergence being given by the magnitude of the largest eigenvalue of the matrix $I - R(U, X) L(U, X)$, where I is the identity matrix.

In design optimisation [13, 174, 219], the grid coordinates X depend on a set of design parameters α, and one wishes to minimise a scalar objective function $J(U, X)$ by varying α. To achieve this using gradient-based optimisation methods [222], at each step in the optimisation process one determines the linear sensitivity of the objective function to changes in α.

For a single design parameter, linearising the objective function gives

$$\frac{\mathrm{d}J}{\mathrm{d}\alpha} = \frac{\partial J}{\partial U}u + \frac{\partial J}{\partial \alpha}. \tag{16.1}$$

Here $u \equiv \mathrm{d}U/\mathrm{d}\alpha$ is the linear sensitivity of the flow variables to changes in the design parameter, which is obtained by linearising the flow equations to give

$$Lu = f, \tag{16.2}$$

where

$$f = -\frac{\partial N}{\partial X}\frac{\mathrm{d}X}{\mathrm{d}\alpha},$$

and L and f are both functions of the nonlinear solution (U, X) for the current value of the design parameter α.

Using forward mode automatic differentiation tools such as ADIFOR [78] or Odyssée [186] (treating $R(U, X)$ as a constant for maximum efficiency since its linearisation is unnecessary because it is multiplied by $N(U, X) = 0$ [456]) one can automatically generate a code for the iterative solution of these linear flow equations. This uses the same iterative procedure as the nonlinear equations and corresponds to the iteration

$$u^{n+1} = u^n - R(Lu^n - f), \tag{16.3}$$

starting from zero initial conditions. R is again a function of the nonlinear solution (U, X), and the linear convergence rate for this will be exactly the same as the asymptotic convergence rate of the nonlinear iteration as it is controlled by the same iteration matrix $I - RL$. However, if there are many design parameters, each one gives rise to a different vector f and linear flow perturbation u. Thus, the computational cost increases linearly with the number of design variables.

To avoid this increasing cost, one can use adjoint methods. The evaluation of the second term on the r.h.s. of Equation (16.1),

$$\frac{\partial J}{\partial \alpha} = \frac{\partial J}{\partial X}\frac{\mathrm{d}X}{\mathrm{d}\alpha},$$

is straightforward and inexpensive, so we focus attention on the first term, which we choose to write as an inner product $(\bar{u}, u) \equiv \bar{u}^T u$, by defining

$$\bar{u} \equiv \left(\frac{\partial J}{\partial U}\right)^T.$$

Since $Lu - f = 0$, simple algebraic manipulation yields

$$(\bar{u}, u) = (\bar{u}, u) - (\bar{f}, Lu - f) = (\bar{f}, f) - (L^T\bar{f} - \bar{u}, u) = (\bar{f}, f)$$

when $\overline{f}$ satisfies the adjoint equation

$$L^T \overline{f} = \overline{u}.$$

The advantage of the adjoint approach is that the calculation of F and the evaluation of the inner product $(\overline{f}, f)$ for each design variable is negligible compared to the cost of determining the single adjoint solution $\overline{f}$, and so the total cost is approximately independent of the number of design variables.

The issue to be addressed in this chapter is how to obtain the adjoint solution $\overline{f}$ as the limit of a fixed point iteration which is the natural counterpart to that used for the nonlinear and linear equations, and which therefore has exactly the same rate of iterative convergence. It will be shown that the naive application of adjoint methodology to the iterative solution of the linear equations results in an algorithm in which the working variables do not correspond to the adjoint variables $\overline{f}$. However, with a slight reformulation it can be cast into the desired form.

The benefit of the adjoint calculation being formulated as a fixed point iteration is that one can obtain very significant computational savings if one can provide a good initialisation for the adjoint variables. This is possible in nonlinear design optimisation, since the adjoint variables computed for one step in the design optimisation can be used to initialise the computation of the adjoint variables for the next step. Indeed, it is usually found that the computational cost of the entire optimisation process is minimised by not fully converging the nonlinear and adjoint flow calculations at each design step, and instead letting the nonlinear and adjoint flow variables as well as the design parameters all evolve towards the optimum solution [306].

This issue of the iterative solution of adjoint equations has been investigated previously by Christianson [126, 127], but the context for his work is more abstract; the references should be consulted for further information.

16.2 Continuous Equations

The iterative solution methods used in CFD are often based on an unsteady evolution towards the solution of a steady system of equations. Therefore, we begin by considering the unsteady solution $u(t)$ of the coupled system of differential equations

$$\frac{du}{dt} = -P(Lu - f),$$

for some constant preconditioning matrix P, subject to the initial conditions $u(0) = 0$. The functional of interest is the inner product $(\overline{u}, u(T))$, with the final time T chosen to be sufficiently large that du/dt is very small and therefore $u(T)$ is very close to being the solution of the steady equations.

Introducing the unsteady adjoint variable $\bar{u}_u(t)$, and using integration by parts, the unsteady adjoint formulation is given by

$$(\bar{u}, u(T)) = (\bar{u}, u(T)) - \int_0^T \left(\bar{u}_u, \frac{du}{dt} + P\left(L\,u - f\right)\right) dt$$

$$= (\bar{u} - \bar{u}_u(T), u(T))$$

$$- \int_0^T \left(-\frac{d\bar{u}_u}{dt} + L^T P^T \bar{u}_u, u\right) - (P^T \bar{u}_u, f) \, dt$$

$$= \int_0^T (P^T \bar{u}_u, f) \, dt,$$

where $\bar{u}_u(t)$ satisfies the differential equation

$$\frac{d\bar{u}_u}{dt} = L^T P^T \bar{u}_u,$$

which is solved backwards in time subject to the final condition $\bar{u}_u(T) = \bar{u}$.

With the equations in this form, one would obtain the correct value for the functional, exactly the same value as one would obtain from the unsteady linear equations over the same time interval, but it is not immediately clear how the working variables $\bar{u}_u(t)$ are related to the steady adjoint solution $\bar{f}$.

To obtain the link with the steady adjoint equation, we define

$$\bar{f}(t) = \int_t^T P^T \bar{u}_u \, dt,$$

so that the functional is $(\bar{f}(0), f)$, and $\bar{f}(t)$ satisfies the differential equation

$$-\frac{d\bar{f}}{dt} = P^T \bar{u}_u$$

$$= P^T \left(\bar{u} - \int_t^T \frac{d\bar{u}_u}{dt} \, dt\right)$$

$$= P^T \left(\bar{u} - \int_t^T L^T P^T \bar{u}_u \, dt\right)$$

$$= -P^T \left(L^T \bar{f} - \bar{u}\right),$$

subject to the final condition $\bar{f}(T) = 0$.

In this form, the connection with the iterative solution of the steady adjoint equations becomes apparent. $\bar{f}(t)$ evolves towards the steady adjoint solution, and if T is very large then $\bar{f}(0)$ will be almost equal to the steady adjoint solution.

16.3   Discrete Equations

Having considered the continuous equations to gain insight into the issue, we now consider the discrete equations and their iterative solution. As described in §16.1, many standard iterative algorithms for solving the linearised equations can be expressed as

$$u^{n+1} = u^n - R(Lu^n - f).$$

After performing N iterations starting from the initial condition $u^0 = 0$, the functional $(\bar{u}, u^N)$ is evaluated using the final value u^N.

Proceeding as before to find the discrete adjoint algorithm yields

$$
\begin{aligned}
(\bar{u}, u^N) &= (\bar{u}, u^N) - \sum_{n=0}^{N-1} \left(\bar{u}_u^{n+1}, u^{n+1} - u^n + R(Lu^n - f) \right) \\
&= (\bar{u} - \bar{u}_u^N, u^N) \\
&\quad - \sum_{n=0}^{N-1} \left\{ -(\bar{u}_u^{n+1} - \bar{u}_u^n, u^n) + (L^T R^T \bar{u}_u^{n+1}, u^n) - (R^T \bar{u}_u^{n+1}, f) \right\} \\
&= (\bar{u} - \bar{u}_u^N, u^N) \\
&\quad + \sum_{n=0}^{N-1} \left\{ (\bar{u}_u^{n+1} - \bar{u}_u^n - L^T R^T \bar{u}_u^{n+1}, u^n) + (R^T \bar{u}_u^{n+1}, f) \right\},
\end{aligned}
$$

in which we have used the following identity which is the discrete equivalent of integration by parts,

$$\sum_{n=0}^{N-1} a^{n+1} (b^{n+1} - b^n) = a^N b^N - a^0 b^0 - \sum_{n=0}^{N-1} (a^{n+1} - a^n) b^n.$$

Consequently, if $\bar{u}_u$ satisfies the difference equation

$$\bar{u}_u^n = \bar{u}_u^{n+1} - L^T R^T \bar{u}_u^{n+1},$$

subject to the final condition $\bar{u}_u^N = \bar{u}$, then the functional is equal to $(\bar{f}^0, f)$, where $\bar{f}^0$ is defined to be the accumulated sum

$$\bar{f}^0 = \sum_{m=0}^{N-1} R^T \bar{u}_u^{m+1}.$$

The above description of the discrete adjoint algorithm corresponds to what would be generated by reverse mode automatic differentiation tools such as Odyssée [179], ADJIFOR [114], or TAMC [212]. As it stands, it is not clear what the connection is between the adjoint solution $\bar{f}$ and either the sum $\bar{f}^0$ or the working variable $\bar{u}_u^n$.

As with the continuous equations, it is preferable to cast the problem as a fixed point iteration towards the solution of the discrete adjoint equations. To do this we define $\overline{f}^n$ for $0 \leq n < N$ to be

$$\overline{f}^n = \sum_{m=n}^{N-1} R^T \overline{u}_u^{m+1},$$

with $\overline{f}^N = 0$. The difference equation for $\overline{f}^n$ is

$$\begin{aligned}
\overline{f}^n - \overline{f}^{n+1} &= R^T \overline{u}_u^{n+1} \\
&= R^T \left(\overline{u} - \sum_{m=n+1}^{N-1} (\overline{u}_u^{m+1} - \overline{u}_u^m) \right) \\
&= R^T \left(\overline{u} - \sum_{m=n+1}^{N-1} L^T R^T \overline{u}_u^{m+1} \right) \\
&= -R^T \left(L^T \overline{f}^{n+1} - \overline{u} \right),
\end{aligned}$$

showing that the new working variable $\overline{f}^n$ evolves towards the solution of the adjoint equations. The rate of convergence is exactly the same as for the linear iteration since it is governed by the matrix $I - R^T L^T$ whose eigenvalues are the same as its transpose $I - LR$ and hence also $I - RL$, since if v is an eigenvector of the former then $L^{-1}v$ is an eigenvector of the latter with the same eigenvalue.

16.4 Applications

In applying the theory presented above to formulate adjoint algorithms, the key is to first express the nonlinear and linear iterative method in the correct form to determine the matrix R, and thereby determine the matrix R^T for the adjoint iterative scheme. Not all algorithms can be expressed in this way with a constant matrix R. In the conjugate gradient algorithm, for example, the matrix R changes from one iteration to the next.

Reference [217] applies the theory to two kinds of iterative solver. The first is a quite general class of Runge-Kutta methods which is used extensively in CFD, and includes both preconditioning and partial updates for viscous and smoothing fluxes. Putting the linear iterative solver into the correct form requires a number of manipulations, and having then determined R^T further manipulations are necessary to express the adjoint algorithm in a convenient form for programming implementation. The correctness of the analysis has been tested with a simple MATLAB program which can solve either a simple scalar o.d.e. or an upwind approximation to the convection equation with a harmonic source term. In the latter case, the theory presented in this chapter has been extended to include problems with complex variables, by replacing all vector and matrix transposes by their complex conjugates. In either case, it is verified that an identical number of iterations of either the linear problem or its adjoint yields identical values for the functional of interest.

Reference [217] also briefly considers the application of the theory to preconditioned multigrid methods [418], in which a sequence of coarser grids is used to accelerate the iterative convergence on the finest grid. Provided the smoothing algorithm used on each grid level within the multigrid solver is of the form given in Equation (16.3), it shown that all is well if the restriction operator for the adjoint solver is the transpose of the prolongation operator for the linear solver, and *vice versa*. This feature has been tested in unpublished research in developing a three-dimensional adjoint Navier-Stokes code using unstructured grids. Again, identical values have been obtained for the functional of interest after equal number of multigrid cycles with either the linear solver or its adjoint counterpart.

16.5   Conclusions

In this chapter we have shown that the naive application of adjoint methods to the iterative solution of a linear system of equations produces an algorithm which does not correspond to the iterative solution of the corresponding adjoint system of equations. However, with some algebraic manipulations it can be transformed into an algorithm in which the working variables do converge to the solution of the adjoint equations.

Mathematically, the two approaches produce identical results if the second calculation starts from zero initial conditions. The advantage of the second formulation is that the computational costs can be greatly reduced if one has a good initial estimate for the solution. This happens in nonlinear design optimisation in which the adjoint solution for one step in the optimisation can be used as the initial conditions for the adjoint calculation in the following step.

This has implications for the use of automatic differentiation software in generating adjoint programs. The AD tools can still be used to generate the subroutines which construct the adjoint system of linear equations, but to achieve the maximum computational efficiency it appears it is necessary to manually program the higher-level fixed point iterative solver.

Acknowledgments: The author is very grateful to Andreas Griewank for his considerable help in the writing of this chapter. This research was supported by EPSRC under grant GR/L95700, and by Rolls-Royce plc and BAe Systems plc.

17

Aerofoil Optimisation via AD of a Multigrid Cell-Vertex Euler Flow Solver

Shaun A. Forth and Trevor P. Evans

ABSTRACT We report preliminary results in the use of ADIFOR 2.0 to determine aerodynamic sensitivities of a 2-D airfoil with respect to geometrical variables. Meshes are produced with a hyperbolic interpolation technique. The flow field is calculated using the cell-vertex method of Hall, which incorporates local time-stepping, mesh sequencing and multigrid. We present results and timings using both Finite Differences (FD) and Automatic Differentiation (AD). We investigate the effect of starting the perturbed calculation for FD and the derivative calculation for AD from either the current or freestream conditions and highlight the need for careful implementation of convergence criteria.

We attempt to make a comparative study of AD and FD gradients in an aerofoil optimisation, using the DERA CODAS method from the perspective of DERA's eventual aim, 3D viscous optimisation of wing-body configurations.

17.1 CFD Design Process

We consider a generic CFD solution algorithm for given design parameters $\mathbf{p}$. We use the simultaneous update of flow and objective variables since the far-field boundary condition is set using the value of the lift [332].

Algorithm 1: A generic CFD algorithm:

1. A mesh generator $\mathbf{X}$ gives a set of discrete mesh points $\mathbf{x} = \mathbf{X}(\mathbf{p})$.

2. Initialise discrete flow-field variables $\mathbf{u}$ and design objectives $\mathbf{c}$.

3. Update $\mathbf{u}$ and $\mathbf{c}$ via an iteration, labelled n, of a numerical solution procedure $\mathbf{U}$ and $\mathbf{C}$ for the Euler or Navier-Stokes equations,

$$\mathbf{u}^{n+1} = \mathbf{U}\left(\mathbf{p}, \mathbf{x}, \mathbf{u}^n, \mathbf{c}^n\right), \qquad (17.1)$$

$$\mathbf{c}^{n+1} = \mathbf{C}\left(\mathbf{p}, \mathbf{x}, \mathbf{u}^{n+1}, \mathbf{c}^n\right). \qquad (17.2)$$

4. Repeat step 3 until the procedure converges.

The design optimisation process proceeds by calculating the gradient $\partial c/\partial p$ of the objective variables with respect to the design variables (we have used up to 17 design variables to date) and then using an SQP optimisation algorithm to modify the design variables. Accurate and efficient calculation of the gradient is important in developing an effective design algorithm.

17.1.1   Calculating gradients of the objectives

Obtaining the gradient $\partial c/\partial p$ is complicated by the implicit definition of the flow field variables as the solution of a discretised PDE via the iterative process of equations 17.1 and 17.2. In this chapter we consider four alternative ways to determine these gradients.

Piggy-Back AD (PB-AD)

In the *Piggy-Back AD (PB-AD)* approach [238] (also termed fully differentiated [29]), we simply differentiate through Algorithm 1 and obtain derivatives with respect to **p** of all variables as they are calculated. It has been shown [29] that, under mild conditions on the iteration functions **U** and **C**, such a scheme converges to the correct derivatives $\partial c/\partial p$.

Two-Phase AD (2P-AD)

Post-differentiation as advocated in [29] involves: first converging the flow solver (Algorithm 1), then performing one iteration involving the derivatives, then finally directly solving the linear system associated with the differentiated flow solver for $\partial u/\partial p$ and $\partial c/\partial p$. This approach is generally impractical in the context of CFD due to the large size of the linear system. In other applications it can be very efficient, see for instance the hand-coded post-differentiation used Chapter 7 [109] of this volume. Instead we adopt an iterative approach termed *Two-Phase AD* (2P-AD) [238], first converging the flow field and objective variables and then switching on the PB-AD algorithm to obtain the derivatives from the converged flow solution. This has the advantage of not performing expensive AD iteration early in the calculation when the flow field is far from converged. We continue to update the flow field and objective variables when calculating the derivatives.

Freestream Start FD (FS-FD)

We use one-sided finite differences (divided differences) to approximate $\partial c/\partial p$ with both the flow solution and its finite-difference perturbation obtained from free-stream starts via Algorithm 1.

Converged Start FD (CS-FD)

Again we use one-sided finite differences and obtain the flow solution from a freestream start, but the calculation of the perturbed solution is initialised with the converged unperturbed flow solution.

17.1.2   Convergence criteria

For two-phase differentiation of the flow solver and finite differencing, we used the convergence condition

$$\left\| \mathbf{c}^{n+1} - \mathbf{c}^n \right\|_\infty < \mathrm{tol}_c, \tag{17.3}$$

where tol_c is the *solver tolerance*. In addition for AD derivatives we used

$$\left\| \frac{\partial \mathbf{c}}{\partial \mathbf{p}}^{n+1} - \frac{\partial \mathbf{c}}{\partial \mathbf{p}}^{n} \right\|_\infty < \mathrm{tol}_{\partial \mathbf{c}/\partial \mathbf{p}}, \tag{17.4}$$

where $\mathrm{tol}_{\partial \mathbf{c}/\partial \mathbf{p}}$ is the *derivative tolerance*. We did not apply convergence conditions on the flow field variables $\mathbf{u}$ or their derivatives $\partial \mathbf{u}/\partial \mathbf{p}$.

17.2   ADIFOR Applied to Cell Vertex Flow Solver

We have applied the four strategies of §17.1.1 to the cell-vertex finite-volume compressible Euler flow solver mheuler of Hall [266]. The solver uses explicit Lax-Wendroff time-stepping and features local time-stepping, mesh sequencing, and multigrid to accelerate convergence to steady state.

Automatic differentiation of the flow solver was performed using ADIFOR version 2.0 revision D [78]. ADIFOR is a source text translation tool that takes existing Fortran 77 code and produces new code consisting of the original source code augmented with code to calculate directional derivatives. Before using ADIFOR we had to perform some minor rewriting of the flow solver so that the nonlinear iteration process abstractly described by equations 17.1 and 17.2 could be placed in a separate subroutine. ADIFOR can be used to calculate an essentially arbitrary number of directional derivatives, the user merely setting the maximum to be used. However in the present work, we used the option AD_SCALAR_GRADIENTS=true to force ADIFOR to produce code for a single, scalar directional derivative. This ensures that we have no short loops within the differentiated code that the compiler cannot *unroll*. In any industrial scale computer program, there are some lines of code that are not globally differentiable. In mheuler the Fortran intrinsics MAX and ABS are used within the second order numerical dissipation required for stability of the flow solver. ADIFOR can log the number of times such a function is called at a point of potential non-differentiability, and the user has control over how this is done. We used the option AD_EXCEPTION_FLAVOR=reportonce to produce a list of such *exceptions* at the end of the calculation. In the future we shall use AD_EXCEPTION_FLAVOR=performance to improve performance once we are satisfied that points of non-differentiability encountered are unimportant.

17.2.1   Comparison of AD and FD directional derivatives

To compare the relative accuracy of the AD and FD directional derivatives, we calculated the inviscid flow about a NACA 64A airfoil section at a Mach number of 0.63 and zero incidence on a "C"-mesh of 25 points normal to the airfoil and 193 around the airfoil and wake cut. We determined the directional derivative associated with a perturbation to a spline controlling the shape on the upper airfoil surface near the leading edge. Assuming that the PB-AD results are correct, then Figure 17.1 shows how

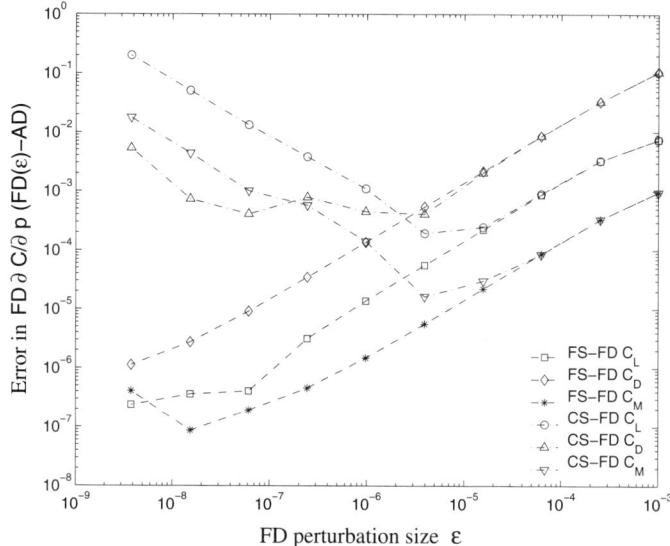

FIGURE 17.1. Error in Finite Difference (FS-FD and CS-FD) approximations to directional derivatives of the objective functions C_L, C_D and C_M as compared with AD (PB-AD).

the error (difference from PB-AD) for the two finite-difference approximations to this single directional derivative vary with the finite-difference perturbation size. Here the design objectives are: lift coefficient C_L, drag coefficient C_D, and pitching moment C_M. The convergence tolerances used were $\text{tol}_c = \text{tol}_{\partial c/\partial \mathbf{p}} = 1 \times 10^{-10}$. The FS-FD results converge linearly to the PB-AD results as the FD step size is reduced until, at very small perturbation size, the error arising from the finite convergence criteria degrades the expected accuracy. This result validates the intrinsic accuracy of automatic differentiation by showing that the numerically approximate finite-difference gradient approaches the AD gradient as the FD step size is reduced. Similar convergence results may be seen in [109] elsewhere in this volume. The CS-FD results initially follow the FS-FD results but are then seen to diverge from the expected behaviour. This, and the slight dif-

TABLE 17.1. Directional derivatives of C_L, C_D and C_M computed by Piggy-Back AD and Two-Phase AD

| | Directional Derivatives of | | |
Method	C_L	C_D	C_M
PB-AD	0.0916389175107	-0.204563844953	0.0105117778674
2P-AD	0.0916389172678	-0.204563844468	0.0105117778962

TABLE 17.2. CPU times corresponding to the calculations of Figure 17.1. 2P-AD and both FD times include 213.2s to obtain a converged flow-field.

| | AD | FD perturbation size | | |
Method		1×10^{-3}	1×10^{-6}	3.8×10^{-9}
PB-AD	1641.5			
2P-AD	1051.2			
FS-FD		420.5	422.7	422.6
CS-FD		398.5	269.2	262.5

ference between the 2P-AD and PB-AD derivatives (as seen in Table 17.1) is thought to be due to the lack of convergence criteria on the flow field variables (and their derivatives). In Table 17.2 we give CPU timings obtained on a COMPAQ Alpha 255-300 for the calculations of Figure 17.1 and Table 17.1. The second column gives the CPU times for the AD results, and columns 3 to 5 give the times for the FD results at 3 different perturbation sizes. For a single directional derivative FD appears more efficient than forward AD, though of inherently limited accuracy.

It is possible to obtain AD gradients with derivative tolerance $tol_{\partial c/\partial p} = 1 \times 10^{-5}$ in CPU times of 368s and 262.4s for PB-AD and 2P-AD respectively. These results are of a similar level of accuracy to the FD results shown in figure 17.1 and for a comparable CPU cost (c.f. Table 17.2).

17.3 AD within an Optimisation Design Method

We used the Euler solver mheuler of §17.2 in an aerodynamic optimisation test case making use of DERA's CODAS package [162]. We compared AD results with those obtained using FD. All the constrained optimisations performed with CODAS used a sequential quadratic algorithm [460].

17.3.1 Aerofoil drag minimisation

In this problem angle of incidence α and 16 cubic spline coefficients were used as design variables. The initial aerofoil geometry was that of an RAE2822 aerofoil. For fixed freestream Mach number of 0.8 and with lift coefficient C_L constrained at 0.9, a minimal drag coefficient C_D solution was sought. Figure 17.2 shows the final pressure coefficient distribution C_P (top left), aerofoil shape (bottom left) with horizontal coordinates X and airfoil displacement Z scaled with the airfoil chord length (here denoted C). The right hand side of Figure 17.2 shows optimisation convergence histories (here ALPHA denotes α the airfoil angle of incidence) plotted against number of calls to the CFD flow solver. In this calculation all gradients were obtained using the Freestream Start FD approach. We see that the

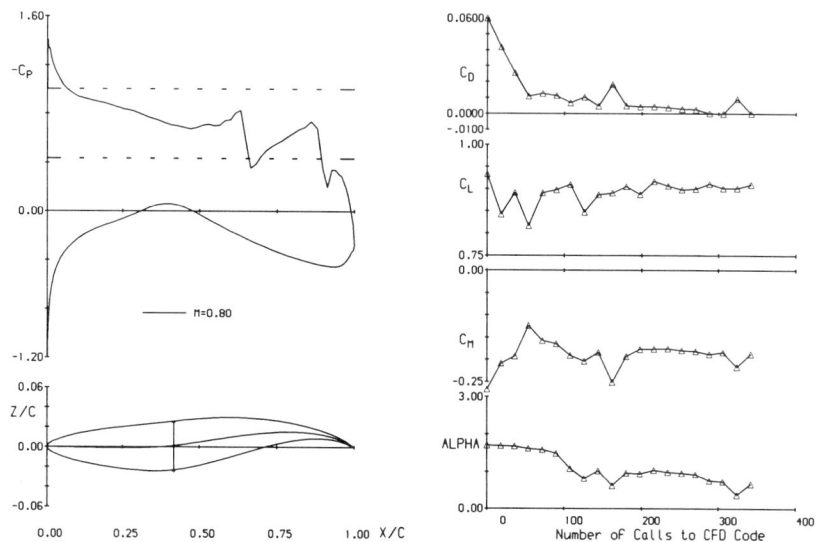

FIGURE 17.2. Finite difference drag optimisation of an RAE2822 aerofoil

drag C_D has been minimised and, after some initial transient behaviour, the lift C_L is close to its constrained value.

In Figure 17.3 we perform the same task using Two-Phase AD to obtain the gradients. Both techniques have converged to essentially the same solution.

The number of flow code iterations used here for FD was that currently employed by DERA in 2D optimisation studies. The number of iterations used for 2P-AD was chosen to give run times similar to those for FD. It was observed that gradient accuracy, compared to an earlier well-converged AD run, was similar for the illustrated FD and AD runs; and the expected resulting similarity in optimiser progress is clear from figures 17.2 and 17.3.

It seems then, that AD and FD have similar success in tasks such as this test case, when run times acceptable to industry are used. It was pleasing, though, that attention to both flow solver and grid generator convergence produced AD gradients with greater accuracy than that possible using FD, giving smoother optimiser performance late in the design process. It may be then that for more challenging search spaces, only AD will be able to

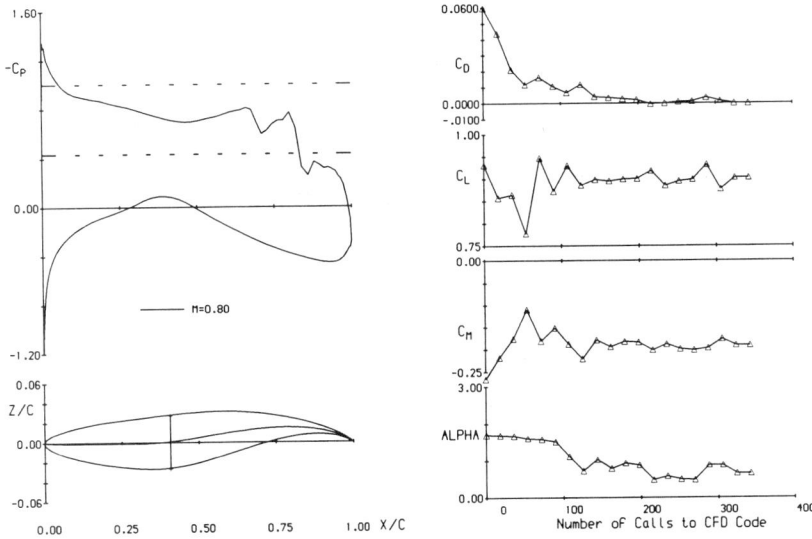

FIGURE 17.3. AD drag optimisation of RAE2822 aerofoil

produce a good design.

17.4   Conclusions

We have confirmed that the forward mode of AD obtained using ADIFOR applied to our cell-vertex Euler flow solver obtains gradients consistent with those from finite differencing, but with greater accuracy. For the forward mode this greater accuracy comes at the expense of greater run times. However we have found that by adjusting the convergence criteria in the iterative solver for the AD gradients we obtain similar accuracy to finite differencing in similar run times. Although we only require objective values such as lift and drag coefficients within our optimisation techniques, we believe that careful control of the convergence of the flow-field itself is important for accuracy of both FD and AD gradients.

We have successfully used ADIFOR within an optimisation of an aerofoil shape to minimise drag for constant lift. For completeness we note that problems were encountered in the inverse design problem of matching a given pressure distribution and full details will be published elsewhere.

In our optimisation problem we have a large number of design variables (which control the shape of our airfoil) and only a small number of outputs (lift, drag, pitching moment). Consequently gradients should be calculated using the reverse or adjoint mode of AD since then run times will be proportional to the small number of outputs rather than to the large number of inputs (as in forward mode). Even for the 2-D optimisation problems we have been considering with 17 design variables and 3 outputs we would expect a 3 to 5 fold reduction in run times, allowing us to calculate more accurate gradients (if we need them) using AD in equivalent run times to FD. For the 3-D wing design problems we are ultimately interested in we will be using in excess of 50 design variables and then reverse mode AD could be in excess of 10 times faster than FD. We shall be attacking such problems with the reverse mode of AD using INRIA's AD tool Odyssée [186] in the EU project AEROSHAPE.

Acknowledgments: The authors would like to thank the UK's Department of Trade and Industry for their funding of this work under the CARAD project.

Additionally the authors thank Mr. A. Gould of BAe Systems (SRC) for both technical support and his liaison with all parties, Dr. P. Betten of ANL for allowing Cranfield University use of ADIFOR and technical contributions from Dr. J. Pryce of Cranfield University (RMCS Shrivenham) and Dr. M. Bartholomew-Biggs, Prof. B. Christianson and Prof. A. Holdo all of the University of Hertfordshire.

18

Automatic Differentiation and the Adjoint State Method

Mark S. Gockenbach, Daniel R. Reynolds and William W. Symes

ABSTRACT   The C++ class `fdtd` uses automatic differentiation techniques to implement an abstract time stepping scheme in an object-oriented fashion, making it possible to use the resulting simulator to solve inverse or control problems. The class takes a complete specification of a *single step* of the scheme, and assembles from it a complete simulator, along with the linearized and adjoint simulations. The result is a (nonlinear) operator in the sense of the Hilbert Class Library, a C++ package for optimization. Performance is equivalent to that of optimized Fortran implementations.

18.1   Introduction

There are two well-known obstacles to the use of time stepping schemes (based on finite difference or finite element discretization) in solving inverse and control problems. First, while the code for the linearized scheme is relatively easy to write, the code for its adjoint is much more difficult, and naive implementations can easily lead to gross inefficiency. Second, when the simulator is coded in procedural style, linkage to optimization code is problematic, often requiring the construction of elaborate, fragile, and non-reusable interfaces.

We attempt to resolve both of these problems through an *abstract time stepping scheme*, implemented as a C++ class `fdtd` ("**F**inite **D**ifference **T**ime **D**omain"). The class implements the structure of an explicit time-stepping scheme, as well as its linearization and two variants of the adjoint scheme, one appropriate for conservative problems, the other for dissipative problems. The dissipative adjoint scheme encodes Griewank's optimal checkpointing scheme [235]. A user is expected to supply (to `fdtd`) a *stencil operator* defining a single time step, along with its linearization and adjoint; `fdtd` then assembles the entire simulation.

Furthermore, the `fdtd` class defines an operator in the sense of the Hilbert Class Library (HCL), a C++ software package for optimization [224]. Since `fdtd` is an operator in the HCL sense, it can link directly to any optimization algorithms implemented in HCL.

18.2   Abstract Formulation of Time-Stepping Schemes

A primary reason for implementing time stepping schemes in a C++ class is to ease the use of optimization in data-fitting problems. For example, suppose that a coefficient in a PDE is to be estimated by solving

$$\min_c J(c) = \frac{1}{2}\|G(c) - D^{obs}\|^2, \tag{18.1}$$

where $c \in C$ denotes an unknown parameter, D^{obs} is observed data, and G is the *forward map* (mathematically, the solution operator of the PDE). The map G involves simulation of space-time fields, followed by sampling. In a typical application, only parts of the fields are observable. We therefore assume that

$$G(c) = SU = \sum_{n=0}^{N} S_n U^n,$$

where $U^n \in \mathcal{U}$ is (related to) the nth time level of the simulated field, $S : \mathcal{U} \to \mathcal{D}$ is the *sampling operator*, and $\mathcal{D}$ is the data space.

Most algorithms for solving (18.1) require the gradient of J,

$$\nabla J(c) = DG(c)^*(G(c) - D^{obs}),$$

so we require the adjoint of the derivative (linearization) $DG(c)$.

In [226], we present a formulation of explicit marching methods that eases the computation of this adjoint. Here we just summarize the resulting formulas, identifying the operations that are common to all marching schemes and those that are application-dependent.

Any marching scheme can be considered to be formally two-level, by concatenating several time levels if necessary. Therefore, we write

$$U^{n+1} = H_n(c, U^n), \quad n = 0, 1, \ldots, N - 1,$$

and call $H_n : C \times \mathcal{U} \to \mathcal{U}$ the *stencil operator*.

The *linearization* of the map $c \mapsto G(c)$ is obtained by linearizing the time-stepping equations: $DG(c)\delta c = S\delta U$, where

$$\delta U^0 = 0, \quad \delta U^{n+1} - A_n \delta U^n = \delta F^{n+1}, \quad n = 0, 1, \ldots, N - 1,$$

$\delta U^n = (DU(c)\delta c)^n$, $A_n = D_U H_n(c, U^n)$, and $\delta F^{n+1} = D_c H_n(c, U^n)\delta c$ ($\delta F^0 = 0$). We can also write this as $M\delta U = \delta F$, where $M : \mathcal{U}^{N+1} \to \mathcal{U}^{N+1}$ is a block lower bidiagonal linear operator. The explicit time-stepping scheme is equivalent to solving $M\delta U = \delta F$ by forward substitution.

Write B for the operator mapping δc to δF. Then $DG(c) = SM^{-1}B$, and so $DG(c)^* = B^*M^{-*}S^*$.

We assume that S_n and S_n^* are supplied by the user along with the stencil operator H_n and its derivatives and adjoints $D_c H_n(c, U)$, $D_U H_n(c, U)$, $D_c H_n(c, U)^*$, and $D_U H_n(c, U)^*$. We now show how to compute $DG(c)^*\delta D$.

Write $\delta V = S^* \delta D$. Straightforward calculations show that

$$(S^* \delta D)^n = S_n^* \delta D, \quad B^* \delta W = \sum_{n=1}^{N} D_c H_{n-1}(c, U^{n-1})^* \delta W^n.$$

Next we need $M^{-*} \delta V$. The operator M^* is a block upper bidiagonal linear operator, and $\delta W = M^{-*} \delta V$ can be computed by back substitution (reverse time-stepping):

$$\delta W^N = \delta V^N, \quad \delta W^{n-1} = A_{n-1}^* \delta W^n + \delta V^{n-1}, \quad n = N, N-1, \ldots, 1.$$

We refer to δW as the *adjoint state* and to the equation $M^* \delta W = \delta V$ as the *adjoint state equation*.

Thus the procedure for computing $DG(c)^* \delta D$, for $\delta D \in \mathcal{D}$, is:

1. Solve the simulation problem to produce the field U.

2. For $n = N, N-1, \ldots, 1$:

   (a) Compute $\delta V^n = S_n^* \delta D$.

   (b) Compute δW^n from δW^{n+1} $(\delta W^N = \delta V^N)$.

   (c) Add $D_c H_{n-1}(c, U^{n-1})^* \delta W^n$ to the output vector δc.

A logistical problem immediately asserts itself: U is produced by stepping *forward* in time, δW by stepping *backwards*. Unless the state space has small dimension, storage of the reference field U is out of the question.

For conservative problems, backwards time stepping is stable, so the final value of U^n can be used as initial data to run the scheme for U backwards simultaneously with the adjoint state scheme for δW. The sum of the terms $D_c H_{n-1}(c, U^{n-1})^* \delta W^n$ is accumulated, and when $n = 0$ is reached the adjoint has been computed.

If the underlying continuous system is dissipative, this simple approach can not be used. In that case, a *checkpointing scheme* due to Griewank [235], extended in [86], is employed. The idea is to combine saving various time levels U^n to use as intermediate initial data to restart the computation of U during the solution of the adjoint state system.

18.3   The Class `fdtd`

18.3.1   The Hilbert Class Library

The Hilbert Class Library (HCL) [224] is a library of C++ classes representing the basic mathematical objects (vectors, functions, operators, etc.) used in optimization problems. HCL allows algorithms to be implemented in an abstract fashion. This means that the optimization code does not deal directly with the data structures or interfaces used by application code. Therefore, optimization algorithms implemented using HCL classes can be used with applications of arbitrary complexity.

In order to solve problem (18.1) using an HCL optimization algorithm, it is necessary to represent the forward map G in a class derived from HCL_Op, which is the abstract base class for nonlinear operators in the HCL framework. Given the forward map G as an operator in this sense, HCL makes it easy to assemble the least-squares functional and invoke an optimization routine. Therefore, the class fdtd is derived from HCL_Op.

In order to create an instance of fdtd implementing a specific marching scheme, one must implement the stencil operator H_n, as well as its derivatives and their adjoints, as an operator in the HCL framework. In addition, the sampling operator must be implemented as an HCL linear operator. We have created two base classes to facilitate the implementation of these operators:

1. StencilOp, derived from HCL_OpProductDomain. The base class represents operators depending on more than one independent variable (therefore, for instance, partial derivatives are meaningful).

2. SampleOp, derived from HCL_LinearOp.

These are both abstract base classes, meaning that they provide a template for creating the necessary derived classes. The actual work involved in creating a derived class is limited to the implementation of methods that perform the calculations mentioned above: $H_n(c, U^n)$, $D_U H_n(c, U^n)\delta U^n$, $D_U H_n(c, U^n)^*\delta W^{n+1}$, etc.

The design of fdtd and the helper classes StencilOp and SampleOp ensures that the user must provide only those calculations that are special to his or her problem. Moreover, this design—namely, the separate implementation of the (problem specific) stencil operator and the (abstract) time-stepping scheme—introduces the possibility of the use of current Automatic Differentiation (AD) tools by the user. Present AD tools, such as TAMC [211], are not able to produce efficient adjoint code for the full time-stepping simulation (at least, not automatically), because of the need to either save or recompute the reference field U for use during solution of the adjoint state equation. However, TAMC and other such tools *can* produce efficient code for the adjoint linearization of a single time step.

It is therefore possible to use modern AD tools to implement the stencil operator, that is, to produce code implementing the calculations

$$D_U H_n(c, U^n)\delta U^n, D_U H_n(c, U^n)^*\delta W^{n+1}, \text{ etc.}$$

Therefore, the user may only be required to write code for the computation of $H_n(c, U^n)$. This was the approach followed in our example, described below in §18.4.

18.3.2 The implementation of fdtd

The major methods of the class fdtd are: the constructor, Image (performs $D \leftarrow G(c)$), DerivImage (performs $\delta D \leftarrow DG(c)\delta c$), and DerivAdjImage (performs $\delta c \leftarrow (DG(c))^*\delta D$).

The constructor must be provided with objects representing the stencil operator H_n and the sampling operator S. These are user-defined and form the interface between fdtd and a specific example, as discussed above.

The computational routines Image, DerivImage, and DerivAdjImage have fairly simple implementations, because the object-oriented structure of HCL allows the algorithms to be coded at a high level. The only complication arises in the adjoint code for dissipative schemes.

The solution of the adjoint state equation requires the following computation:

$$\delta W^{n-1} = D_c H_{n-1} \left(c, U^{n-1}\right)^* \delta W^n + \delta V^{n-1}, \ n = N, N-1, \ldots, 1;$$

the value U^{n-1}, which cannot be computed from U^n (by backward time-stepping) because the original scheme is dissipative, must be available. The two obvious ways of producing U^{n-1}, recomputing it from U^0 by forward time-stepping and saving it during the original forward simulation, are too costly in terms of time and memory, respectively.

The class fdtd therefore uses the AD technique of checkpointing—saving certain time levels U^k during the forward simulation, and then recomputing U^{n-1} by forward time-stepping from the nearest checkpoint. As Griewank [235] showed, a properly implemented checkpointing scheme can reduce the growth in time and memory requirements to only a factor of $\log(N)$ (over the requirements for the forward simulation), where N is the number of time steps.

18.4 A Concrete Example

To test the fdtd class, we implemented the standard (2,4) (leapfrog) finite difference scheme for the 2D acoustic wave equation with a point source. The parameter c in this example is the acoustic wave speed, and the state variable U is the pressure field. Since the leapfrog scheme is three-level, we define U^n to be (p^{n-1}, p^n), where p is the pressure.

The Fortran subroutine a2cptsrc, which applies the finite difference stencil (i.e. evaluates $H_n(c, U^n)$), is the basis of our implementation. The AD package TAMC [211] was used to produce four further Fortran subroutines, implementing the linearizations of a2c24ptsrc with respect to the velocity and the pair (p^{n-1}, p^n), and the adjoints of these linearizations. A small amount of hand-editing of the TAMC output was necessary because TAMC does not always generate code for the desired operation (for example, all TAMC adjoint code adds the computed result to the output

TABLE 18.1. Times on a Sun SPARC Ultra 60 workstation.

grid	FDTD (s)	Fortran (s)	ratio
161×161	8.244	7.960	1.0357
321×321	86.4233	85.9967	1.0050

variable, which is not necessarily the desired action). See [223] for details on using TAMC-generated code.

The class a2c24StencilOp was then defined to represent H_n. All significant arithmetic in our implementation is confined to subprograms written in Fortran. The C++ code defining a2c24StencilOp merely extracts pointers to the data and grid parameters from the vector class representing the velocity and pressure fields.

The sampling operator for this application is based on a C++ class TimeSample, written to sample at arbitrary locations functions defined on regular rectangular grids.

Details about this example and its implementation can be found in [226]. The question we address here is whether the use of C++ and object-oriented programming imposes significant overhead compared to a purely procedural (Fortran) implementation.

For comparison, we created a purely Fortran implementation (a2c24sim) of the leapfrog scheme. We then compared the total time required for a simulation, using both codes (fdtd and a2c24sim). The results (on a Sun SPARC Ultra 60), which are given in Table 18.1, suggest that the overhead is not more than a few percent.

We also profiled the execution of the main methods of fdtd, namely Image, DerivImage, and DerivAdjImage (that is, the computations of $G(c)$, $DG(c)\delta c$, and $DG(c)^*\delta D$). This was done using the GNU tool gprof with the code compiled under g++ and g77. The results (for an 81×81 grid) showed that in every case, 98% or more of the total execution time was spent in the Fortran subroutines. We believe that the combination of direct timing and profiling presented here provides conclusive evidence that the use of C++ (and specifically HCL) to manage high-level manipulations does not imply a performance penalty.

Acknowledgments: This work was supported in part by NSF grants DMS-9627355, DMS-9973423, and DMS-9973308, by the Los Alamos National Laboratory Computer Science Institute (LACSI) through LANL contract number 03891-99-23, and by The Rice Inversion Project.

19

Efficient Operator Overloading AD for Solving Nonlinear PDEs

Engelbert Tijskens, Herman Ramon and Josse De Baerdemaeker

ABSTRACT By employing automatic differentiation (AD), solvers for nonlinear systems of PDEs can be developed which relieve the user from the extra work of linearising a nonlinear PDE system and at the same time improve performance. This is achieved by extending common AD techniques using operator overloading to take advantage of the fact that in a FEM/FD/FV framework, a limited number of functions and their partial derivatives with respect to the unknowns have to be evaluated many times. The extension is implemented in C++ for both forward and reverse modes, and compared to hand coded evaluation of derivatives and two state-of-the-art AD implementations, ADIC [84] and ADOL-C [242, 243]. An application is discussed which dramatically reduces the cost of solver development.

19.1 Introduction

The maturity of the mathematical and numerical aspects of Finite Element, Finite Difference and Finite Volume Methods (FEM/FD/FV) and the recent advances in object-oriented software engineering has lead to the point where the construction of solvers for general systems of partial differential equations (PDE) has become feasible for users without a strong mathematical, numerical or programming background in a time efficient manner. General packages, e.g. Diffpack® [344, 406], have dramatically reduced the efforts required from the user to build such solvers. Ideally, the user would only have to enter the PDE system in a symbolic form together with initial and boundary conditions and perhaps some constraining algebraic equations. This is still beyond reach. In practice, the user has to derive discretised equations and provide details about the grid and the elements used. This is generally not a major obstacle, but for nonlinear systems, the user also must linearise the PDE system or provide the Jacobian of the discretised system. Although a trivial task in principle, this is time-consuming, error-prone and can potentially be very hard or even prohibitive, depending on the constitutive equations for the model parameters in the system [232]. The situation certainly does not encourage the exploration of a large number of models. In the last decade automatic differentiation (AD) has become mature and can in principle be used to relieve the user from the task of providing accurate Jacobians. AD comes in forward and reverse modes, as well as hybrids of the two [238], which can be implemented as a source preprocessor, e.g. ADIFOR [79], ADIC [84], Odyssée [186], or using operator overloading, e.g. KDPACK [499], FastDer [480], ADOL-C [242].

19.2   Multiple Point Operator Overloading AD

Consider a function f depending on p variables

$$f(u_1, ..., u_p) : \mathbb{R}^p \to \mathbb{R}, \tag{19.1}$$

which is to be evaluated in n points. The function f and the arguments $\mathbf{u}^i = \begin{bmatrix} u_1^i & u_2^i & \cdots & u_p^i \end{bmatrix}$ ($i = 1, ..., n$) are mapped onto the following structures, respectively:

$$\mathbf{f}(u_1, ..., u_p) = \begin{bmatrix} f\left(\mathbf{u}^1\right) & \nabla f\left(\mathbf{u}^1\right) \\ f\left(\mathbf{u}^2\right) & \nabla f\left(\mathbf{u}^2\right) \\ \vdots & \vdots \\ f\left(\mathbf{u}^n\right) & \nabla f\left(\mathbf{u}^n\right) \end{bmatrix}, \qquad \mathbf{u}_j = \begin{bmatrix} u_j^1 & \mathbf{e}_j \\ u_j^2 & \mathbf{e}_j \\ \vdots & \vdots \\ u_j^n & \mathbf{e}_j \end{bmatrix}, \tag{19.2}$$

where $\mathbf{e}_j = \nabla u_j$ is the j-th unit vector in $\mathbb{R}^p$. Note that gradients are taken with respect to $\mathbf{u} \in \mathbb{R}^p$. A general k-ary algebraic operator K is correspondingly mapped onto the operator

$$\mathbf{K}(\mathbf{f}_1, \mathbf{f}_2, ..., \mathbf{f}_k) = \begin{bmatrix} K\left(f_1\left(\mathbf{u}^1\right), ..., f_k\left(\mathbf{u}^1\right)\right) & \sum_{j=1}^{k} \frac{\partial K}{\partial f_j} \nabla f_j\left(\mathbf{u}^1\right) \\ K\left(f_1\left(\mathbf{u}^2\right), ..., f_k\left(\mathbf{u}^2\right)\right) & \sum_{j=1}^{k} \frac{\partial K}{\partial f_j} \nabla f_j\left(\mathbf{u}^2\right) \\ \vdots & \vdots \\ K\left(f_1\left(\mathbf{u}^n\right), ..., f_k\left(\mathbf{u}^n\right)\right) & \sum_{j=1}^{k} \frac{\partial K}{\partial f_j} \nabla f_j\left(\mathbf{u}^n\right) \end{bmatrix}.$$
$$\tag{19.3}$$

In the usual operator overloading approach n equals 1. In the current implementation the structures in equation (19.2) are represented by instances of a particular C++ class DaElement, and the operators are the usual algebraic operators overloaded to act on DaElement objects. A major concern in the implementation was the realisation of a class interface as natural as possible. Therefore DaElement objects should behave just as doubles, and all common arithmetic C++-operators acting on doubles should have overloaded equivalents acting on DaElement objects. A data storage manager object (DaSession) takes care of efficient memory management, allows the operators to return results by reference and to deal efficiently with intermediate results. This solution allows for maximal freedom in writing expressions over DaElements.

Can operator overloading AD also be computationally efficient? We show that the extension to multiple point AD reduces all sources of overhead to a minimum in both forward and reverse modes. In single point operator overloading AD there are three sources of overhead. An obvious source of overhead is the function call overhead resulting from replacing the usual operators with their overloaded equivalents. In addition, there is the need to build dynamical structures for storing the computational graph in the reverse mode. In the forward mode many unneeded computations with 0 an 1 will be performed. Clearly, in single point operators ($n = 1$) the two last sources of overhead will be incurred in every function call, making it proportional to the number of required evaluations.

The extension to multiple point AD ($n > 1$) reduces the number of function calls by a factor n as one function call handles n points at a time. Hence, the overhead of building dynamical structures for storing the computational graph is incurred only once. The extension also offers other interesting perspectives. The data storage of structure (19.2) and the computations in the operators (19.3) can be organised columnwise (columns of length n). This allows columns with a constant value (most often 0 or 1) to be replaced by a scalar. A single test will then suffice to check whether multiplication with a gradient column is necessary. Unnecessary computations in the forward mode can thereby be eliminated.

Thus, multiple point operator overloaded AD reduces all sources of overhead to a minimum. As usual, this comes at the cost of increased memory requirements, as the memory needed by a multiple point version scales with n. For the forward mode the memory needed is proportional to $n(p + 1)w_{max}$, with w_{max} the maximal number of living vertices in the computational graph. For the reverse mode it is approximately $n \sum_{i=1}^{N} (1 + d_i)$, where N is the number of nodes in the tree and d_i is the number of daughter nodes of the i-th node. Here, the advantage is with the forward mode. Finally, the overloaded operators can be expressed in terms of operations between columns which can be executed very fast on pipelined architectures and can be parallellised easily.

Table 19.1 presents a quantitative comparison of the computational efficiency for obtaining derivatives in different ways (hand coded, our implementation of multiple point operator overloading implementations of reverse mode and forward mode (FastDer++), ADIC [84] and ADOL-C [242] for a simple test function $y(\mathbf{u}) = r^2 \exp(-r^2)$, $r^2 = \sum_{i=1}^{P} u_i^2$. Although the simplicity of this function is in extreme contrast with the complexity of the cost functions to be differentiated in optimisation problems, it is typical of the kind of nonlinear parameters generally encountered in quasilinear PDEs, as these parameters typically represent empirical relationships of a few independent variables. Hence the performance results are believed to be representative for the situation typically encountered in systems of quasilinear PDEs. More detailed results can be found in [478].

In this multiple point setting FastDer++ is competitive with the single point AD tools ADOL-C and ADIC, showing that AD tools need specialisation with respect to optimisation problems and nonlinear PDE systems. It also shows that when the computational graph is not too big – which is actually the case for a large class of nonlinear PDE models arising in engineering – the forward mode is not to be ruled out a priori on the basis of its worse complexity result [238].

TABLE 19.1. Comparison between AD implementations. Average run times per point on a 200 MHz Pentium II processor for the test function $y(\mathbf{u}) = r^2 \exp(-r^2)$, $r^2 = \sum_{i=1}^{p} u_i^2$. Run times were recorded for array lengths of $n = 2^m (m = 2, 3, \ldots, 15)$.

	$p = 2$	$p = 10$
Hand coded, not optimised	2.74	14.9
Hand coded, optimised	1.11	3.95
FastDer++, Reverse mode	9.2	37.3
FastDer++, Forward mode	3.64	19.1
ADIC	8.06	22.9
ADOL-C	54.4	101

19.3   Application in a FEM Framework

From the beginning the FastDer++ system was developed for application in nonlinear partial differential equation (PDE) solvers in a finite element or finite differences framework [480]. In recent years general FEM/FD/FV packages, e.g., Diffpack® [344, 406], have become available which dramatically reduce the required efforts to build PDE solvers. The treatment of nonlinear systems generally implies a linearisation either at the PDE level or at the algebraic level – an error-prone procedure that requires extra work in terms of symbolic manipulations and programming. As Newton-Raphson (NR) linearisation naturally implies taking partial derivatives, automatic differentiation can help to reduce the extra work and the chance of introducing errors. This strategy can be demonstrated best by considering a solver for a simple problem, e.g., the quasilinear diffusion equation:

$$-\nabla \cdot [\lambda(u)\nabla u] = f. \tag{19.4}$$

This can be linearised in a Newton-Raphson scheme by putting $u^{k+1} = u^k + \delta u$:

$$-\nabla \cdot \left[\lambda(u^k)\nabla \delta u + \lambda'(u^k)\nabla u^k \delta u\right] = f - \nabla \cdot \left[\lambda(u^k)\nabla u^k\right]. \tag{19.5}$$

The problem is solved iteratively by first taking an initial guess for $u^{k=0}$, solving for δu, and restarting the computation with $u^{k+1} = u^k + \delta u$. Discretising equation (19.5) in a FEM framework leads to the element matrix and vector:

$$\mathbf{A}_{ij}^{(e)} = \int_{\Omega^{(e)}} \lambda(u^k)\nabla N_i \cdot \nabla N_j d\Omega + \int_{\Omega^{(e)}} \nabla N_i \cdot \left(\lambda'(u^k)\nabla \phi^k\right) N_j d\Omega,$$
$$\mathbf{b}_i^{(e)} = \int_{\Omega^{(e)}} f \nabla N_i d\Omega - \int_{\Omega^{(e)}} \nabla N_i \cdot \left(\lambda(u^k)\nabla u^k\right) d\Omega,$$

(19.6)

where N_i are the shape functions, and $\Omega^{(e)}$ is the e-th element. This requires the evaluation of the function $\lambda(u)$ and its derivative $\lambda'(u) = \partial\lambda/\partial u$ at least once in each element. The FastDer++ system is ideally suited for this task. On the other hand, successive substitution linearisation leads to the equation

$$-\nabla \cdot \left[\lambda(u^k)\nabla u^{k+1}\right] = f.$$

(19.7)

It is solved iteratively by first taking an initial guess for $u^{k=0}$, solving for u^{k+1}, and restarting the computation with $u^{k+1} = u^k$. Here, the element matrix and vector

$$\mathbf{A}_{ij}^{(e)} = \int_{\Omega^{(e)}} \lambda(u^k)\nabla N_i \cdot \nabla N_j d\Omega, \quad \mathbf{b}_i^{(e)} = \int_{\Omega^{(e)}} f \nabla N_i d\Omega,$$

(19.8)

only require the evaluation of the function $\lambda(u)$. The element equations resulting from equations (19.6) and (19.8) are typically implemented by the developer of the PDE solver, while the nonlinear function $\lambda(u)$ and its derivative $\lambda'(u)$ are typically provided by the user. From the standpoint of the user, the linearised equations (19.5) or (19.7) and the derivative $\lambda'(u)$ are a nuisance, which distract attention from the real problem, equation (19.4) and the function $\lambda(u)$. Programming $\lambda(u)$ is in fact as far as the user really wants to go. This requirement can be met by integrating the FastDer++ tool in the solver. This allows us to fully hide the details of linearisation since both linearisation schemes are fully specified by the function $\lambda(u)$ since the derivative $\lambda'(u)$ can be obtained automatically. The user does not even have to be aware of the derivative computation, but simply selects one of the linearisation schemes, and provides a function to evaluate $\lambda(u)$ in terms of DaElement objects, just as if they were doubles. In the eyes of the user, linearisation is carried out automatically by the solver.

A very general solver based on the Least-Squares FEM (LSFEM) method has been developed using this strategy by Tijskens et al. [479]. The LSFEM is a method for solving linear and quasilinear systems of first order PDEs. A detailed analysis of the method is presented by [100, 101, 310]. The standard matrix form of a LSFEM system for a vector field $\mathbf{u}$ of N_{un} unknowns is expressed as:

$$\sum_{k=1}^{d} \mathbf{A}_k \frac{\partial \mathbf{u}}{\partial x_k} + \mathbf{A}_0 \mathbf{u} = \mathbf{f} \text{ in } \Omega \subset \mathbb{R}^d, \quad \mathbf{Bu} = \mathbf{0} \text{ on } \partial\Omega,$$

(19.9)

where $\mathbf{A}_k \in \mathbb{R}^{N_{eq} \times N_{un}}$ ($k = 0, ..., d$) and the boundary operator $\mathbf{B}$ is algebraic. The number of equations N_{eq} satisfies $N_{eq} \geq N_{un}$. Since almost any

PDE-based multiple physics problem can expressed in the form (19.9), the potential scope of this method is very wide. In addition, the method guarantees convergence irrespective of the mathematical type of the problem, and most problems of scientific and engineering interest can be cast in a form that guarantees optimal convergence. Consequently, the method allows for a uniform formulation of any problem that can be expressed as a system of first-order PDEs and is ideally suited for model exploration and rapid solver development. The LSFEM solver discussed here consists of a hierarchy of C++ classes for solving linear, quasilinear and time-dependent problems. It builds extensively on the Diffpack® functionality (see [344, 406]). The class for quasilinear problems supports Newton-Raphson and successive substitution linearisation schemes with manual and automatic linearisation. In the former case the partial derivatives $\partial \mathbf{A}/\partial \mathbf{u}\,(k = 0, ..., d)$ are provided by the user, whereas in the automatic case they are evaluated behind the scenes by FastDer++, with n equal to the number of elements in the grid. As a test case a 2D quasilinear Poisson equation has been solved by the LSFEM method using these classes ([479]). On a square grid of 100×100 elements, the run time to build the linear system (excluding the time spent in the linear solver) using automatic linearisation was decreased by 5%, compared to manual linearisation.

The major advantage of building nonlinear PDE solvers with integrated AD, however, is a reduced cost of solver development. The user is relieved entirely from the linearisation procedure and is allowed to directly program the nonlinear system rather than its linearised version. This is advantageous since the terms to be programmed for the nonlinear system are typically smaller in number and less involved than for the linearised system. This reduces the programming work and the amount of debugging needed. On average a reduction between 30% and 60% is expected. Also, the programmed problem is now much closer to the real problem so that the user can rely on an understanding of the problem during debugging. An additional advantage is that two linearisation schemes are obtained with the same (already reduced) effort. This is an additional asset in view of the complementarity of both linearisation schemes. With traditional methods, extra programming work was required to implement different linearisation schemes. Finally, the method is computationally efficient.

In conclusion we can state that although the difference between automatic and manual linearisation is not dramatic, the solvers built with these classes will probably not be optimal in terms of *run* time efficiency, but they are certainly near-optimal in terms of *programmer's* time efficiency.

20

Integrating AD with Object-Oriented Toolkits for High-Performance Scientific Computing

Jason Abate, Steve Benson, Lisa Grignon, Paul Hovland, Lois McInnes and Boyana Norris

ABSTRACT Often the most robust and efficient algorithms for the solution of large-scale problems involving nonlinear PDEs and optimization require the computation of derivatives. We examine the use of automatic differentiation (AD) for computing first and second derivatives in conjunction with two parallel toolkits, the Portable, Extensible Toolkit for Scientific Computing (PETSc) and the Toolkit for Advanced Optimization (TAO). We discuss how the use of mathematical abstractions in PETSc and TAO facilitates the use of AD to automatically generate derivative codes and present performance data demonstrating the suitability of this approach.

20.1 Introduction

As the complexity of advanced computational science applications has increased, the use of object-oriented software methods for the development of both applications and numerical toolkits has also increased. The migration toward this approach can be attributed in part to code reusability and encapsulation of data and algorithms. Code reuse justifies expending significant effort in the development of highly optimized object toolkits encapsulating expert knowledge. Furthermore, well-designed toolkits enable developers to focus on a small component of a complex system, rather than attempting to develop and maintain a monolithic application.

Many high-performance numerical toolkits include components designed to be combined with an application-specific nonlinear function. Examples include optimization components, nonlinear equation solvers, and differential algebraic equation solvers. Often the numerical methods implemented by these components also require first and possibly second derivatives of the function. Frequently, the toolkit is able to approximate these derivatives by using finite differences (FD); however, the convergence rate and robustness are often improved if the derivatives are computed analytically.

This represents an ideal situation for using automatic differentiation (AD) [232, 238]. Developing correct parallel code for computing the derivatives of a complicated nonlinear function can be an onerous task, making an automated alternative very attractive. In addition, the well-defined interfaces used by object toolkits simplify AD by removing the need for the programmer to identify the independent and dependent variables and/or write code for the initialization of seed matrices.

We examine the use of AD to provide code for computing first and second derivatives in conjunction with two numerical toolkits, the Portable, Extensible Toolkit for Scientific Computing (PETSc) and the Toolkit for Advanced Optimization (TAO). We describe how AD can be used with these toolkits to generate the code for computing the derivatives automatically. We present results demonstrating the suitability of AD and PETSc for the parallel solution of nonlinear PDEs and mention preliminary results from the use of AD with TAO.

20.2   Portable, Extensible Toolkit for Scientific Computing

PETSc [25, 26] is a suite of data structures and routines for the scalable solution of scientific applications modeled by partial differential equations. The software integrates a hierarchy of components that range from low-level distributed data structures for vectors and matrices through high-level linear, nonlinear, and timestepping solvers. The algorithmic source code is written in high-level abstractions so that it can be easily understood and modified. This approach promotes code reuse and flexibility and, in many cases, helps to decouple issues of parallelism from algorithm choices.

Newton-based methods (see, e.g., [405]), which offer the advantage of rapid convergence when an iterate is near to a problem's solution, form the algorithmic core of the nonlinear solvers within PETSc. The methods employ line search, trust region, and pseudo-transient continuation strategies to extend the radius of convergence of the Newton techniques, and often solve the linearized systems inexactly with preconditioned Krylov methods. The basic Newton method requires the Jacobian matrix, $J = F'(u)$, of a nonlinear function $F(u)$. Matrix-free Newton-Krylov methods require Jacobian-vector products, $F'(u)v$, and may require an approximate Jacobian for preconditioning.

20.3   Toolkit for Advanced Optimization

TAO [42, 43] focuses on scalable optimization software, including nonlinear least squares, unconstrained minimization, bound constrained optimization, and general nonlinear optimization. The TAO optimization algorithms use high-level abstractions for matrices and vectors and emphasize the reuse of external tools where appropriate, including support for using the linear algebra components provided by PETSc and related tools.

Many of the algorithms employed by TAO require first and sometimes second derivatives. For example, unconstrained minimization solvers that require the gradient, $f'(u)$, of an objective function, $f(u)$, include a limited-memory variable metric method and a conjugate gradient method, while solvers that require both the gradient, $f'(u)$, and Hessian, $f''(u)$, (or Hessian-vector products) include line search and trust region variants of Newton methods. In addition, algorithms for nonlinear least squares and constrained optimization often require the Jacobian of the constraint functions.

20.4   Experimental Results

The AD tools used in this research are the ADIFOR [79] and ADIC [84] systems. Given code for a function $f(u)$ in Fortran 77 or ANSI C, respectively, these tools generate code for $f'(u)$ and, if desired, $f''(u)$.

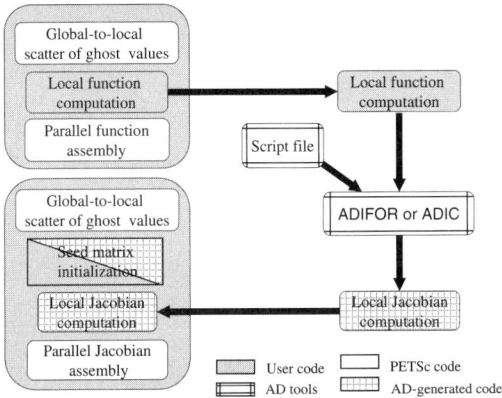

FIGURE 20.1. Schematic diagram of the use of AD tools to generate the Jacobian routine for a nonlinear PDE computation.

Other work [190, 210, 352, 354] has demonstrated the benefits of well-defined interfaces for automating the AD process. The object-oriented designs of PETSc and TAO lead to such well-specified interfaces. However, rather than exploit this feature directly, we have chosen to take advantage of the structure of a typical PETSc/TAO nonlinear function evaluation to simplify the AD process. The parallel nonlinear function code usually includes several calls to PETSc/TAO routines for generalized vector scatters before and after the actual local function computation. These calls take care of data structure setup and communication, enabling a completely local function computation. In the current semi-automatic approach, illustrated in Figure 20.1, we differentiate this local function using AD. This produces code for local derivative computation, which is coupled with code for assembling the gradient, Jacobian, or Hessian. While in principle it is possible to differentiate through the parallel scatter and assembly routines [110, 183, 288, 289], currently the corresponding seed-matrix initialization and assembly code are generated manually. Future development will automate this process.

We present experimental results for nonlinear PDEs and unconstrained minimization problems that demonstrate the utility of AD in conjunction with parallel numerical libraries. These computations were run on an IBM SP with 120 MHz P2SC nodes with two 128 MB memory cards each and a TB3 switch. We have observed analogous qualitative behavior on a range of other current parallel architectures.

20.4.1   Using AD and PETSc for nonlinear PDEs

We used AD and PETSc to solve the steady-state, three-dimensional compressible Euler equations on mapped, structured meshes using a second-order, Roe-type, finite-volume discretization. In particular, we solved in parallel a nonlinear system, using matrix-free Newton-Krylov-Schwarz algorithms with pseudo-transient continuation to model transonic flow over an ONERA M6 airplane wing. See [258] for details about the problem formulation and algorithmic approach. The linearized Newton correction equations were solved by using restarted GMRES preconditioned with the restricted additive Schwarz method with one degree of overlap.

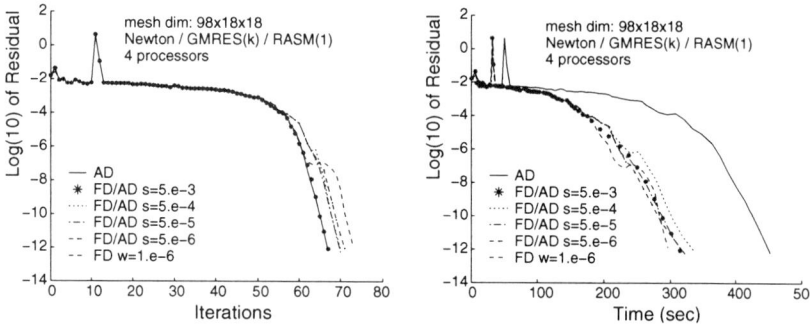

FIGURE 20.2. Comparison of convergence using matrix-free Newton-Krylov--Schwarz methods with finite differencing (FD), AD, and hybrid variants (FD/AD) with various switching parameters, s. Iterations (left) and time (right).

As discussed in depth in [291] and summarized in Figure 20.2, our re-

TABLE 20.1. Execution times (sec) for various stages of the elastic-plastic torsion minimization problem.

Number of Processors	1		16		32	
Number of Vertices	10,000		160,000		320,000	
Hessian Computation Method	FD	AD	FD	AD	FD	AD
Linear System Solution	1.19	1.20	7.10	7.04	9.16	9.04
Compute Function	0.01	0.01	0.04	0.02	0.04	0.02
Compute Gradient	0.21	0.20	0.22	0.24	0.23	0.24
Compute Hessian	1.89	1.48	3.86	1.58	6.11	1.65
Total Time	5.20	4.86	14.13	11.87	18.73	14.32

sults indicate that, for matrix-free Newton-Krylov methods, AD offers significantly greater robustness and converges in fewer iterations than FD. This figure plots the residual norm $\|F(u)\|_2$ versus both nonlinear iteration number and computation time on four processors for a model problem of dimension 158,760. The runtime for AD is slightly higher than for FD (when an appropriate step size is used), due to the higher cost of computing Jacobian-vector products using AD. However, coupling AD with FD in a hybrid scheme provides the robustness of AD with the lower computation time of FD, without needing to identify the proper step size for FD. Additional experiments show that this hybrid technique scales well [291].

20.4.2   Using AD and TAO for unconstrained minimization

We evaluated the preliminary performance of AD in conjunction with TAO using a two-dimensional elastic-plastic torsion model from the MINPACK-2 test problem collection [21]. This model uses a finite element discretization to compute the stress field on an infinitely long cylindrical bar to which a fixed angle of twist per unit length has been applied. The resulting unconstrained minimization problem can be expressed as $min\ f(u)$, where $f : \Re^n \to \Re$. All of the following numerical experiments use a line search Newton method with a preconditioned conjugate gradient linear solver.

We compared two approaches for computing the full Hessian of $f(u)$. First, we applied AD to the hand-coded routine computing the analytic gradient of $f(u)$. Second, we used FD to approximate the Hessian, again using the analytic gradient of $f(u)$. In both cases we used graph coloring techniques to exploit the sparsity of the Hessian computation [136, 227].

The graphs in Figure 20.3 show the scaling of the complete minimization problem using either no preconditioning or a block Jacobi preconditioner. The block Jacobi preconditioner uses a subdomain solver of ILU(0) on each block, with one block per processor. For this relatively simple problem, both AD and FD exhibit rapid convergence in terms of number of iterations. However, AD outperforms FD in terms of total time to solution, mainly because of the good scalability of the AD Hessian computation. Overall,

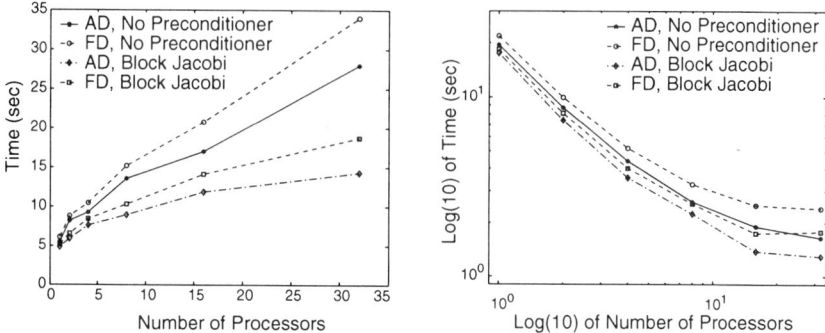

FIGURE 20.3. Comparison of total execution times for the elastic-plastic torsion minimization problem using a line search Newton method. Fixed local (left) and fixed global (right) problem sizes 10,000 and 22,500, respectively.

the results for a fixed local 100×100 mesh size indicate that the problem does not scale well for either AD or FD (although using a block Jacobi preconditioner helps somewhat). This situation is due in part to the poor performance scaling of the linear system solution (see Table 20.1). We are currently exploring the causes of this poor scalability.

20.5 Summary and Future Work

We have presented a methodology for using AD to compute first and second derivatives for use in the parallel solution of nonlinear PDEs and optimization. The robustness and, in some cases, performance of the resulting code are superior to results with finite difference approximations. Also, in contrast to hand-coding, AD can easily be re-applied if the function changes.

Future work includes moving from a semi-automatic approach to an approach, in which code for seed matrix initialization and derivative matrix assembly are completely automated. This task will be aided by the existence of well-defined interfaces for the nonlinear function component and will leverage work on developing a differentiated version of PETSc [293].

Acknowledgments: This work was supported by the Mathematical, Information, and Computational Sciences Division subprogram of the Office of Advanced Scientific Computing Research, U.S. Department of Energy, under Contract W-31-109-Eng-38.

We thank Jorge Moré and Barry Smith for stimulating discussions about issues in derivative computations with PDE and optimization software and Gail Pieper for proofreading an early draft of this manuscript.

Part V

Applications in Science and Engineering

21

Optimal Sizing of Industrial Structural Mechanics Problems Using AD

Gundolf Haase, Ulrich Langer, Ewald Lindner and Wolfram Mühlhuber

ABSTRACT We consider minimizing the mass of the frame of an injection moulding machine as an example of optimal sizing. The deformation of the frame is described by a generalized plane stress state with an elasticity modulus scaled by the thickness. The resulting constrained nonlinear optimization problem is solved by sequential quadratic programming (SQP), which requires gradients of the objective and the constraints with respect to the design parameters. As long as the number of design parameters is small, finite differences may be used. For several hundreds of varying thickness parameters, we use the reverse mode of automatic differentiation (AD). AD works fine but requires huge memory and disk capabilities and limits the use of iterative solvers for the governing state equations. Therefore, we combine AD with the adjoint method to get a fast and flexible hybrid gradient evaluation procedure. Numerical results show the potential of this approach and imply that this method can also be used for finding an initial guess for a shape optimization.

21.1 Introduction

The design of mechanical structures in many industrial applications has to fulfill various constraints. In most cases, engineers designing a machine component desire an optimal design subject to several constraints, but deadlines force them to truncate the design process after a few iterations and accept a suboptimal design. Therefore, tools supporting such a design process must 1) be flexible enough to handle the various requirements, requiring only a little work when the requirements change, and 2) be fast.

As an example for a typical optimal sizing problem, we minimize the mass of the frame of an injection moulding machine. We focus on the efficient and flexible calculation of the gradients of the objective and the constraints for use in standard optimization procedures (such as SQP). Since implementing analytic derivatives is infeasible, we present a very flexible approach using automatic differentiation (AD), coupled with a well-known adjoint method approach from shape optimization. Numerical results show the strength of

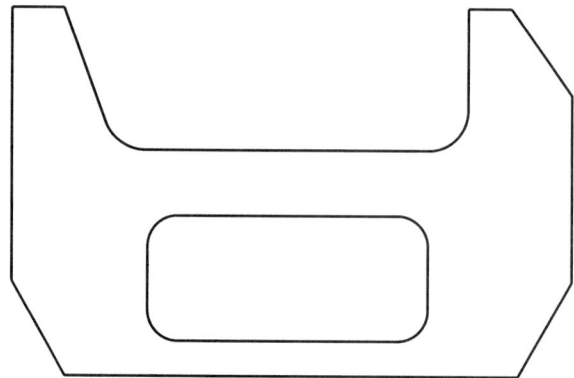

FIGURE 21.1. Cross section of the original shape

this approach.

There are extensive reviews of the various methods for structural optimization using finite elements [272] or for topology optimization [40]. Mahmoud [363] focuses on an optimal sizing approach similar to the one in this chapter, but he uses approximate representations of the objective and the constraints to reduce the overall costs of the method. Stangl [469] presents a generalization of that approach to a class of nonlinearly elastic materials. For approaches using a topology optimization for getting an initial guess of the topology used in a shape optimization afterwards, see e.g. [375, 434]. In this volume, Forth and Evans [199] and Kim and Hovland [323] apply AD to similar problems in fluid dynamics and inverse problems, respectively. Tadjouddine et al. [471] describe an efficient combination of AD and hand-coded parts of the derivative whose concept is similar to the approach in this chapter.

21.2   Modeling of the Problem

The frame of an injection moulding machine is shown by its 2D-cut Ω (Figure 21.1). For a frame of homogeneous thickness typical dimensions are:

thickness of one plate	180 mm
mass of one plate	3.8 tons
clumping force (surface force)	300 tons $\hat{\approx}$ 16 N/mm^2
length	2.8 m
height	1.7 m
supporting areas	2

The primary goal of the design phase is to minimize the mass of the frame, subject to several other requirements.

Let $V_0 = \{v \in H^1(\Omega) \mid v = 0 \text{ on } \Gamma_D, \text{meas } \Gamma_D > 0\}$ denote the set of admissible displacements, where $\partial\Omega = \Gamma_D \cup \Gamma_N$, $\Gamma_D \cap \Gamma_N = \emptyset$. For a fixed thickness $\rho(x)$, the displacement field $u \in V_0$, satisfies

$$a(\rho; u, v) = F(v) \qquad \text{for all } v \in V_0 \tag{21.1}$$

with

$$a(\rho; u, v) = \int_\Omega \rho \frac{\partial u_i}{\partial x_j} E_{ijkl} \frac{\partial v_k}{\partial x_l} \, dx, \quad F(v) = \int_\Omega \langle f, v \rangle \, dx + \int_{\Gamma_N} g v \, ds \ ,$$

where E_{ijkl} denotes the elasticity tensor, f the volume force density and g the surface force density on a part Γ_N of the boundary. The design problem is

$$\int_\Omega \rho \, dx \to \min_{u,\rho}$$

$$\begin{aligned}
\text{subject to} \quad & a(\rho; u, v) = F(v) & & \text{for all } v \in V_0 \\
& 0 < \underline{\rho} \le \rho \le \overline{\rho}, & & \text{a.e. in } \Omega \\
& \sigma^{\text{vM}}(u) \le \sigma^{\text{vM}}_{\max}, \quad \sigma^{\text{ten}}(u) \le \sigma^{\text{ten}}_{\max} & & \text{a.e. in } \Omega \\
& \alpha(u) \le \alpha_{\max}
\end{aligned} \tag{21.2}$$

where $\sigma^{\text{vM}}(u)$ denotes the v. Mises stress, and $\sigma^{\text{ten}}(u)$ denotes the tensile stress in the frame. The change in the shrinking angle of the clumping unit (vertical edges on top, called *wings*) is denoted by $\alpha(u)$.

For discretizing the problem, we use triangular finite elements with piecewise constant shape functions for approximating ρ and piecewise quadratic ones for approximating u. We denote the discrete approximation of ρ and u by ρ and u, respectively. In our application, the upper limits on the angle and the stresses are treated either as constraints or as soft limits, which can be violated to some extent, if the mass would be significantly smaller. Furthermore, the pointwise constraints on σ^{vM} and σ^{ten} are replaced using a higher order ℓ^p norm. Treating the upper limits as soft constraints leads to the reformulation:

$$\text{mass}(\rho) + w_1 \left(\max \left(\|\sigma^{\text{vM}}\|_p - \sigma^{\text{vM}}_{\max}, 0 \right) \right)^2 + w_2 \left(\max \left(\|\sigma^{\text{ten}}\|_p - \sigma^{\text{ten}}_{\max}, 0 \right) \right)^2$$

$$+ w_3 \left(\max \left(\alpha - \alpha_{\max}, 0 \right) \right)^2 \quad \to \min_{u,\rho}$$

$$K(\rho) u = F \quad \text{and} \quad \underline{\rho} \le \rho \le \overline{\rho} \tag{21.3}$$

21.3   Sketch of the Optimization Strategy

Problem (21.3) is a special case of

$$J(u, \rho) \to \min_{u,\rho} \quad \text{subject to } K(\rho) u = f(\rho) \text{ and } \underline{\rho} \le \rho \le \overline{\rho}, \tag{21.4}$$

where ρ denotes the vector of design parameters and u the solution of the governing finite element (FE) state equation. The splitting of the parameter vector into design parameters ρ and the solution of the FE state equation u is typical for problems in shape optimization. From the optimization's point of view, the discretized state equation can be interpreted as equality constraints. In our case it is linear with respect to u, and $K(\rho)$ is symmetric and positive definite for all admissible parameters ρ. Therefore, u can be eliminated:

$$\tilde{J}(\rho) = J(K^{-1}(\rho)\, f(\rho),\, \rho) \to \min_{\rho}$$

$$\text{subject to} \qquad \underline{\rho} \le \rho \le \overline{\rho}$$

$$(21.5)$$

For a standard SQP optimization method, (21.5) is preferred to (21.4) because it has many fewer parameters (see [222, 405]). The optimizer used in our code is based on a Quasi-Newton approximation of the Hessian using a modified BFGS update formula of Powell [422] to avoid the need for Hessian information of the objective.

21.4   Calculating Gradients

Quasi-Newton strategy and update formulas need gradients for the objective and the constraints. Hand coded analytic derivatives are usually too complicated, error prone, and time consuming. Furthermore, hand coding is not well suited to the design process, as we would loose completely flexibility. Hence, we consider black box methods such as finite differences or automatic differentiation (AD) [238] or methods exploiting the special structure of the state equation such as the direct or the adjoint methods [272]. None of these methods is well suited for our problem, so we developed a hybrid method combining AD and the adjoint method. Since our finite element code is completely written in C++ and heavily uses virtual inheritance, we use ADOL-C [242]. We applied the reverse mode within our calculations, as we have at most a few nonlinear constraints (c.f. (21.2) and (21.3)), while the dimension of the design space is usually large.

21.4.1   Direct and adjoint methods

Both methods are well-known in shape optimization and take into account the special structure of the FE state equation. Differentiating the discretized state equation with respect to a design parameter ρ_i leads to

$$K\frac{\partial u}{\partial \rho_i} = \frac{\partial f}{\partial \rho_i} - \frac{\partial K}{\partial \rho_i}u.$$

$$(21.6)$$

For the direct method, (21.6) is solved numerically. Then the gradient of the objective can be calculated by (c.f. (21.5))

$$\frac{\mathrm{d}\tilde{J}}{\mathrm{d}\rho_i} = \frac{\partial J}{\partial \rho_i} + \left\langle \frac{\partial J}{\partial u}, \frac{\partial u}{\partial \rho_i} \right\rangle. \tag{21.7}$$

The adjoint method solves (21.6) formally and inserts the result in (21.7), which leads to

$$\frac{\mathrm{d}\tilde{J}}{\mathrm{d}\rho_i} = \frac{\partial J}{\partial \rho_i} + \left\langle K^{-T} \frac{\partial J}{\partial u}, \frac{\partial f}{\partial \rho_i} - \frac{\partial K}{\partial \rho_i} u \right\rangle. \tag{21.8}$$

21.4.2 Comparison

ADOL-C needs a file containing the evaluation graph of the function in a symbolic form which is generated at run time. Structural optimization problems require huge memory and disk capabilities for that purpose, e.g., about 1 GB is needed to store the evaluation graph for a problem with about 450 design parameters and about 7500 degrees of freedom. The flexibility of ADOL-C with respect to changes in the objective is similar to finite differences. For the reverse mode, the calculation time of the gradient is independent of the number of design parameters and takes the time of about 15 - 20 native C++ function evaluations as long as the evaluation graph can be stored in the main memory of the computer. Compared to the use of finite differences, this is a tremendous speedup, even for problems with only 10 - 20 design parameters. The coupling of AD with iterative solvers is a topic of current research (see e.g. [238] and references therein). Since the use of iterative methods (e.g., multilevel methods) is important for solving fine discretisations of the state equation efficiently, the applicability is limited to problems where direct solvers can be used.

The direct method needs the solution of one state equation per design parameter, whereas the adjoint method needs the solution of one adjoint problem for the objective and in principle for each constraint. Depending on the number of design parameters and constraints, the better suited method can be chosen. As analytic partial derivatives of J with respect to ρ and u are needed, both methods can only be applied to simple objectives, where this can be done easily. Furthermore, the flexibility of the method suffers from the need of hand-coded gradient routines. Compared to finite differences or the use of AD for the whole function, this approach is much faster. Finite differences need many more solutions of the FE state problem, while this approach avoids the huge evaluation graph from the solution of the state equation required by AD. For both methods, any solver can be used for solving the state problem, including iterative solvers such as conjugate gradient methods with multilevel preconditioning.

TABLE 21.1. Comparison of the run time for various differentiation strategies

		Finite Diff.	Pure AD	Hybrid M.
Problem dim.	Nr. of design par.	24	24	24
	Nr. of elements	3981	3981	3981
	DOFs of state equ.	16690	16690	16690
Optimizer statistics	Iterations	83	100	100
	gradient eval.	12.40 h	6.00 h	0.18 h
	function eval.	0.23 h	2.36 h	0.20 h
Runtime	Total CPU time	12.4 h	4.88 h	0.39 h
	Total elapsed time	12.6 h	8.42 h	0.40 h

21.4.3   Hybrid Method

The strengths of the direct and the adjoint method and of AD lie in completely different areas. AD provides very high flexibility with respect to the objective, but has severe space and time requirements and cannot easily be used with iterative solvers for the state equation. In contrast, the direct and adjoint methods can be combined easily with iterative solvers and provide a fast way for calculating the needed gradients, but they lack the needed flexibility. Hence, we combine both approaches into a new hybrid method.

The main drawback of the direct or the adjoint method is the need for analytic partial derivatives of the objective and the constraints with respect to ρ and u, which can be provided easily by AD. Then $\frac{\partial K}{\partial \rho_i}$ and $\frac{\partial f}{\partial \rho_i}$ can be hand-coded or produced by AD. For optimal sizing problems, routines for $\frac{\partial K}{\partial \rho_i}$ and $\frac{\partial f}{\partial \rho_i}$ can be hand-coded easily. Furthermore, they only depend on the type of state problem considered and not on the used objective, justifying the additional effort of coding even for more complex problems.

21.5   Numerical Results

The following numerical results for the problem stated in §21.2 were calculated on an SGI Origin 2000 at 300 MHz. At the beginning, we tried to use only few design parameters. Therefore, we divided our domain into a number of sub-domains and approximated the thickness with a constant function per sub-domain. For evaluating the gradient, either finite differences, an AD approach for (21.5), or the hybrid method were used. For a better comparison (see Table 21.1), the calculation was terminated after a fixed number of steps. The run using finite differences terminated earlier because the search direction was not a descent direction.

All three methods lead to a similar design with about 5% reduction of the mass compared to the starting configuration (which is the current design of the frame). Compared to finite differences and the pure AD approach, the hybrid method is significantly faster, combining a fast function evaluation

TABLE 21.2. Comparison of the run time for many design parameters

		Pure AD	Hybrid M.	Hybrid M.
Problem	Nr. of design par.	449	449	449
dim.	Nr. of elements	1796	1796	7184
	DOFs of state equ.	7518	7518	29402
Evaluation	Operations	45521797	1399910	5578270
graph	Total file size	953 MB	32.4 MB	129.2 MB
Optimizer	Iterations	800	800	800
statistics	gradient eval.	18.0 h	0.54 h	3.35 h
	function eval.	16.5 h	1.29 h	8.13 h
Runtime	Total CPU time	32.3 h	3.73 h	14.01 h
	Total elapsed time	38.5 h	3.76 h	14.12 h

and a fast gradient evaluation. The gradient evaluation is the main drawback for finite differences. For the pure AD approach we implemented additional safeguards. To detect when a regeneration of the evaluation graph was necessary, we compared the value of the objective using the evaluation graph and the value using a native C++ implementation, which explains the longer run time of the function evaluation.

In the following we used the coarsest grid of our FE triangulation for discretizing the thickness. For solving the state problem, each coarse grid element was subdivided into 16 elements using 2 levels of uniform refinement. The state equation was discretized on this refined triangulation . Table 21.2 contains results for the pure AD approach and the hybrid method and results for an even finer discretization of the state equation.

Analyzing the run time behaviour we see that the pure AD approach is no longer competitive due to the large file containing the evaluation graph. Furthermore, it can be seen that for the hybrid approach the optimizer represents a considerable portion of the total run time, a portion that grows when using more design parameters as the complexity of one optimization step is proportional to $(\dim \rho)^3$ (due to the use of dense matrix linear algebra), whereas the complexity of solving one state equation is proportional to $\dim u$ (if solvers with optimal complexity e.g., conjugate gradients with multigrid or multilevel preconditioning are used).

21.6   Remarks and Conclusions

In this chapter, we presented differentiation strategies needed to solve a real optimal sizing problem. We focused our attention on the flexibility of the gradient routine and on the possibility of combining the gradient module with iterative solvers for the state equation. A hybrid method combines the strengths of AD and the adjoint method. This method preserves the flexibility of the pure AD approach at the cost of a completely hand-coded gradient routine. Furthermore, the huge memory and disk requirements of the pure AD approach are reduced significantly.

Our current implementation of the optimizer is based on dense matrix linear algebra. Therefore, it is only well suited for small to medium size optimization problems. To close the gap with topology optimization, which is of high practical importance, new optimization methods for large scale problems have to be developed such as an approach using multigrid methods proposed by Maar and Schulz [361].

Acknowledgments: This work was supported by the Austrian Science Fund - "Fonds zur Förderung der wissenschaftlichen Forschung (FWF)" - SFB F013 "Numerical and Symbolic Scientific Computing."

22

Second Order Exact Derivatives to Perform Optimization on Self-Consistent Integral Equations Problems

Isabelle Charpentier, Noël Jakse and Fabrice Veersé

ABSTRACT Great efforts have been made to search for enhanced semi-empirical forms of the bridge function b appearing in disordered condensed matter problems which are solved using the integral equation method. Currently, parameterized forms of the bridge function are manually chosen to fit as closely as possible the thermodynamic self-consistent equations depending on g and its derivatives. We discuss a second-order differentiation of the computer code that enables construction of parameterized bridge functions using optimal control techniques.

22.1 Introduction

In the method of distribution functions of the classical statistical physics, the microscopic properties of dense fluids [269] are described by the pair-correlation function g, which represents the local arrangement of atoms or molecules. The integral equation theory is one of the most powerful semi-analytic methods to obtain the pair-correlation function of a fluid. Even though its favorite field is the case of simple liquids, often taken as a benchmark, it can be applied to various complex situations including liquid metals, noble gases or polymer hard-chains.

The integral equation method solves the set of equations formed by the Ornstein-Zernike integral equation [407] and a closure relation involving g. Even if a formal exact expression for the closure exists [390], it includes a particular function b involving the pair-correlation function g itself. Therefore, considerable efforts have been made since the pioneering work of Rowlinson [444] to search for enhanced semi-empirical forms of the bridge function that satisfy thermodynamic self-consistency conditions.

The procedure of thermodynamic self-consistency involves an explicit numerical calculation of the density and temperature derivatives of the pair-correlation function [204]. Besides the classical finite difference method,

Vompe and Martynov [492] obtain these derivatives as solutions of differentiated integral equations. Both methods are approximate. Currently, parameterized forms of the bridge function have been manually tuned to fit as closely as possible a thermodynamic self-consistent condition. The inaccuracy of these derivatives is probably the main reason why optimal control methods were not in use in the liquid physics.

An alternate solution is to use algorithmic differentiation [238] to obtain exact derivatives of g as discussed in [122]. The ability of automatic differentiation (AD) tools ADIFOR [78] and Odyssée [186] makes derivatives of g available to the search for bridge functions using optimal control techniques. The aim of this chapter is to show that it is easy to construct codes that enable optimization for determining the parameterized function b, even though it requires second-order derivatives.

In this chapter, §22.2 briefly describes integral equation problems and objective consistency functions. Differentiation steps and numerical results are discussed in §22.3. Some research directions conclude the presentation.

22.2   Liquid State Theory

Let us consider a fluid of density ρ and temperature T whose molecules, separated by a distance r, interact via the Lennard-Jones pair potential

$$\forall r > 0, \qquad u(r) = 4\varepsilon \left\{ (\sigma/r)^{12} - (\sigma/r)^6 \right\}, \tag{22.1}$$

where σ is the position of the node, and ε is the well-depth of the pair potential. The latter is generally used to model real fluids such as pure gases considered as homogeneous and isotropic [269]. Beside the simulation methods, such as molecular dynamics or Monte-Carlo methods, the integral equation theory has proven to be a powerful semi-analytic method to calculate the pair-correlation function g from which it is possible to predict the structural and thermodynamic properties of fluids.

22.2.1   Integral equation problem

Due to the homogeneity and isotropy of the fluid, all the correlation functions can be considered as spherically symmetric and expanded only on r, the modulus of vector $\vec{r}$. In this framework, Morita and Hiroike [390] derived a one-dimensional formal exact expression for the pair-correlation g by means of the Ornstein-Zernike integral equation plus a closure relation:

$$\begin{cases} g(r) = 1 + c(r) + \rho \int (g(r') - 1) c(|\vec{r} - \vec{r'}|) dr', & \forall r > 0 \\ g(r) = \exp\left(-u(r)/k_B T + g(r) - 1 - c(r) + b(r)\right). \end{cases} \tag{22.2}$$

In the former system, c is the direct correlation function, and k_B is Boltzmann's constant. The bridge function b can be written as an infinite sum of irreducible bridge diagrams [374]. The direct calculation of this expansion is too slowly convergent, so great efforts were devoted to the design of semi-empirical bridge functions, implicit functions of g usually written explicitly in terms of the renormalized indirect correlation function

$$\gamma^*(r) = g(r) - c(r) - 1 - u_2(r)/(k_B T), \quad (22.3)$$

where u_2 (= ε if $r \leq 2^{1/6}\sigma$ and $u(r)$ otherwise) is the attractive part of the potential u according to [497].

As a result, b contains parameters chosen to ensure the thermodynamic self-consistency (§22.2.2) of the resulting approximate integral equations and to satisfy the zeroth separation theorems [350]. Here, $b \equiv b_p$ contains p parameters defining a vector $\alpha_p \in IR^p$. From a physical point of view, we discuss three bridge functions depending on one [506], three [350], and five [168] parameters, respectively. For example, function b_3 takes the form

$$b_3(\alpha_3, r) = -\frac{\alpha_3^1 \gamma^{*2}(r)}{2} \left(1 - \frac{\alpha_3^2 \alpha_3^3 \gamma^*(r)}{1 + \alpha_3^3 \gamma^*(r)} \right). \quad (22.4)$$

22.2.2   Objective thermodynamic consistency

Given ρ and T and an expression b_p, a unique bridge function is obtained by fixing the values of α_p to satisfy an objective thermodynamic consistency accounting for as many conditions as the number of parameters. Since one is interested in b_3 three conditions are needed: the Gibbs-Helmholtz I, the virial-compressibility J and the Gibbs-Duhem K [350]. They require the explicit numerical calculation of the pair-correlation function derivatives with respect to the density and temperature. For the sake of simplicity only I is given here:

$$I(\alpha_p, \rho, T) = \int \left[u(r) \left\{ g(r) + \rho \frac{\partial g(r)}{\partial \rho} \right\} + r \frac{du(r)}{dr} \left\{ g(r) - T \frac{\partial g(r)}{\partial T} \right\} \right] r^2 dr.$$

Such derivatives were exactly computed in a previous study [122] using the forward mode of differentiation of Odyssée. The objective thermodynamic consistency S is then computed as proposed in [340]

$$S(\alpha_p, \rho, T) = I^2(\alpha_p, \rho, T) + J^2(\alpha_p, \rho, T). + K^2(\alpha_p, \rho, T) \quad (22.5)$$

If one wants $S = 0$, which amounts to satisfying the thermodynamic self-consistency by an optimization process, its derivatives with respect to α_p have to be obtained. This implies computing second order derivatives of g, namely $\partial^2 g/\partial \alpha_p \partial T$ and $\partial^2 g/\partial \alpha_p \partial \rho$. As a consequence, one needs to differentiate twice the code that computes g.

22.3   Differentiation Steps and Numerical Results

The original code `paircorrel` is made of 7 routines (Figure 22.1). Among them, `lmv` and `newton` are iterative loops that solve nonlinear systems.

```
paircorrel +- lmv +- fft
                +- cfg
                +- newton +- gauss
                +- test
                +- fft
```

FIGURE 22.1. Graph of the original code for the computation of g.

```
thermo +- odydg +- paircorreltl +- lmvtl +- ...
        +- fft
        +- simpson
```

FIGURE 22.2. Graph of thermodynamic code for the computation of S (or R).

Given a b, the routine lmv [340] calls the routine newton to solve system (22.2) through a Newton-Raphson method. The resulting functions g and c then allow us to deduce a new γ (22.3) to update b (22.4) in the routine cfg. The routine lmv loops until the convergence of the process is achieved according to the routine test. Before the Newton-Raphson iterations, a fast Fourier transform routine (fft) is used to calculate solutions of (22.2) in the dual space. These are then computed using a linear solver gauss. The appearance of iterative solvers fft and gauss makes the original code pathological as far as the differentiation in reverse mode is considered. In this chapter we are only faced with the direct mode.

For a given thermodynamic state defined by ρ and T, only cross derivatives $\partial^2 g / \partial \alpha_p \partial \rho$ and $\partial^2 g / \partial \alpha_p \partial T$ are needed. The correctness of the gradients has been systematically checked.

22.3.1 Differentiate the original code with respect to T and ρ

The differentiation in direct mode is straightforward [122] using either Odyssée or ADIFOR, but the resulting codes differ.

Odyssée generates differentiated variables of the same dimension as the original ones. This allows propagation of perturbation in only one direction. Since two derivatives of g cannot be computed simultaneously, one introduces the additional routine odydg for those calculations. The routine odydg then consists of a global loop that calls paircorreltl twice. Figure 22.2 shows the graph of the thermodynamic code thermo that uses the routines generated by Odyssée (with suffix tl) to compute S (22.5).

Consider an n-dimensional variable. ADIFOR generates a differentiated variable with $(n+1)$ dimensions giving the possibility of propagating several directions of perturbation. This implies the writing of a loop (of length n) for each computation of a derivative statement. A unique run of the original code then provides a trajectory for the computation of all the derivatives under interest. The graph of the code is similar to that of the original code

TABLE 22.1. Characteristics of the thermo codes (2 derivatives).

	Run time (in s.)			Memory		
p	pair correl	ADIFOR	Odyssée	pair correl	ADIFOR	Odyssée
1	0.58	3.37	2.33	347 044	877 176	710 452
3	0.69	3.68	2.57	363 444	942 888	759 628
5	0.76	3.73	2.61	363 460	942 992	759 660

(Figure 22.1), the names of the differentiated routines having a prefix g_.

Routine thermo uses either odydg or g_pairrcorrel results to compute S. Table 22.1 shows run times as well as the memory (number of scalar data) required for the original code paircorrel and the different thermodynamic codes. In both cases the differentiation is performed with respect to two scalar parameters: the temperature and the density. It is then natural to obtain similar run times for similar bridges function b_p ($p \in \{1, 3, 5\}$). However one observes that the codes generated by Odyssée are faster than those produced using ADIFOR even if one runs the original code two times when using Odyssée. This can be partly explained by the fact that ADIFOR generated codes use more local variables (uninitialized data) to carry out the differentiation. Indeed a glance at the generated codes shows that the sequences of statements generated for the computation of the derivatives differ from one AD tool to the other. ADIFOR moreover links the differentiated routines to specific libraries. Nevertheless, for the fast code we deal with, paircorrel lasts less than 1 second, does not allow for exact measurements of these contributions.

22.3.2   Differentiation of the thermodynamic code with respect to the p parameters of b_p

Neither Odyssée nor ADIFOR 2.0 directly provide second-order derivatives, but one can use these AD tools repeatedly to get them. The cross derivatives are obtained by a differentiation with respect to α_p ($p \in \{1, 3, 5\}$), of the thermodynamic code thermo (Figure 22.2) generated using Odyssée. Since Odyssée does not allow computing the p derivatives $\partial S / \partial \alpha_i$ ($i = 1, .., p$) in one pass, the p calls to routine thermot1 are grouped in an additional routine called gradient.

Table 22.2 presents time and memory requirements of the gradient codes. From the Odyssée results, one observes that the computation of each directional derivative lasts approximately 4.1 seconds. This result is simply obtained by dividing the run time by the number p of directional derivatives. One also observes that the required memory is quite fixed. Such behavior is natural since the computations are performed by routine gradient, which uses a global loop to call p times the differentiated thermodynamic routine

TABLE 22.2. Characteristics of the gradient codes.

p	Run time			Memory		
	thermo	ADIFOR	Odyssée	thermo	ADIFOR	Odyssée
1	2.33	9.23	4.08	710 452	1 265 940	1 255 912
3	2.57	20.95	14.89	759 628	2 504 944	1 321 476
5	2.61	25.67	26.15	759 660	3 677 904	1 321 504

thermotl. The codes produced by Odyssée are cheaper in terms of run time and memory than those generated by ADIFOR for bridge functions using up to three parameters.

The trend of the run times for second-order codes, as a function of p, suggests that ADIFOR is more efficient for larger values of p. Indeed the time necessary for the computation of a directional derivative reduces on average when p increases because ADIFOR generated codes contain intermediate calculations that are outside the loops that propagate derivatives (§22.3.1). On the other hand, memory usage is an affine function of p. This can be observed by subtracting the memory requirement used for thermo from the one used to run the ADIFOR second order code. It then appears that one directional derivative uses about 550 000 scalar data. Finally, time and memory consumption of differentiated codes differ from one tool to the other. This is mainly due to the manner in which loops are used to propagate several directions of perturbation.

22.4   Conclusion

ADIFOR and Odyssée do not yet generate high order derivative codes, although they may be used to obtain the cross-derivatives required for the optimal tune of the parameterized bridge functions. We have shown that the resulting codes have different behaviors depending on the chosen AD tool and on the number of directions of perturbation under study. For a small number of parameters, Odyssée is better than ADIFOR in terms of run time and memory. ADIFOR 3.0 enables a direct computation of the Hessian and will probably reduce the run times even further.

Improvements accuracy and run time allow us to solve many new liquid state theory problems using the integral equation method. In a first study [122], the simulation using the code thermo generated by Odyssée improves the physical results systematically. For stiff thermodynamic states (high density and a low temperature) the accuracy with respect to experiments is sometimes twice as good. Moreover, for the first time, we can use the density derivative of the chemical potential to complete the objective thermodynamic self-consistency.

The ultimate goal in classical statistical physics is to determine a general bridge function. We have described the codes for optimization processes based on a objective thermodynamic self-consistency. Such processes that require the computation of second order derivatives were never achieved for bridge functions containing more than one parameter. It is now possible to consider more sophisticated bridge functions such as those described in §22.2.1. Bridge functions written as an expansion in terms of γ are also attainable. Physical results of optimization process based on objective thermodynamic self-consistency will be presented soon.

One may also use the reverse mode of Odyssée applied to the code `paircorrel` as an efficient alternative. Such a code has been generated, and first results look promising for optimizing with respect to observed data coming from experiments or molecular dynamics simulation.

23

Accurate Gear Tooth Contact and Sensitivity Computation for Hypoid Bevel Gears

Olaf Vogel

ABSTRACT

We investigate a new method for gear tooth contact analysis based on a mathematical model of the generating process. The approach allows the computation of the paths of contact as the solution set of a nonlinear operator equation using a minimally augmented defining system for certain first order singularities. As a byproduct we obtain curvature properties of the gear tooth flanks, the reduced curvatures at the contact points, as well as the boundaries of the paths of contact in a convenient way. Using automatic differentiation, all the geometric quantities are computable with the machine accuracy of the computer. For the first time, we provide analytical sensitivities of the determined contact properties with respect to arbitrary machine tool settings such as additional motion parameters. Thus the systematic optimization of the gear tooth contact as well as the evaluation of the impact of perturbations on the contact pattern become available in a straightforward fashion. In order to obtain the desired sensitivity information, we compute derivative tensors of an implicit function up to order three using the software tool ADOL-C.

23.1 Introduction

The widespread field of applications of hypoid bevel gears ranges from transportation equipment such as cars, ships, and helicopters to heavy earth-moving and construction equipment. The extremely good characteristics of hypoid bevel gear drives for the power transmission with intersecting or crossing shafts are obtained at a significant cost for its design and optimization caused by the highly complicated shape of the gear tooth flanks. Since the quality of hypoid gear drives strongly depends on the position and size of the contact pattern and on the transmission errors under load, computerized *tooth contact analysis* is common practice to determine the best machine tool settings. Unfortunately, all the known techniques for the investigation of tooth contact properties and tooth strength analysis are based on approximations of the gear tooth flanks obtained by methods of varying power [66, 170, 358, 458, 468]. Moreover, trial design on the basis of engineering knowledge is the current way of "optimization," i.e. the machine settings are altered as long as the features match most design specifications. However, the potential of today's gear-generating machine tools with a continuously growing number of machine settings makes an optimization by hand impossible.

In order to overcome the problems mentioned above we present here a new methodology for the geometric gear tooth contact analysis that applies techniques of singularity theory to characterize the contact of meshing gears analytically. The constructed defining system for the path of contact can be solved efficiently and with high accuracy. Our method is *direct* in the sense that the paths of contact as well as the transmission error are obtained approximation-free without determining the tooth flanks first. Furthermore, the approach naturally allows the computation of the sensitivities of the determined contact properties with respect to arbitrary machine settings of modern machine tools including *additional motion parameters* which enter nonlinearly into the model. Thus, for the first time, analytical sensitivity information is available. We consider this result to be a large step towards a systematic computational optimization and a reasonable evaluation of the gear tooth contact. The application of automatic differentiation (AD) to provide all the needed derivatives is fundamental to make the presented method workable, efficient, and flexible.

23.2 Modelling Gear Tooth Flanks and Their Meshing Based on a Virtual Machine Tool

Modern methods for computing spiral and hypoid bevel gear tooth flanks are based on mathematical models of their generating process. Various models of known gear-generating machine tools can be generalized to a *virtual machine tool* [298], here represented by an abstract vector function $F_\mathbf{w} : \mathcal{D}_\mathbf{w} \subseteq \mathbb{R}^3 \to \mathbb{R}^3 : w_\mathbf{w} \mapsto x_\mathbf{w} = F_\mathbf{w}(w_\mathbf{w})$ that describes, in a coordinate system fixed with respect to the gear work blank, a family of surfaces swept out by the cutter blade profile while manufacturing an arbitrary *gear wheel* (index $\mathbf{w}$). The generated gear tooth flank is contained in the image of the *singular set* $\mathcal{S}_\mathbf{w}$ that consists of those points $w_\mathbf{w}^*$ where $F_\mathbf{w}'(w_\mathbf{w}^*)$ suffers a rank drop by one. As shown in [246] the singular set can be obtained constructively as the regular solution set of some scalar equation. This approach is used in [298] to accurately compute arbitrary tooth flank points using automatic differentiation and in [490] to investigate the tooth contact of meshing gears, for the first time, without any approximation of the tooth flanks as for example by splines.

In order to model the meshing of *gear* and *pinion* in a gear drive, we represent both gear wheels with respect to a common coordinate system fixed with the gear drive [491]. Let the angle $\psi_\mathbf{w} \in \mathcal{I}_\mathbf{w} \equiv [\psi_\mathbf{w}^-, \psi_\mathbf{w}^+] \subseteq \mathbb{R}$ specify the rotation of the gear wheel about its axis. For each fixed angle

$\psi_{\mathbf{w}}$, the virtual motion of the cutter blade with respect to the common coordinate system is described by the parametrized function

$$\tilde{F}_{\mathbf{w}} : \mathcal{D}_{\mathbf{w}} \times \mathcal{I}_{\mathbf{w}} \times \mathcal{P}_{\mathbf{w}} \subseteq \mathbb{R}^3 \times \mathbb{R} \times \mathbb{R}^{n_{\mathbf{w}}} \to \mathbb{R}^3 :$$

$$(w_{\mathbf{w}}, \psi_{\mathbf{w}}; p_{\mathbf{w}}) \mapsto \tilde{x}_{\mathbf{w}} = \tilde{F}_{\mathbf{w}}(w_{\mathbf{w}}, \psi_{\mathbf{w}}; p_{\mathbf{w}}) \equiv Q(\psi_{\mathbf{w}}; p_{\mathbf{w}}) \, F_{\mathbf{w}}(w_{\mathbf{w}}; p_{\mathbf{w}}) + q(\psi_{\mathbf{w}}; p_{\mathbf{w}})$$

with an orthogonal rotation matrix $Q(\psi_{\mathbf{w}}; p_{\mathbf{w}}) \in \mathbb{R}^{3 \times 3}$ and a translation vector $q(\psi_{\mathbf{w}}; p_{\mathbf{w}}) \in \mathbb{R}^3$. Throughout the parameter vector $p_{\mathbf{w}}$ consists of all settings of the virtual machine tool and of the gear drive with respect to which we want to compile sensitivities. This vector $p_{\mathbf{w}}$ is considered to be fixed during the machining process. The transformed function $\tilde{F}_{\mathbf{w}}(\psi_{\mathbf{w}}; p_{\mathbf{w}})$ diffeomorphically inherits the properties of $F_{\mathbf{w}}$ such that the computation of gear tooth flanks is possible as sketched above. However, our goal is to determine the *paths of contact* without computing the contacting tooth flanks first.

Replacing the general index w by the indices g and p for gear and pinion, respectively, and using the notation $w^{\top} \equiv [w_{\mathbf{g}}^{\top}, w_{\mathbf{p}}^{\top}]$, $\psi^{\top} \equiv [\psi_{\mathbf{g}}, \psi_{\mathbf{p}}]$, the paths of contact are determined by the set $\mathcal{S}^*$ consisting of those vector pairs (w^*, ψ^*) that determine a *tooth contact* between meshing gear and pinion in the sense that a) w^* must specify flank points, where both tooth flank surfaces get in contact with a common tangent plane; b) the tooth flanks may not intersect locally. In case of hypoid bevel gears, we assume point contact between the tooth flanks. In fact, there is nearly line contact, making the exact computation of gear tooth contact a nontrivial task.

23.3   Direct Analytical Characterization and Computation of Gear Tooth Contact

In [491] a method is proposed for characterizing and computing gear tooth contact directly exploiting the properties of the functions $\tilde{F}_{\mathbf{g}}$ and $\tilde{F}_{\mathbf{p}}$ modelling the underlying virtual machine tools. Consider the function

$$G : \mathcal{D} \times \mathcal{I} \times \mathcal{P} \subseteq \mathbb{R}^6 \times \mathbb{R}^2 \times \mathbb{R}^n \to \mathbb{R}^3 :$$

$$(w, \psi; p) \mapsto d = G(w, \psi; p) \equiv \tilde{F}_{\mathbf{g}}(w_{\mathbf{g}}, \psi_{\mathbf{g}}; p_{\mathbf{g}}) - \tilde{F}_{\mathbf{p}}(w_{\mathbf{p}}, \psi_{\mathbf{p}}; p_{\mathbf{p}})$$

with $\mathcal{D} \equiv \mathcal{D}_{\mathbf{g}} \times \mathcal{D}_{\mathbf{p}}$, $\mathcal{I} \equiv \mathcal{I}_{\mathbf{g}} \times \mathcal{I}_{\mathbf{p}}$, and $\mathcal{P} \equiv \mathcal{P}_{\mathbf{g}} \times \mathcal{P}_{\mathbf{p}}$ at some tooth contact $(w^*, \psi^*) \in \mathcal{S}^*$. Immediately, the necessary conditions $G(w^*, \psi^*; p) = 0_3$ and $\mathrm{def}(\partial_w G(w^*, \psi^*; p)) = 1$, i.e. the rank drop of $\partial_w G(w^*, \psi^*; p)$, follow from condition (a). The *singular set* $\mathcal{S}(\psi) \equiv \{w^* | \mathrm{def}(\partial_w G(w^*, \psi; p)) = 1\}$ can be obtained constructively as a regular solution set of a nonlinear equation system $g(w^*, \psi; p) = 0_4$ [246]. Since the individual motion of each tooth flank in the contact point is transversal to the common tangent plane at that point, the resulting defining system for $\mathcal{S}^*$

$$G(w, \psi; p) = 0_3, \qquad g(w, \psi; p) = 0_4 \tag{23.1}$$

has a regular solution curve. Using a parametrization in $\tau \equiv \psi_{\mathrm{p}} - \psi_{\mathrm{p}}^-$ the system can be solved point-wise by Newton's method to determine solution points $w^* = w^*(\tau, p)$ and $\psi^* = \psi^*(\tau, p)$ for $\tau \in \left[0, \psi_{\mathrm{p}}^+ - \psi_{\mathrm{p}}^-\right]$.

To improve geometrical understanding and to develop sufficient tooth contact conditions, we investigate for each tooth contact (w^*, ψ^*) the regularly unfolded and augmented equation system

$$G(w, \psi^*; p) = \lambda r, \qquad Tw = s$$

with $\lambda \in \mathbb{R}$, $r \in \mathbb{R}^3$, $T \in \mathbb{R}^{4 \times 6}$, and $s \in \mathbb{R}^4$. The system implicitly defines the functions $\lambda = \lambda(s, \psi^*; p)$ and $w = w(s, \psi^*; p)$ near $s^* \equiv Tw^*$ and thus parametrizes pairs of points w_{g} and w_{p} in s whose images $\tilde{F}_{\mathrm{g}}(w_{\mathrm{g}}, \psi_{\mathrm{g}}^*; p_{\mathrm{g}})$ and $\tilde{F}_{\mathrm{p}}(w_{\mathrm{p}}, \psi_{\mathrm{p}}^*; p_{\mathrm{p}})$ have the distance λ along the direction r. The necessary tooth contact conditions require $\lambda(s^*, \psi^*; p) = 0$ and the *reduced function* λ to be stationary as a function of the *reduced variables* s in the sense that $\partial_s \lambda(s^*, \psi^*; p) \equiv g^\top(w^*, \psi^*; p) = 0_4^\top$. Furthermore, the sufficiency condition (b) is satisfied if and only if λ attains a saddle point with $\partial_s^2 \lambda(s^*, \psi^*; p) \equiv H(w^*, \psi^*; p) \in \mathbb{R}^{4 \times 4}$ having two positive and two negative eigenvalues [490]. The key ingredient in showing this is the construction of local parametrizations near the tooth contact point for both gear and pinion flank as well as for their distance in direction r in the *same* two parameters. Differentiating these parametrizations allows the estimation of the principal curvatures and their directions of the *normal curvature* of both tooth flanks and of the *reduced curvature* in each of the determined contact points.

As indicated by the formulas above, the tooth contact conditions derived for the reduced function λ operating on the reduced variables s can be expressed directly in terms of the original function G operating on w. All the required possibly directional derivatives up to order two of the model functions G, $\tilde{F}_{\mathrm{g}}$, and $\tilde{F}_{\mathrm{p}}$ are calculated using the automatic differentiation tool ADOL-C [241] to compute all geometric quantities of interest with machine accuracy of the computer. Figure 23.1 shows the computed path of contact on a pinion flank with Dupin's indicatrices of the reduced curvature which can be used to approximate the bearing area under load. Comparisons with the *nested* approach used in [491] have shown that the new direct method is approximately five times faster and delivers qualitatively better results in the sense that the common normal direction in the contact points is reflected more accurately.

23.4 Computation of Sensitivities

A major step in the optimization and evaluation of the gear tooth contact is the computation of analytical sensitivities with respect to the parameter vector p. Sometimes p may consist of a suitable subset of over a hundred design and manufacturing parameters of a gear pair. In other cases p may

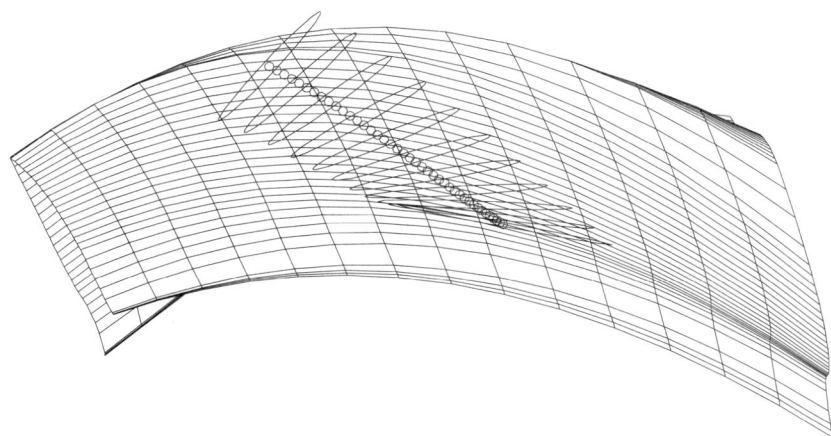

FIGURE 23.1. Computed path of contact with reduced Dupin's indicatrices

contain arbitrary model parameters introduced to investigate the influence
of perturbations, such as misalignments due to deformations of the gear
box while meshing under load, onto the contact pattern.

Applying the implicit function theorem to the extended defining system

$$G_\oplus(w, \psi, \tau; p) \equiv \begin{bmatrix} G(w, \psi; p) \\ g(w, \psi; p) \\ \psi_p - \tau - \psi_p^-(p) \end{bmatrix} = 0_8$$

at some solution point enables the simple calculation of the sensitivities
$\partial_p w^* = \partial_p w^*(\tau, p)$ and $\partial_p \psi^* = \partial_p \psi^*(\tau, p)$, the first of which determines the
impact of machine settings on the position of the path of contact on the gear
flanks. Here one has to take into account that the boundaries of the path of
contact, where the tooth contact "jumps" from one pair of tooth flanks to
the next, depend on the parameters p as well. In [490] it is shown how the
intervals $\left[\psi_p^-(p), \psi_p^+(p)\right]$ and $\left[\psi_g^-(p), \psi_g^+(p)\right]$ as well as their dependence
on p can be evaluated exactly coupling two equation systems of the form
(23.1).

Figure 23.2 visualizes the computed sensitivity of the path of contact with
respect to a particular design parameter p_{p0}^{hm} on a projected gear flank. It
shows the direction and the expected amount of the displacement of each
contact point for small changes $\Delta p_{p0}^{hm} = \pm 0.35$ of the parameter around
$p_{p0}^{hm} = 0$. Thus one can observe how the influence of the parameter p_{p0}^{hm} is
distributed over the whole path of contact. A main objective in optimizing
gear pairs is the minimization of the *transmission error*, which measures
the variation in the ratio between the angular velocities of gear and pinion, respectively. Figure 23.3 shows how small changes of p_{p0}^{hm} affect the
transmission error, which is of order 10^{-5}.

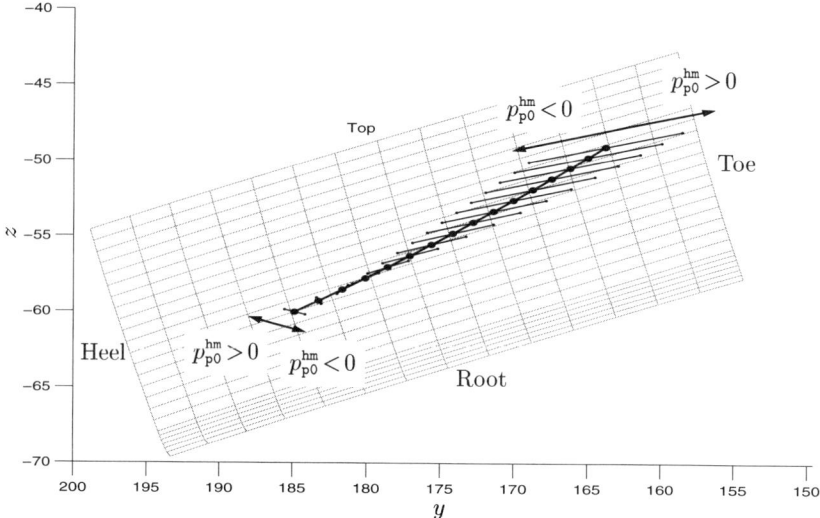

FIGURE 23.2. Sensitivity of path of contact with respect to $p_{\mathrm{p0}}^{\mathrm{hm}}$

Typical derivatives needed for the computation of sensitivities of the principal curvatures and their directions of the reduced curvature at a contact point are given by

$$\partial_s^3 \lambda \left(s^*, \psi^*; p\right) \partial_p s^*, \quad \partial_\psi \partial_s^2 \lambda \left(s^*, \psi^*; p\right) \partial_p \psi^*, \quad \partial_p \partial_s^2 \lambda \left(s^*, \psi^*; p\right),$$
$$\partial_s^2 w \left(s^*, \psi^*; p\right) \partial_p s^*, \quad \partial_\psi \partial_s w \left(s^*, \psi^*; p\right) \partial_p \psi^*, \quad \partial_p \partial_s w \left(s^*, \psi^*; p\right)$$

with $\partial_p s^* = T \partial_p w^*$. These more complex derivative tensors of the implicitly defined functions $\lambda = \lambda(s, \psi^*; p)$ and $w = w(s, \psi^*; p)$ can be obtained by applying the new driver routine `inverse_tensor_eval(...)` of the ADOL-C library [241] to the invertible vector function

$$G_{\mathrm{AD}} : \mathbb{R}^6 \times \mathbb{R} \times \mathbb{R}^4 \times \mathbb{R}^n \to \mathbb{R}^3 \times \mathbb{R}^4 \times \mathbb{R}^4 \times \mathbb{R}^n$$

defined by

$$G_{\mathrm{AD}}(w, \lambda; s, p)^\top \equiv \left[\left(G(w, \psi^*; p) - \lambda r\right)^\top, \ (Tw - s)^\top, \ s^\top, \ p^\top\right]$$

at the solution point $(w^*, 0; s^*, p)$. The driver routine uses the propagation of univariate Taylor polynomials as proposed in [249] and thus facilitates the computation of directional derivatives of arbitrary high order. A preliminary comparison of this approach with derivative code generated by Maple has shown its efficiency and flexibility.

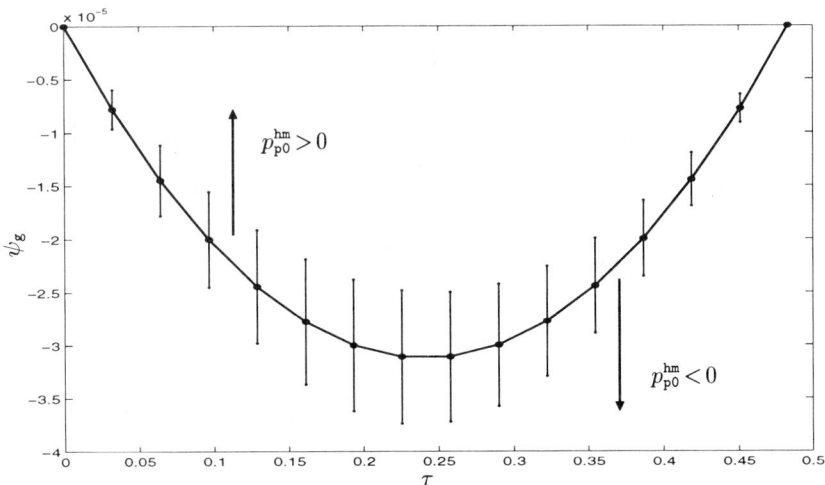

FIGURE 23.3. Sensitivity of transmission error with respect to p_{p0}^{hm}

23.5   Higher Derivative Tensors from Univariate Taylor Series vs. Maple

In order to compare the univariate approach of calculating higher derivative tensors employed by ADOL-C with the multivariate approach used by Maple generated code, in [496] first the new ADOL-C routine `tensor_eval` (. . .) was applied to compute the directional derivatives

$$\nabla_S^k F_w (u_w) \equiv \frac{\partial^k}{\partial v^k} F_w (u_w + Sv) \Big|_{v=0} \in \mathbb{R}^{3 \times q^k}$$

for $S = [s_1, s_2, \ldots, s_q] \in \mathbb{R}^{(3+n_w) \times q}$. We made a slight modification of the model function F_w with $u_w^\top = [w_w^\top, p_w^\top]$, and the number of parameters $p_w \in \mathbb{R}^{n_w}$ was increased by varying the degree of the polynomials modelling additional motions. Second, a Maple procedure equivalent to the about 180 C statements needed to evaluate F_w was differentiated applying the Maple command `diff( . . . )` to produce full multivariate derivative tensors from which we obtained the desired directional derivatives. The resulting derivative procedures were automatically transformed to C code allowing us to compare the run times of both techniques. The results achieved for $q \in \{1, 2, 3\}$ and $n_w \in \{0, 6, 12, 18\}$ are shown in Table 23.1.

The missing values indicate that it was impossible to produce the C code for calculating the directional derivatives by Maple in an acceptable time. The code generation was interrupted after several days. This problem is caused by the fact that the symbolic application of the multivariate chain rule leads to humongous code if the number of parameters is high. Since in the relevant case of tooth contact optimization the number will be

TABLE 23.1. Higher derivative tensors from univariate Taylor series vs. Maple

$n_{\mathsf{w}} = 0$

q	Maple	ADOL-C	Ratio
1	0.12 ms.	0.20 ms.	0.60
2	0.13 ms.	0.45 ms.	0.29
3	0.14 ms.	0.99 ms.	0.14

$n_{\mathsf{w}} = 6$

q	Maple	ADOL-C	Ratio
1	1.02 ms.	0.25 ms.	4.1
2	1.20 ms.	0.60 ms.	2.0
3	—	1.28 ms.	—

$n_{\mathsf{w}} = 12$

q	Maple	ADOL-C	Ratio
1	—	0.28 ms.	—
2	—	0.67 ms.	—
3	—	1.49 ms.	—

$n_{\mathsf{w}} = 18$

q	Maple	ADOL-C	Ratio
1	—	0.32 ms.	—
2	—	0.82 ms.	—
3	—	1.49 ms.	—

even higher, i.e. $n_{\mathsf{w}} \approx 50$, the univariate approach is obviously preferable. Its flexibility is especially attractive when one wishes to vary the set of parameters of interest. In any case, having only three dependent variables and many parameters suggests applying the reverse mode for compiling the desired sensitivities, which is planned for future developments.

23.6   Conclusions

We have presented a promising approach for the approximation-free computation of the gear tooth contact of hypoid bevel gear drives. Besides the run time gain and the higher accuracy, the main advantage of the novel approach lies in the possibility to compute the sensitivities of the determined geometric quantities with respect to the machine settings, which enter into the model of a virtual machine tool. This sensitivity information generated by automatic differentiation may be used in modern optimization algorithms to improve the contact properties of a gear drive. We believe automatic differentiation to be the key to further progress towards systematic gear tooth contact optimization by numerical simulation. Furthermore, we want to extend its application to tooth strength analysis.

Acknowledgments: The author would like to thank Ulf Hutschenreiter for the helpful discussions, Andrea Walther for performing the run time comparison, and the members of the DFG funded research group "Identifikation und Optimierung komplexer Modelle auf der Basis analytischer Sensitivitätsberechnungen" for their comments.

24

Optimal Laser Control of Chemical Reactions Using AD

Adel Ben-Haj-Yedder, Eric Cances and Claude Le Bris

ABSTRACT This chapter presents an application of automatic differentiation to a control problem from computational quantum chemistry. The goal is to control the orientation of a linear molecule by using a designed laser pulse. In order to optimize the shape of the pulse we experiment with a nonlinear conjugate gradient algorithm as well as various stochastic procedures. Work in progress on robust control is also mentioned.

24.1 Introduction

We consider a linear triatomic molecule HCN [159] subjected to a laser field $\overrightarrow{\mathcal{E}(t)}$. Our purpose is to use the laser as a control of the molecular evolution (see [348] for the general background), and more precisely as a control of the *orientation* of the molecular system. On the basis of experiment, it is believed that controlling the alignment of a molecular system is a significant step towards controlling the chemical reaction the system experiences.

An isolated molecular system is governed by the time-dependent Schrödinger equation (TDSE)

$$\begin{cases} i\hbar \frac{\partial \psi}{\partial t}(t) = H\,\psi(t), \\ \psi(0) = \psi^0, \end{cases} \qquad (24.1)$$

where $\psi(t)$ denotes the wave function of the molecule, and H is the Hamiltonian of the free molecular system. The Hamiltonian H can be written as $H = H_0 + V(x)$, where H_0 is a second order elliptic operator corresponding to the kinetic energy (typically $H_0 = -\Delta$), and $V(x)$ denotes a multiplicative operator accounting for the potential to which the molecule is subjected. When a laser field $\overrightarrow{\mathcal{E}(t)}$ is turned on, the dynamics of the molecular system is governed by:

$$\begin{cases} i\hbar \frac{\partial \psi}{\partial t} = H\psi + \overrightarrow{\mathcal{E}(t)} \cdot \mathcal{D}(t)\,\psi, \\ \psi(0) = \psi^0, \end{cases} \qquad (24.2)$$

where $\mathcal{D}(t)$ denotes the electric dipolar momentum operator. In our problem, $\mathcal{D}(t)$ is approximated at the second order by: $\mathcal{D}(t)\,\psi = (\overrightarrow{\mu_0} + \alpha \cdot \overrightarrow{\mathcal{E}(t)})\psi$,

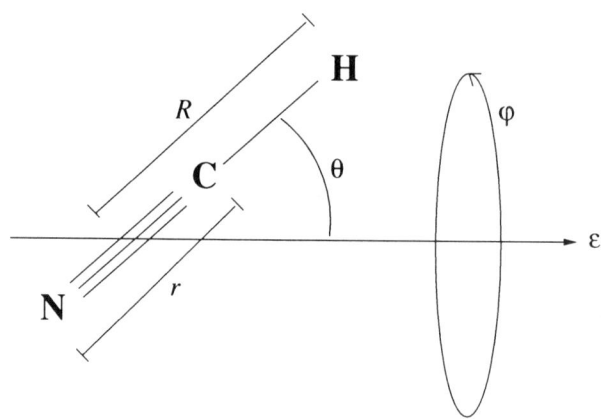

FIGURE 24.1. Model for HCN molecule

where $\vec{\mu_0}$ and α denote the permanent dipolar momentum and the polarisability tensor, respectively.

For the molecular system under study (the HCN molecule), the Hamiltonian can be written in a very convenient way by resorting to the Jacobi coordinates $(\mathbf{R}, \theta, \varphi)$ (the notation $\mathbf{R}$ refers to the pair (R, r), see Figure 24.1)

$$H(\mathbf{R}, \theta, \varphi, t) = T_{\mathbf{R}} + H_{rot}(\mathbf{R}, \theta, \varphi) + V(\mathbf{R}) + H_{laser}(\mathbf{R}, \theta, \varphi, t), \quad (24.3)$$

where

$$T_{\mathbf{R}} = -\frac{\hbar^2}{2\mu_{HCN}} \frac{1}{R^2} \frac{\partial}{\partial R} \left(R^2 \frac{\partial}{\partial R} \right) - \frac{\hbar^2}{2\mu_{CN}} \frac{1}{r^2} \frac{\partial}{\partial r} \left(r^2 \frac{\partial}{\partial r} \right),$$

$$H_{rot}(\mathbf{R}, \theta, \varphi) =$$
$$-\frac{\hbar^2}{2(\mu_{HCN} R^2 + \mu_{CN} r^2)} \left[\frac{1}{\sin\theta} \frac{\partial}{\partial \theta} \left(\sin\theta \frac{\partial}{\partial \theta} \right) + \frac{1}{\sin^2\theta} \frac{\partial^2}{\partial \varphi^2} \right],$$

$$H_{laser}(\mathbf{R}, \theta, \varphi, t) = -\mu_0(R, r)\mathcal{E}(t) \cos\theta$$
$$-\frac{\mathcal{E}^2(t)}{2} \left[\alpha_\|(R, r) \cos^2\theta + \alpha_\perp(R, r) \sin^2\theta \right].$$

24.2   Models and Results

Our goal is to control the orientation of the molecular system with the laser beam direction. We optimize the objective functional (the criterion) J, a measure of the rate of orientation given by

$$J = \frac{1}{T} \int_0^T \left[\left(\int_0^{\frac{\pi}{2}} - \int_{\frac{\pi}{2}}^{\pi} \right) \mathcal{P}(\theta, t) \sin\theta \, d\theta \right] dt,$$

where $\mathcal{P}(\theta, t)$ is the angular distribution of the molecule given by:

$$\mathcal{P}(\theta, t) = \frac{\langle \Psi(\theta, t) | \Psi(\theta, t) \rangle_R}{\langle \Psi(t) | \Psi(t) \rangle}$$

with

$$\langle f | g \rangle_R = \int_{r_{min}}^{r_{max}} \int_{R_{min}}^{R_{max}} f^*(R, r) g(R, r) R^2 \, dR \, r^2 \, dr,$$

and

$$\langle f | g \rangle = \int_0^\pi \int_{r_{min}}^{r_{max}} \int_{R_{min}}^{R_{max}} f^*(R, r, \theta) g(R, r, \theta) R^2 \, dR \, r^2 \, dr \sin \theta \, d\theta.$$

The criterion J takes its values in the range $[-1, 1]$, the values -1 and 1 corresponding respectively to a molecule pointing in the desired direction and in the opposite direction. Our goal is therefore to minimize J. As a first step towards the treatment of the more sophisticated model (24.3), we consider here the case of a rigid rotor: the problem depends only on the variable θ. The equation (24.2) depending only on the variable θ is numerically solved by a Fortran program written by Dion [159] which uses a operator splitting method coupled with a FFT (for the kinetic part).

The laser field is the superposition of 10 laser pulses, each of the form

$$\mathcal{E}(t) = f(t)\mathcal{E}_0 \cos(\omega t + \phi), \tag{24.4}$$

where:

$$f(t) = \begin{cases} 0 & \text{if} & 0 < t < t_0 \\ \sin^2 \left[\frac{t - t_0}{t_1 - t_0} \frac{\pi}{2} \right] & \text{if} & t_0 < t < t_1 \\ 1 & \text{if} & t_1 < t < t_2 \\ \sin^2 \left[\frac{t - t_3}{t_2 - t_3} \frac{\pi}{2} \right] & \text{if} & t_2 < t < t_3 \\ 0 & \text{if} & t > t_3 \, . \end{cases} \tag{24.5}$$

Let us emphasize at this point that we do not pretend that such a superposition of laser beams is feasible experimentally: we are just testing here the mathematical attack of the problem.

We first search for a local minimum of the criterion by means of a deterministic optimization algorithm (namely a nonlinear conjugate gradient procedure). The gradient is computed by an adjoint code automatically generated by Odyssée [186]. In the present calculations, 70 parameters have to be optimized, namely $t_0^i, t_1^i, t_2^i, t_3^i, \mathcal{E}^i, \omega^i, \phi^i$ for $i = 1, 10$. As in the direct program we have 50,000 iterations, the adjoint program needs a lot of memory to run. To reduce the size of memory needed, the adjoint program was modified by deleting the temporary variables in the linear parts of the program. Table 24.1 gives an idea of the size of the direct code and the adjoint code. The reduction of the memory we have obtained by optimising the generated code by hand is coherent with the reduction obtained by an

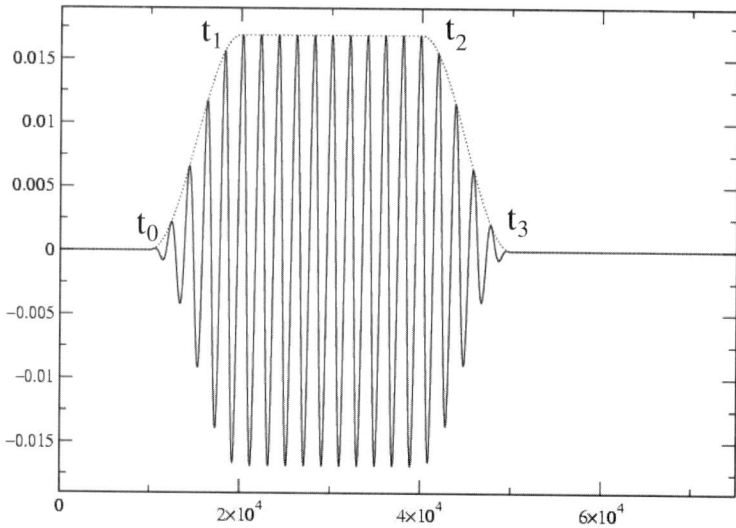

FIGURE 24.2. One laser field amplitude $\mathcal{E}_i(t)$

TABLE 24.1. Some data about the program

	direct code	standard adjoint	optimised adjoint
Size (lines)	433	2075	1190
Memory needed	12 Ko	520 Mo	103 Mo
Time (CPU)	60s	—	141s

automatic optimisation as shown in [185]. The calculations are done on a Pentium II, 466 Mhz Celeron with 128 Mb RAM and running with Linux.

Numerical results show that the gradient values are most important for the variables ω^i and ϕ^i (which can thus be considered as the most significant ones from a physical viewpoint). When running the optimization program with a sample of representative initial guesses, it appears that the program always converges after a few iterations toward a *local* minimum generally located in the neighborhood of the chosen initial guess. This observation leads us to also turn to stochastic algorithms for searching a global minimum (see §24.3 below). Two cases of initial guesses were investigated: the case of laser pulses that approximately have the same frequencies, and the case of laser pulses with significantly different frequencies. In the first case, despite the number of local minima, it is nevertheless possible to reach quite a satisfactory local minimum ($J = -0.67$) by starting from an initial guess where all the 10 laser pulses have the same characteristics (see Table 24.2). At this minimum, the values of the pulsations ω_i are very close to each other (for example $\omega_1 = 1060.52256$ and $\omega_2 = 1060.83702$). Unfortunately, a further analysis demonstrates that the so-obtained results are due

TABLE 24.2. Laser field characteristics for a first starting point for $i = 1, 10$

$\mathcal{E}$	ω	ϕ	t_0	t_1	t_2	t_3
10^{13}	1060	0	0	14000	56000	75000

TABLE 24.3. Laser field characteristics for a second starting point for $i = 1, 10$

$\mathcal{E}$	ω	ϕ	t_0	t_1	t_2	t_3
10^{12}	$500 \times i$	0	0	14000	56000	75000

to numerical instabilities: refining the discretization grid in both time and space makes the criterion go up to the value -0.0044 (a by far less good result!). With such close pulsations, a very fine grid is needed to avoid the numerical instabilities.

In the second case (see Table 24.3) the laser field found by the optimization program is stable with respect to a refinement of the grid. In this case, we have remarked that the laser field presents several very sharp peaks (see Figure 24.3 (a)). For this reason, consolidated by theoretical arguments, we have chosen next to perform optimization with laser pulses consisting not of functions of type (24.4) but rather in superpositions of Dirac functions. The Dirac functions are numerically approximated by Gaussian functions of the form: $\mathcal{E}_i \frac{1}{\sqrt{\delta t \pi}} \exp(-\frac{(t-t_i)^2}{\delta t})$. After some modifications of the direct code and a new automatic differentiation, we have used the new optimization program to minimize the criterion. The parameters are the instants t_i when the Dirac functions appear and the intensities $\mathcal{E}_i$ of the peaks. The calculations show that the gradient of the criterion with respect to the parameters t_i is indeed very small, and that the main control parameters are the intensities $\mathcal{E}_i$. The shape of the laser field given by the optimization program depends also on the number of Dirac functions. For 75 Dirac functions, the criterion $J = -0.56$, and the corresponding pulse shape is given in Figure 24.3 (b).

24.3 Comparison with Stochastic Algorithms

As already mentioned, the problem under investigation has many *local minima*. Stochastic procedures therefore seem appropriate to search for a *global minimum* in an effective way.

Our first trial consisted of using a simulated annealing algorithm with different temperature programs, which yielded disappointing results. We next combined the simulated annealing procedure with a nonlinear conjugate gradient algorithm, but the improvements were small compared to the cost in CPU time.

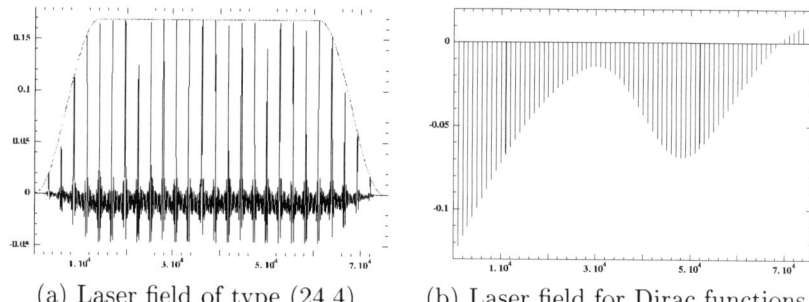

(a) Laser field of type (24.4).   (b) Laser field for Dirac functions.

FIGURE 24.3. Laser field given by the optimization program.

Using a genetic algorithm program given by [115], we got results better than those given by simulated annealing (although slightly worst than those given by the nonlinear conjugate gradient algorithm). However, we expect that coupling a genetic algorithm with conjugate gradient techniques will improve the search for a global minimum. Further work in this direction is in progress.

24.4   Robustness

To avoid instabilities (small variations of the laser pulse leading to large variations of the criterion) a very fine grid in space and in time is needed. However, such a fine grid dramatically increases the computational cost in both memory and CPU time resources and will not be tractable in practice. Another way to avoid theses instabilities is to consider the new criterion $J = J(u, w)$, where u represents the same 70 parameters as above and where w represents a disturbance of the parameters [507]. The robust control problem consists of finding the solution of the minimax problem:

$$\min_{u} \max_{w} J(u, w).$$

Apart from the numerical instabilities mentioned above, robust control also enables us to take into account uncertainties due to the model itself. We have slightly modified the direct program and then used automatic differentiation to obtain the adjoint code. With a saddle point search algorithm the local optimization was slightly improved. This work is in progress and will be reported on in a future publication.

24.5   Conclusion

The results obtained so far by automatic differentiation for the problem of laser control of molecular systems seem to us very promising. It clearly brings some new contribution to the tools used by the community of chemists on the subject. For the different cases we have studied so far the optimization program provides a laser pulse which corresponds to a local minimum. Our work is now continuing along two directions: (a) implement *robust* control and (b) improve the search for a *global minimum.*

The use of AD tools in our work allows us to obtain better results (smaller objective function values), but less robust results, than results obtained by using stochastic algorithms. The combination of the two methods should make the optimization even more efficient. When we treat the complete problem (depending on all the molecular degrees of freedom R, r and θ, see Figure 24.1), checkpointing methods are likely to be necessary.

In parallel with the numerical work presented here, some experimental work is in progress.

Acknowledgments: Special thanks are due to C. Dion for his help in understanding the direct code and for making the necessary modifications before automatic differentiation. We thank the Odyssée development team for putting this software in our disposition for this work. We also wish to thank M. Barrault for his help and for useful discussions on optimization algorithms.

Part VI

Maintaining and Enhancing Parallelism

25

Automatically Differentiating MPI-1 Datatypes: The Complete Story

Alan Carle and Mike Fagan

ABSTRACT   This chapter describes the problems met while extending ADIFOR 3.0 to handle programs coming from the MPI datatypes. To give a complete picture of the design space, this chapter first explains why creating a differentiated interface for MPI is preferable to differentiating MPI source itself. Next, we identify the two main problems that MPI datatypes present and present two potential solution techniques. In conclusion, we give some preliminary implementation results.

25.1   Introduction

A popular way to add parallelism to a sequential programming language is to use a library that defines routines for communication of data via "message-passing." The MPI-1 (Message Passing Interface Version 1.0, 1.1) [463] library provides a standardized message-passing system for writing portable message-passing programs. MPI-1, released in 1994, provides a wide-range of facilities including point-to-point communication, collective communication and arithmetic, and process synchronization for programs written in Fortran 77 or C. MPI provides primitive datatypes representing floating point and integer data, and a collection of constructor operations for hierarchically creating derived datatypes.

There is a clear need to provide automatic differentiation for programs written using the MPI message-passing library. Odyssée and ADIFOR 3.0 both provide minimal support for MPI. Odyssée [183, 184] provides forward and reverse mode support for synchronous SEND, RECV, and REDUCE operations for primitive MPI datatypes. ADIFOR 3.0 [112] additionally supports the asynchronous versions of SEND and RECV, and a few other operations, also limited to the primitive datatypes. In this chapter, we discuss methods to extend ADIFOR 3.0 to support general MPI-1 datatypes. Furthermore, in an effort to give a "complete" story, we discuss two possible methods, as well as one method we rejected.

25.2   An Idea We Rejected

One possible strategy for providing an AD facility for MPI involves differentiating the source code for the MPI library itself and then coupling the differentiated library with an appropriately differentiated version of the user's source program. By definition, if the user's source code and the MPI library are correctly differentiated, then this process will result in a correct, but probably unportable, differentiated version of the user's program. In fact, source code for a public domain reference implementation of the MPI (MPICH) library is available, though the library contains source code written in both Fortran and C and makes fairly extensive use of the C preprocessor. Developing a pair of AD tools for Fortran and C that could generate compatible code for Fortran and C programs should be possible – the two AD tools would need to agree on a strategy for associating derivatives with variables in the user's program, and they would need to agree on a convention for inter-language procedure invocations. Although developing a pair of compatible AD tools for Fortran and C has merit, it seems unlikely that an AD tool for C will be able to satisfactorily differentiate the MPI source code (from the standpoints of both software engineering and code performance) given that it has more in common with an operating system device driver than a numerical library[1]. In light of the clear portability problem, and likely engineering and performance problems, we rejected this strategy.

A more realistic approach to differentiating programs that use the MPI library requires the creation of a differentiated interface capable of being used with any standard-conforming MPI library. This is the approach taken by both Odyssée and ADIFOR 3.0.

25.3   Synopsis of Relevant MPI Datatype Concepts

To understand the impact of MPI datatypes on AD processing, we define and summarize some of the important properties of these objects. According to the MPI reference manual [463], an MPI datatype is an opaque data structure that describes a collection of primitive data items. A primitive data item is one of the fundamental Fortran 77 types: real, double, integer, logical, or character[2]. MPI datatypes use *offsets* from a base address to indicate the location of the relevant data items. The address of an item is computed by adding the base address and the offset. The data structure is "opaque" because the internal type and offset components of the MPI

---

[1] This should be considered to be a challenge to AD tool developers for C.

[2] Some Fortran implementations support additional "primitive" types (such as extended precision integers or floats). The MPI implementation may support additional "primitive" types as well.

datatype are not accessible by the MPI client[3].

The MPI reference denotes an MPI datatype as a set of (type, offset) pairs, usually listed in order of increasing offsets. This presentation is called a **type map**. As an example:

```
T1 = {(float,0),(integer,12),(float,20),(float,28)}
```

This example describes a datatype that has a `float` (i.e. F77 `real`) data at offsets 0, 20, and 28 from a given base address, and an `integer` datum at offset 12 from the base. We could create this datatype using this pseudo F77 code fragment:

```
common /CC/ r1(3),i1(2),z(2),r2(2)
real r1,r2,z   ! z inactive
integer i1
integer d(4),bb(4),t(4),NT ! MPI datatype variables

d = (1,12,20,28)            ! offset array init
t = (FLOAT,INTEGER,FLOAT,FLOAT) ! type array init

call MPI_type_create_struct(4,d,t,NT,...)
call MPI_type_commit(NT)
```

For *any* set of primitive types and offsets, MPI can define a datatype with those primitive types and offsets. Since *any* offset can be inserted into an MPI datatype, then *any* address in the address space can be referenced by an MPI datatype. This effectively adds a "pointer" notion to Fortran 77 programs using MPI datatypes.

25.4   Problems for ADIFOR 3.0 Caused by Datatypes

To see what potential problems can arise, we attempt the most straightforward possible solution on a simple problem instance. We examine a simple send/recv pair in forward mode.

```
Send Proc: call mpi_send(recv_proc,SBuf,DT,...)
Recv Proc: call mpi_recv(send_proc,RBuf,DT,...)
```

Considering only forward mode for the moment, the AD tool must produce code that accomplishes the following:

Derivatives of `RBuf` = Derivatives of `SBuf`

This could seemingly be accomplished by ADIFOR-style augmentation

```
Send Proc: call mpi_send(recv_proc,g_SBuf,DT,...)
Recv Proc: call mpi_recv(send_proc,g_RBuf,DT,...)
```

---

[3]The data structure is opaque to the *programmer, not* to the implementation.

However, careful consideration of this straightforward attempt reveals the difficulty. Since MPI datatype offsets can be arbitrary, the locations referenced by the SBuf (or RBuf) can be any conceivable program location. There is no reason to expect that derivatives associated with some arbitrary location can be found using the same type map at a different offset. The pointer-like nature of the MPI datatype implies that we must do something non-trivial to locate the derivative storage associated with an arbitrary location. This same problem was noted by Lee and Hovland [352] for the C language AD processor ADIC.

The reverse mode introduces one additional concern for the AD process. To see this, we note that reverse mode update for our simple send/recv pair would be

adjoints of SBuf += adjoints of RBuf

Note that the updating operation for adjoints is +=. This means that

```
mpi_send(send,a_RBuf,DT,...)
mpi_recv(recv,a_SBuf,DT)
```

does *not* implement the adjoint update operation, as it mimics assignment. The effect of the proposed send/recv pair would be

adjoints of SBuf = adjoints of RBuf

This is clearly wrong. The problem is that there is no += operation provided by MPI.

To summarize, the set of problems for AD of Fortran 77 using MPI datatypes are:

Location: Given a location, what is the location of the associated derivative values?

Reverse update: The adjoint propagation rule uses +=, not =. MPI does not provide such an operation. Hence, how do we update adjoint values in reverse mode?

25.5   Description of Two Possible Solution Methods

To solve the problems discussed in the previous section, we devised two different solution methods, **Derivative Aggregation** and **Active Lookup**, which we summarize in this section. The subsequent sections will show, for each solution method, how each of the two problems of the previous section can be solved.

25.5.1   Overview of derivative aggregation

The derivative aggregation technique changes representation for derivatives. Each active scalar location is transformed into an array of values: the first element of the array holds the original value of the scalar. All other array elements hold derivative values[4].

To illustrate the concept, suppose the program variable to be augmented is declared as:

```
real A(4,5)
```

This corresponds to 20 scalars. We would augment this variable to hold five derivative values as follows:

```
real g_A(6,4,5)
```

In this case, g_A(1,3,4) corresponds to the original A(3,4). The values g_A(2,3,4) thru g_A(6,3,4) are the storage for the 5 derivative values for the A(3,4) scalar.

Using this derivative aggregation scheme, MPI datatypes referencing locations that have derivatives can be easily associated with MPI datatypes that reference the derivatives. To illustrate, let

```
T1 = {(integer,0),(float,16),(float,40)}
```

Further assume that the middle (float,16) item is augmented with four derivative values. The integer item and the last float are not augmented[5]. Then

```
g_T1 = {(float,20),(float,24),(float,28),(float,32)}
```

will correctly reference the derivatives associated with any T1 datatype, no matter what the base. In an analogous manner, for any datatype, an associate datatype for the derivatives can always be constructed that will always correctly reference the derivatives.

To solve the reverse update problem using this technique, we resorted to MPI's reduce operation. The reduce operation is part of MPI's *collective* operations, and requires additional system data structures. The += operation can be implemented using the following steps:

1. Build a new MPI communicator object having only two group members: the sender and the receiver

2. Make the sender the root process for the communicator

3. Use MPI_reduce ("in place" variant) with the MPI_ADD built-in operation. Use the datatype associated the derivatives.

---

[4]This method essentially simulates the C++ overloading technique for AD. Since Fortran 77 does not have overloading, we must generate the additional code necessary to accomplish the same effect.

[5]For the last float to be "inactive," ADIFOR would have to determine that *all* uses of the element are "inactive."

By definition, an MPI reduce operation that is done "in place" leaves the sum of all buffers in the root buffer. Since we have arranged for the derivatives to be transferred in the message buffers, the sum of all the derivatives will be held by the root process. By construction, this means that the sending process buffer holds the sum of the sending process original derivatives and the receiving process derivatives. This is precisely the += operation we were originally seeking.

25.5.2   Overview of active lookup

To solve the location problem, for a given program location, we must determine the location of the derivatives for that location (if there are any). It would seem reasonable that we could answer queries of this sort by storing the derivative location for active scalars in a lookup structure keyed on the location (address). The lookup structure would be populated as part of the derivative initialization process. The differentiated code could then use this lookup structure to obtain the requisite derivative location, and update accordingly.

It would seem that an update structure would have a performance penalty associated with it, so the "complete" story should indicate why we still considered this method. There are three reasons why we considered the active lookup a viable method:

1. (Most important) The update structure allows us to retain the conventional ADIFOR derivative variable declaration strategy. We would not need to alter the tool as much. In addition, it is backward compatible with other ADIFOR-differentiated code. The derivative aggregation method does *not* have this advantage.

2. The active lookup approach will work for other pointer-like Fortran extensions. Hence, we could implement this solution once (as a runtime library), and use it in other contexts

3. The performance penalty might not be too bad.

To illustrate the idea, suppose we have the declarations from the example in §25.3

```
common /CC/ r1(3),i1(2),z(2),r2(2)
real r1,r2,z   ! z inactive
integer i1
```

with the associated declarations of four derivatives

```
common /g_CC/ g_r1(4,3),g_r2(4,2)
```

Then we would initialize the lookup structure as follows:

```
do ii=1,3
   call d_lookup_init(r1(ii),g_r1(1,ii))
enddo
do ii=1,2
   call d_lookup_init(r2(ii),g_r2(1,ii))
enddo
```

The active lookup structure goes a long way towards solving the location problem, but we need one more component technique to complete our solution: we need to find out which locations are referenced by a datatype instance. Since an MPI datatype is "opaque," access to the offset and type information is not provided by MPI[6].

To make the datatype transparent (i.e. "not opaque"), we can augment the MPI code to record the type and offset information for every instance of a datatype construction operation. For example, suppose that the original code uses the datatype construction operation from §25.3:

```
call MPI_type_create_struct(4,d,t,NT,...)
```

To make NT transparent, we would augment this construction thusly:

```
call MPI_type_create_struct(4,d,t,NT)
call g_MPI_type_create_struct(4,bb,d,t,offs_NT,
+                              types_NT,n_NT)
```

where offs_NT and types_NT store the offset and type information for the datatype, and n_NT stores the number of elements of NT in the NT structure.

To illustrate the method, we show augmented pseudocode to recv the derivatives for MPI_recv(send,Base,T1,...).

```
do j = 1,n_T1
   D_loc = D_lookup(Base(offs_T1(j)))
   if (D_loc .ne. 0) then
      call MPI_recv(send,D_loc,types_T1(j),...)
   endif
enddo
```

This technique is easily applied to the reverse mode update problem. The augmented pseudocode for MPI_send(recv,Base,T1,...) would be:

```
do j = 1,n_T1
   D_loc = D_lookup(Base(offs_T1(j)))
   if (D_loc .ne. 0) then
      call MPI_recv(send,tmpbuf,types_T1(j),...)
      D_loc = D_loc + tmpbuf ! array update
```

---

[6]The current *MPI-2* standard [256] does provides accessor functions for type and offset information in an MPI datatype.

```
        endif
     enddo
```

We could have used the `MPI_reduce` operation for this derivative update as well. The reason we did not is that creating and destroying the required communicators takes too much time. There may be cases for which the reduce performance improvement exceeds the communicator performance penalty. This aspect of the problem certainly deserves more study.

25.6 (EXTREMELY) Preliminary Results

Using a small test program we created[7], we examine some of the characteristics of two proposed protocols for differentiating MPI datatype programs. Our test program consisted of about 100 lines of random calculations, and three MPI datatypes, one of which did not involve active variables. We wrapped a loop around the main calculations to make sure that the program took long enough for the system to measure. The system did 4000 sends and 4000 receives. We used two Sun workstations with the freely available MPICH [259] implementation of MPI. There were three independent variables and one dependent variable. Results were collected over an average of ten runs of the test programs. All timings are in seconds, as determined by the Unix function 'time'.

This small test is *not* a performance study. It is more of a proof-of-concept for the strategies suggested. Both strategies pass proof-of-concept. The lack of large variation in the times for each chosen method merely suggests that there is lots of work to be done before there is a definitive answer for automatic differentiation of MPI datatype programs.

TABLE 25.1. CPU times in seconds for calculations in three MPI datatypes

Forward Mode			**Reverse Mode**		
Function	Derivative Aggregation	Active Lookup	Function	Derivative Aggregation	Active Lookup
4.3	7.4	7.9	4.3	88.4	97.6

Acknowledgments: This work was supported in part by NASA Langley Research Center under contract NCC 1-234 and Purchase Order DFN.1155, and by the NSF through the Center for Research on Parallel Computation (CRPC) and the National Computational Science Alliance (NCSA). The authors also wish to thank the NSF for providing the Origin 2000 computing platform (mapy) via SCREMS grant 98-72009. Mapy provided important 2nd MPI platform for software development and algorithm verification.

---

[7]We could not find a suite of "common kernels" test programs in Fortran that use the MPI datatypes operations. Hence, we have no idea what is "typical" for Fortran programs using datatypes, and we were forced to create our own.

26

Sensitivity Analysis Using Parallel ODE Solvers and Automatic Differentiation in C: SensPVODE and ADIC

Steven L. Lee and Paul D. Hovland

ABSTRACT PVODE is a high-performance ordinary differential equation solver for the types of initial value problems (IVPs) that arise in large-scale computational simulations. Often, one wants to compute sensitivities with respect to certain parameters in the IVP. We discuss the use of automatic differentiation (AD) to compute these sensitivities in the context of PVODE. Results on a simple test problem indicate that the use of AD-generated derivative code can reduce the time to solution over finite difference approximations.

26.1 Background

In complicated, large-scale computational simulations, the governing equations can often be spatially discretized and then numerically solved in parallel (using domain decomposition) as an initial value problem for a system of ordinary differential equations (ODEs) or differential-algebraic equations (DAEs). PVODE [97] and IDA [277] are powerful, parallel codes for solving these types of ODEs and DAEs, respectively. The codes are written in C and use MPI to achieve parallelism and portability. Typically, the equations contain parameter values (e.g., chemical reaction rates) that are not precisely known. In analyzing the simulations, the scientist would like to know which parameters are most influential in affecting the behavior of the simulation. Such sensitivity information is useful because it identifies which parameters will require precise measurements if the simulation results are to be made more accurate. This article summarizes preliminary work in which automatic differentiation (AD) is being used with PVODE to create a solver that computes sensitivity information for ODE systems.

In computing sensitivities for ODEs, one is interested in solving

$$y'(t) = f(t, y, p), \ y(t_0) = y_0(p), \ y \in \mathbb{R}^N, \ p \in \mathbb{R}^m, \tag{26.1}$$

where the solution vector $y(t)$ depends upon an additional vector of parameters p, and the *sensitivities* are defined as

$$s_i(t) = \frac{\partial y(t, p)}{\partial p_i}, \quad i = 1, \cdots, m.$$

One approach for computing these sensitivities is to apply AD techniques to the entire PVODE solver. However, PVODE is a variable-stepsize, variable-order solver and, for this situation, Eberhard and Bischof [173] have demonstrated that AD may compute unexpected derivative values unless an *a posteriori* correction is applied. In contrast to such a "black-box" approach, it is often superior to couple the use of AD with some insight into the computational requirements of the problem. To do this, we formally differentiate the original ODE (26.1) with respect to each component p_i of p. Thus, we obtain the sensitivity ODEs

$$s_i'(t) = \frac{\partial f}{\partial y} s_i(t) + \frac{\partial f}{\partial p_i}, \quad s_i(t_0) = \frac{\partial y_0(p)}{\partial p_i}, \quad i = 1, \cdots, m. \qquad (26.2)$$

The initial sensitivity vector $s_i(t_0)$ is either all zeros (if p_i occurs only in f), or has nonzeros according to how $y_0(p)$ depends on p_i. The time integration of $y'(t)$ and each $s_i'(t)$ can be accomplished by solving an ODE system of size $N(m + 1)$, where

$$Y = \begin{pmatrix} y(t) \\ s_1(t) \\ \vdots \\ s_m(t) \end{pmatrix} \quad \text{and} \quad F(t, Y, p) = \begin{pmatrix} f(t, y, p) \\ \frac{\partial f}{\partial y} s_1(t) + \frac{\partial f}{\partial p_1} \\ \vdots \\ \frac{\partial f}{\partial y} s_m(t) + \frac{\partial f}{\partial p_m} \end{pmatrix}.$$

The new ODE-sensitivity IVP to be solved is simply

$$Y'(t) = F(t, Y, p), \quad Y(t_0) = Y_0(p), \qquad (26.3)$$

and each $s_i'(t)$ can be evaluated by computing $\frac{\partial f}{\partial y} s_i(t) + \frac{\partial f}{\partial p_i}$ via AD, or by approximating their sum via finite differences.

In general, the sensitivities of the ODE problem can be solved for in a variety of ways. For example, the ODE problems can be decoupled: compute and store the solution $y(t)$ in advance; then use interpolation, along with AD or finite differences, to evaluate $s_i'(t)$ wherever needed by an ODE solver. If the same ODE solver is used for (26.1) and (26.2), the effort needed to integrate them is often comparable since the ODEs have the same Jacobian matrix $\frac{\partial f}{\partial y}$ and therefore the same stiffness properties. For a comprehensive review of methods for computing sensitivity information in ODE systems, see [429].

SensPVODE [351] is a variant of PVODE that *simultaneously* computes the solution and the sensitivities in the augmented ODE system (26.3). Also, for many large-scale applications, implicit time integration methods are required. Several papers describe how to modify Newton's method for efficiently solving the nonlinear systems that arise at each time step [189, 370]. Also, we note that the sensitivity ODEs (26.2) are linear in $s_i(t)$, even if the original ODE (26.1) is nonlinear. This observation is significant in the next section as we discuss the need to properly scale the sensitivities that we compute.

26.2   Scaled Sensitivities Using Finite Differences

Several observations motivate our modifications to the sensitivity ODEs (26.2). First, the units for the ODE solution, $[y(t)]$, and the units for the sensitivity vectors, $[s_i(t)]$, do not match. This mismatch in units can lead to scaling problems, especially when using finite difference methods. Fortunately, the issue is easily remedied. In particular, the sensitivity vectors have units of $[y]/[p_i]$. For $y(t)$ and the sensitivities to share the same units, the linearity of the sensitivity ODEs (26.2) allows us to multiply the sensitivities by their respective parameter values to obtain the *scaled* sensitivity ODEs

$$w_i'(t) = \frac{\partial f}{\partial y} w_i(t) + \bar{p}_i \frac{\partial f}{\partial p_i}, \tag{26.4}$$

where

$$w_i(t) = \bar{p}_i s_i(t),$$

and $\bar{p}_i$ is a nonzero constant that is dimensionally consistent with p_i. Typically $\bar{p}_i = p_i$. In general, the scale factor $\bar{p}_i$ can be any nonzero multiple of p_i, and this can sometimes be used to create a well-scaled problem for the ODE variables and sensitivities.

To improve the accuracy of estimating the scaled sensitivity derivatives in (26.4), SensPVODE has an option that applies centered differences to each term separately:

$$\frac{\partial f}{\partial y} w_i \approx \frac{f(t, y + \delta_y w_i, p) - f(t, y - \delta_y w_i, p)}{2 \, \delta_y} \tag{26.5}$$

and

$$\bar{p}_i \frac{\partial f}{\partial p_i} \approx \frac{f(t, y, p + \delta_i \bar{p}_i e_i) - f(t, y, p - \delta_i \bar{p}_i e_i)}{2 \, \delta_i}. \tag{26.6}$$

As is typical for finite differences, the proper choice of perturbations δ_y and δ_i is a delicate matter. Our recommended value for δ_y and δ_i takes into account several problem-related features: the relative ODE error tolerance RTOL, the machine unit roundoff $\epsilon_{\text{machine}}$, and the weighted root-mean-square (RMS) norm of the scaled sensitivity w_i. We then define

$$\delta_i = \sqrt{\max(\text{RTOL}, \epsilon_{\text{machine}})} \quad \text{and} \quad \delta_y = \frac{1}{\max(\|w_i\|, 1/\delta_i)}. \tag{26.7}$$

The terms $\epsilon_{\text{machine}}$ and $1/\delta_i$ are included as divide-by-zero safeguards in case RTOL $= 0$ or $||w_i|| = 0$. Roughly speaking (i.e., if the safeguard terms are ignored), δ_i gives a $\sqrt{\text{RTOL}}$ relative perturbation to parameter i, and δ_y gives a unit weighted RMS norm perturbation to y. Of course, the main drawback of this approach is that it requires four evaluations of $f(t, y, p)$.

A less costly technique for estimating scaled sensitivity derivatives is also based on centered differences. However, it uses the formula

$$
w_i' = \frac{\partial f}{\partial y} w_i + \overline{p}_i \frac{\partial f}{\partial p_i} \approx \frac{f(t, y + \delta w_i, p + \delta \overline{p}_i e_i) - f(t, y - \delta w_i, p - \delta \overline{p}_i e_i)}{2\,\delta}
$$

(26.8)

in which

$$
\delta = \min(\delta_i, \delta_y).
$$

If $\delta_i = \delta_y$, a Taylor series analysis shows that the sum of (26.5) and (26.6) and the value of (26.8) agree to within $O(\delta^2)$. However, the latter approach is half as costly, since it requires only two evaluations of $f(t, y, p)$. To take advantage of this savings, it may also be desirable to use the latter formula when $\delta_i \approx \delta_y$. In [351], we explore the possibility of allowing SensPVODE to select the finite difference formula based on how closely δ_i and δ_y agree.

In summary, the sensitivity version of PVODE incorporates a variety of finite difference formulas for approximating scaled sensitivity derivatives. However, for some problems, finite differences do not work. Typically, difficulties arise in applications where the solution components are very badly scaled. In addition to failure or accuracy problems, finite differences may be inefficient for functions $f(t, y, p)$ that are expensive to evaluate. Such shortcomings motivate the need for an efficient, exact, and (preferably) automated process for computing sensitivity derivatives within SensPVODE.

26.3   Scaled Sensitivities Using AD

Automatic differentiation must be nearly as easy to use as finite differences, or it will only be used when finite differences fail, if at all. Previous work [1, 147, 190, 210, 354] has demonstrated that it is possible to automate the AD process by exploiting the existence of well-defined interfaces for the user's function implementing $f(t, y, p)$. This makes it easy to identify the independent and dependent variables and to properly initialize the AD-generated code.

Applying AD is complicated by the fact that the user's function is implemented in C with MPI parallelism [260]. We are therefore adding support for MPI to the ADIC [84] automatic differentiation tool, building on earlier work by Hovland [288, 289]. The use of C poses challenges from the standpoint of automation. PVODE, like many other numerical toolkits, allows the user to pass around application-specific data in a user-defined struct. As part of the AD process, it may be necessary to associate derivatives with some of the variables in this structure. To avoid aliasing problems, this generally implies changing the type of these variables [84]. Thus, all code (not just the function) must be modified to use this new datatype. Our initial approach has been to circumvent the problem through the use of two data structures, one with derivatives and one without, copying data back and forth as necessary. This removes the need for the user to manually modify driver and initialization routines that access the data structures, but introduces the overhead of copying. To eliminate this overhead while preserving ease of use, we plan to use a single data structure, using ADIC to automatically modify all user code to use the new datatype.

26.4 Experimental Results

We applied SensPVODE to a simple test case, a two-species diurnal kinetics advection-diffusion system in two space dimensions. The PDEs can be written as

$$\frac{\partial c_i}{\partial t} = K_h \frac{\partial^2 c_i}{\partial x^2} + V \frac{\partial c_i}{\partial x} + \frac{\partial}{\partial y}\left(K_v(y)\frac{\partial c_i}{\partial y}\right) + R_i(c_1, c_2, t) \quad (i = 1, 2),$$

where the subscripts i are used to distinguish the chemical species. The reaction terms are given by

$$R_1(c_1, c_2, t) = -q_1 c_1 c_3 - q_2 c_1 c_2 + 2q_3(t)c_3 + q_4(t)c_2, \quad \text{and}$$
$$R_2(c_1, c_2, t) = q_1 c_1 c_3 - q_2 c_1 c_2 - q_4(t)c_2;$$

and $K_v(y) = K_0 \exp(y/5)$. The scalar constants for this problem are $K_h = 4.0 \times 10^{-6}$, $V = 10^{-3}$, $K_0 = 10^{-8}$, $q_1 = 1.63 \times 10^{-16}$, $q_2 = 4.66 \times 10^{-16}$, and $c_3 = 3.7 \times 10^{16}$. The diurnal rate constants are

$$q_i(t) = \exp[-a_i/\sin \omega t] \quad \text{for } \sin \omega t > 0,$$
$$q_i(t) = 0 \qquad\qquad\quad \text{for } \sin \omega t \leq 0,$$

where $i = 3$ and 4, $\omega = \pi/43200$, $a_3 = 22.62$, and $a_4 = 7.601$. The time interval of integration is $[0, 86400]$, representing 24 hours measured in seconds.

The problem is posed on the square $0 \leq x \leq 20$, $30 \leq y \leq 50$ (all in km), with homogeneous Neumann boundary conditions. The PDE system is treated by central differences on a uniform mesh, with simple polynomial initial profiles. See [351] for more details. For the purpose of sensitivity analysis, we identify the following 8 parameters associated with this problem: $p_1 = q_1$, $p_2 = q_2$, $p_3 = c_3$, $p_4 = a_3$, $p_5 = a_4$, $p_6 = K_h$, $p_7 = V$, and $p_8 = K_0$. In solving for (say) 5 sensitivities, we are computing the ODE solution together with the scaled sensitivities with respect to the first 5 parameters; that is, $y(t)$ and $w_1(t), \ldots, w_5(t)$.

In the numerical experiments that follow, we allowed the number of sensitivities to vary from 1 to 8. In computing the scaled sensitivity derivatives, we compared the use of AD against the finite difference strategies described in §26.2. Two centered difference strategies were examined: separate evaluations, based on the sum of (26.5) and (26.6); and a combined evaluation, given by (26.8). A forward difference method was also tested in which $\frac{\partial f}{\partial y} w_i(t)$ and $\overline{p}_i \frac{\partial f}{\partial p_i}$ are each approximated by forward differences. The results are summarized in Figures 26.1 and 26.2.

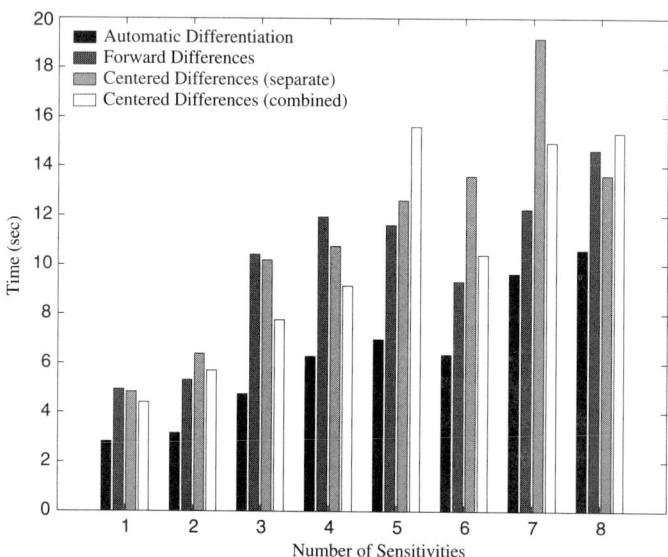

FIGURE 26.1. Performance for various derivative-computation strategies. Results are the average of three runs on 4 processors of an SGI Origin 2000.

Although the present framework for using AD includes some inefficiencies such as the copying of data, Figure 26.1 shows that AD is still markedly faster than each of the three finite difference methods. As shown in Figure 26.2, this advantage can be attributed primarily to the reduced number of time steps. The increased accuracy of the analytic derivatives provided by AD results in larger time steps in the variable-step size, variable-order solver.

26.5   Conclusions and Future Work

SensPVODE provides an efficient and easy-to-use mechanism for computing the sensitivities for simulations that use the PVODE parallel ODE solver. Results for a simple problem indicate that derivatives computed using AD provide performance superior to finite difference approximations. We plan to examine whether this performance advantage holds for more complex problems, and how well this advantage scales with respect to the number of processors used.

Future work also includes developing a mechanism that eliminates the need to copy data from one structure to another, while preserving the

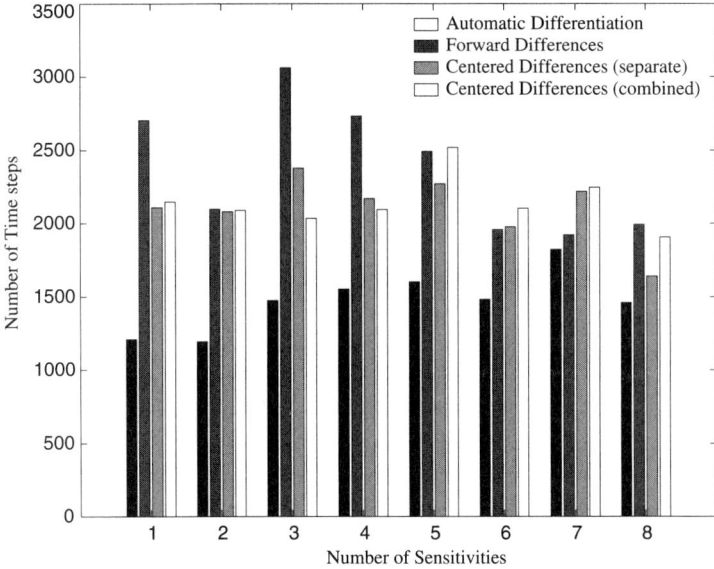

FIGURE 26.2. Number of time steps for derivative-computation strategies.

ease of use of the current implementation. This issue is related to those faced in the use of AD with other numerical toolkits such as PETSc and TAO [1], and we therefore hope to benefit from lessons learned in those projects. In addition, the algorithms used by SensPVODE require the solution of linear systems with multiple right-hand side vectors [351, 370]. A similar situation arises when one differentiates through a linear or non-linear solver [29, 456, 293]. Thus, we expect to leverage other work [75] in the development of block solvers for systems with multiple right-hand sides. All of these developments should increase the efficiency of sensitivity computations using SensPVODE and ADIC.

The SensPVODE package is available for general distribution. Interested users should contact Steven Lee (slee@llnl.gov) and Alan Hindmarsh (alanh@llnl.gov).

Acknowledgments: The work of S. Lee was performed under the auspices of the U.S. Department of Energy by University of California Lawrence Livermore National Laboratory under contract No. W-7405-Eng-48. P. Hovland was supported by the Mathematical, Information, and Computational Sciences Division subprogram of the Office of Advanced Scientific Computing Research, U.S. Department of Energy, under Contract W-31-109-Eng-38.

We thank Gail Pieper for proofreading a draft manuscript and Peter Brown, Alan Hindmarsh, and Linda Petzold for valuable advice regarding sensitivity analysis of ODEs and DAEs.

27

A Parallel Hierarchical Approach for Automatic Differentiation

Marco Mancini

ABSTRACT We evaluate in parallel first-order derivatives given a sequential computer program of the function to be differentiated. Our parallel implementation of an automatic differentiation (AD) algorithm is based on a hierarchical approach. The parallel method is developed by considering as a parallel computational model a shared-memory paradigm. The performance of the derivative codes is evaluated by considering a SGI Origin 2000 and by using the OPENMP standard library. In our computational experiments, we have considered the *Flow in a Driven Cavity* function belonging to the MINPACK-2 test problem collection. The computational results show the performance gain of the parallel approach over both the sequential one and the stripmining technique.

27.1 Introduction

We describe an algorithm for automatic differentiation (AD) that allows us to generate efficiently first-order derivatives by using a hierarchical approach and to perform derivative computation concurrently by taking advantage of parallel technology. In recent years a significant research effort has been devoted to develop parallel automatic differentiation techniques and tools (see [39, 70, 81, 85, 271, 288, 488]). In this chapter, we present an AD algorithm characterized by

- the use of a hierarchical approach to generate derivatives;

- the sparsity detection in derivative computation; and

- the computation of derivatives in parallel by exploiting the associativity of the chain rule.

This work was motivated by the results reported earlier [371], where we found that, in a sequential environment, methods based on a hierarchical approach that exploit the sparsity in derivative computation, are more effective than "standard" approaches such as the pure forward mode.

The rest of the chapter is organized as follows. In §27.2, we present a brief introduction of a hierarchical approach to AD. In §27.3, we describe the implementation details of the method on a shared memory multiprocessor.

In §27.4, we present and discuss the computational results collected on a SGI Origin 2000 with 8 processors.

27.2   The Sequential Hierarchical Approach

The hierarchical approach (HAD, for short) [82] is a powerful methodology to design efficient AD algorithms. The main features of our hierarchical AD algorithm are summarized in what follows.

The program corresponding to the function to be differentiated is partitioned in extended basic blocks (EBB), where each EBB consists of *assignments* and *if statements*. For each block, information about the data flowing in and out is collected at compile time. For each EBB, the HAD propagates derivatives following these two steps:

Step 1: Preaccumulation of local derivatives: We compute the derivatives $\frac{\partial v_o}{\partial v_i}, \forall v_o \in O, \forall v_i \in I$, where I and O are the set of input and output variables of the EBB, respectively. The local derivatives are computed by using the forward-reverse mode at statement level, assuming that the variables belonging to I are independent variables.

Step 2: Accumulation of global derivatives: The global gradients of each v_o are accumulated. We use the forward mode for global propagation of derivatives, that is:

$$\nabla v_o = \sum_{v_i \in I} \frac{\partial v_o}{\partial v_i} \nabla v_i , \quad \forall v_o \in O.$$

Furthermore, the EBB derivative sparsity patterns are exploited to reduce the derivative code complexity by taking into account the flow data dependencies among the variables of each EBB. Due to the presence of *if-statements* in EBBs, conservative assumptions are made during the data dependence analysis. Moreover, we do not perform any array data analysis, since all variables are considered to be atomic elements. See [371] for further details about our hierarchical approach.

27.3   A Parallel Hierarchical Approach

To describe our approach to implement HAD in parallel, we consider the program partitioned at the loop level. We consider simple nested loops, whose body is an EBB. While the HAD is used at the EBB level, parallelism is exploited at the loop level. It is worth emphasizing that the following ideas could be easily generalized to other structures of computer programs.

We can view a loop as a sequence of EBBs, each one corresponding to an iteration of the loop. The idea is to allow the concurrent computation of each EBB derivative block corresponding to different iterations of the

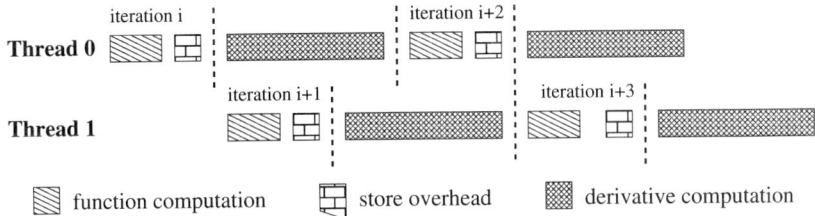

FIGURE 27.1. Parallel computation of derivatives. In this case, granularity is 1, since one loop iteration at a time is assigned to each thread. Precisely, iterations i, $i+2$, ... are assigned to thread 0, while iterations $i+1$, $i+3$, ... are assigned to thread 1.

loop. Each thread performs a certain number of iterations of the loop; after that, it proceeds in computing the corresponding local derivatives of the EBBs and the global derivatives of the output variables of the EBBs. Concurrently, another thread is created to take care of other iterations of the loop (see Figure 27.1).

Two issues must be addressed to fully exploit HAD parallelism:

- Parallelism overhead (thread creation, memory accesses and synchronization).

- Cost to store variables shared by more than one program segment to be differentiated.

The trade-off between parallelism overhead and storage cost can be controlled by varying the computational granularity of the program segments. If we increase granularity, parallelism overhead is decreased, but the cost for storing variables is increased. On the other hand, decreasing granularity has the opposite effect to increase overhead and decrease the storage cost. In particular, we must consider the following issues:

- Active variables used in each iteration of the loop as intermediate variables must be stored to get their correct values during the computation of the derivative block. However, only the variables involved in nonlinear operations must be stored.

- If there are no dependencies between the output and input variables of the EBBs, the update of the global derivatives can be carried out in parallel without any synchronization. Otherwise, we must synchronize the threads for updating the global derivatives of the EBB output variables to get the "correct" derivative values.

- The granularity can be controlled by varying the number of successive iterations of the loop to be assigned to each thread to minimize the overhead introduced by storing variables and thread creation and synchronization.

27.4   Computational Experiments

We have tested the parallel HAD algorithm by considering the *Flow in a Driven Cavity* function belonging to the MINPACK-2 collection [21]. The function is defined on a grid that we set to 512×512. We have carried out computational experiments on a SGI Origin 2000 with 8 processors, by using the OPENMP library. In our computational experiments, we have considered different values of granularity, that is a different number of successive iterations to be assigned to each thread. Moreover the number of directional derivatives considered ranges from 1 to 100.

As we can see from Figure 27.2, if we assign few iterations to each thread, the benefits of the concurrent computation of derivatives is lost. This behavior can be explained by observing that the computational workload assigned to each thread is low, whereas the thread overhead is high with the consequent performance degradation. By increasing the size of granularity (i.e. the number of successive iteration assigned to each thread) parallelism is better exploited. The minimum execution time occurs not at the maximum granularity, probably because the storage cost degrades the performance.

In Figure 27.3 (a), we report the speedup of the parallel hierarchical approach with respect to the sequential version. We obtain a good speedup of almost 5 for eight processors.

Other computational experiments have been carried out by applying the stripmining technique to the derivative code generated by the sequential HAD algorithm. Details on stripmining technique can be found in [81].

The computational results related to the speedup of the stripmining technique are reported in Figure 27.3 (b). Performance is worse than that obtained by our HAD approach because the stripmining technique parallelizes exclusively the derivative code related to the accumulation of global derivatives, whereas the computations of the local derivatives related to each EBB is replicated on each thread.

27.5   Considerations and Further Work

The parallel approach described in §27.3 could be implemented in a source transformation paradigm by using some powerful AD tool infrastructure. A first step towards the development of new AD algorithms is the need of a language-independent intermediate format. AIF [84] is a first attempt at providing an infrastructure in order to make it easier to experiment with new differentiation algorithms. However, for complex AD algorithms, the tool infrastructure should be able to perform typical stages of compilers. We believe that data dependence analysis, data flow analysis and computational models are important for the future of AD in a source transformation paradigm. Only by incorporating sophisticated modules in the infrastruc-

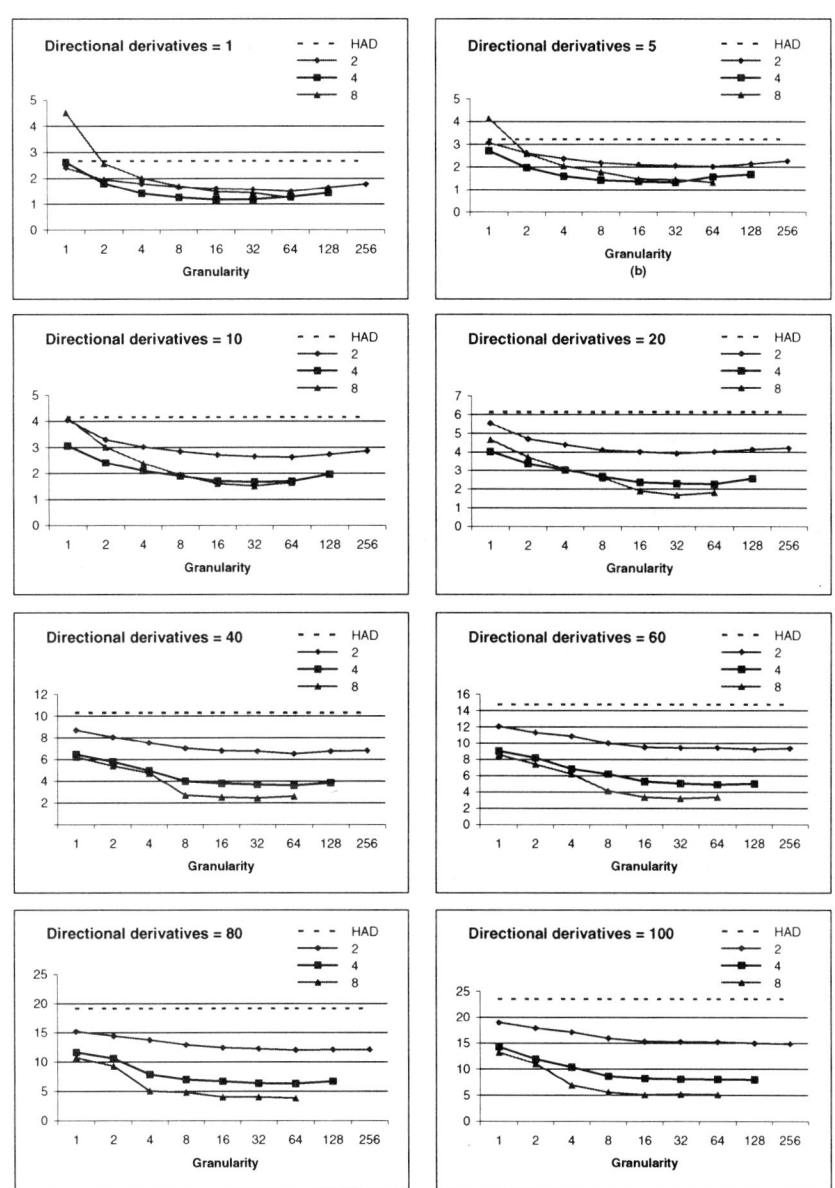

FIGURE 27.2. Jacobian - Flow in a Driven Cavity. Execution time in seconds results of applying the parallel hierarchical approach by varying the granularity and the number of directional derivatives.

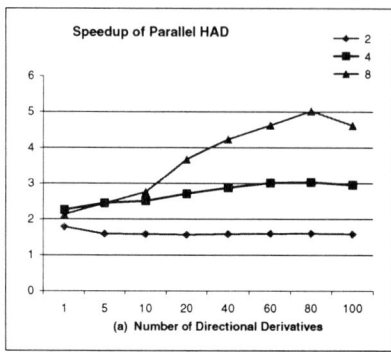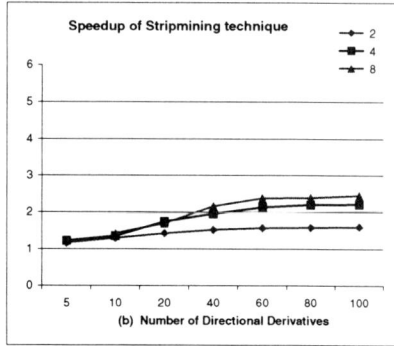

FIGURE 27.3. Jacobian - Flow in a Driven Cavity. Speedup of the parallel hierarchical approach and the stripmining technique.

ture of an AD tool we can hope to develop more complex AD algorithms that go beyond the basic approaches.

Our current work in progress aims to establish a computational model for context-sensitive automatic differentiation strategies in parallel environments. We believe that such a model is very useful to establish the best granularity in order to minimize overhead and to define the best strategy for generating derivatives.

Acknowledgments: I am very grateful to Chris Bischof for helpful comments and suggestions.

28

New Results on Program Reversals

Andrea Walther and Andreas Griewank

ABSTRACT For adjoint calculations, parameter estimation, and similar purposes, one may need to produce all quantities calculated during the execution of a computer program in reverse order. The simplest possible approach is to record a complete execution log and then to read it backwards. This may require massive amounts of storage. Instead one may generate the execution log piecewise by restarting the "forward" calculation repeatedly from suitably placed checkpoints. For such program execution reversals we present parallel reversal schedules that are provably optimal with regards to the number of concurrent processes and the total amount of memory required.

28.1 Introduction

The mathematical specification of many application problems involves nonlinear vector functions $F : \mathbb{R}^n \to \mathbb{R}^m$, $x \mapsto F(x)$, that are evaluated by a computer program. For several purposes one may need to reverse that execution of the evaluation procedure, which we will also denote by F. For example, such reversals are extensively used to calculate adjoints or to adapt parameters in a given model. This is evidenced in several publications on weather data assimilation (e.g. [472]) and the optimization of production processes which use this technique. There, the desired gradients can be obtained with a low temporal complexity by integrating the linear co-state equation backward. This well-known technique is closely related to the reverse mode of algorithmic differentiation (AD) [175]. Similarly, debugging and interactive control may require the reconstruction of previous states by some form of running the program that evaluates F backward. Serious difficulties arise when the simulated process described by F is not invertible or ill conditioned. In these cases one cannot simply apply the inverse function F^{-1}. As a result, the reversal of a program execution has received some, but only perfunctory, attention in the computer science literature (see e.g. [483]).

An obvious way to reverse the evaluation of F is given by first recording the complete *execution log* onto a data structure called *tape* and subsequently reading the tape backward. For each arithmetic operation, the execution log contains a code and the addresses of the arguments. This approach requirements memory proportional to the run time needed to evaluate F. Hence the practical applicability is limited despite the ever increasing size of memory systems.

Another way to reverse the calculation of F employs only a fixed and usually small amount of memory to store suitably placed intermediate states called checkpoints or snapshots during the evaluation process. Then the execution log is generated piecewise by restarting the evaluation repeatedly from the checkpoints. Applying this checkpoint technique, the calculation of F can be reversed even in such cases where the basic approach fails due to excessive memory requirement (see e.g. [235] with respect to AD). This chapter describes our latest research results with respect to the reversal of an evaluation procedure on a multi-processor machine.

28.2   Notation and Assumptions

Throughout this chapter we assume that the evaluation of F comprises the evaluation of subfunctions F_i, $i = 0, \ldots, l - 1$, called physical steps that act on the ith state to calculate the $(i + 1)$th state for $i = 0, \ldots, l - 1$. Hence F describes an *evolutionary system*. The intermediate states of the evolutionary system F represented by the counter i should be thought of as vectors of large dimensions. The physical steps F_i describe mathematical mappings that can in general not be reversed at a reasonable cost. Therefore it is impossible to simply apply the inverses F_i^{-1} in order to run backward from state l to state 0. Also it will be assumed that due to their large size only a certain number of intermediate states can be kept in memory. These saved states form the checkpoints, which can be thought of as pointers to nodes representing the intermediate states i.

Furthermore it is supposed that for each $i \in \{0, \ldots, l - 1\}$ there exist functions $\hat{F}_i$ that in preparation of a subsequent reverse step cause the recording of the intermediate values overwritten during the evaluation of F_i onto the tape. Correspondingly, there must exist functions $\bar{F}_i$ that perform the reversal of the ith physical step using the recorded values. Then all reversal strategies that use checkpoints perform the following *actions* in order to reverse the execution of F:

a: Initialization: Reserve space for c checkpoints and copy the initial state 0 to the first one.

b: Forward sweep: Starting from a state i advance to a state j by performing the physical steps F_h, $h = i, \ldots, j - 1$, without recording the execution log.

c: Recording step: Starting from state i perform recording step(s) to the current final state by writing the execution log onto the tape.

d: Reverse Step: Perform a corresponding number of reverse steps from the currently final state to state i using the tape. Now state i becomes the final state. If the new final state has been saved, free the corresponding checkpoint up.

e: Checkpoint writing: Copy a state into a checkpoint.

f: Checkpoint reading: Read a state from a checkpoint.

It will be supposed that the time needed for writing or reading a checkpoint is negligible in comparison to the execution of one physical step. On a serial machine only one processor is available to perform actions a – f, whereas in the case of a multi-processor machine some actions b, c, e, and f can be carried out on parallel. It is impossible to execute actions d that reverse different physical steps simultaneously because each reverse step $\bar{F}_{i-1}$ for $i = l, l - 1, \ldots, 1$ is based on the results of the previous reverse step $\bar{F}_i$.

For the optimal reversal of the evaluation procedure F one has to take into account four kinds of parameters, namely:

1. the number l of time steps to be reversed,

2. the number p of processors that are available,

3. the number c of checkpoints that can be accommodated,

4. the step costs: $\tau_i = TIME(F_i)$, $\hat{\tau}_i = TIME(\hat{F}_i)$, $\bar{\tau}_i = TIME(\bar{F}_i)$.

In the following sections several aspects of reversal schedules that are optimal with respect to particular criteria are discussed given certain problem parameters 1. – 4. and a constant size of the checkpoints. We always assume that the memory requirement for storing the intermediate states is the same for all i. Otherwise it seems impossible to apply the techniques described below to find optimal reversal schedules. In practical applications nonuniform state sizes might come about, for example, through adaptive grid refinements or through function evaluations that do not conform naturally to our notion of an evolutionary system on a state space of fixed dimension.

28.3   Serial Reversal Schedules

Every optimal reversal schedule performs the functions $\hat{F}_i$ and $\bar{F}_i$ for $i = 0, \ldots, l - 1$ exactly once. Therefore, in the serial case only the step costs τ_i, $i = 0, \ldots, l - 1$, must be taken into account. Time-minimal reversal schedules have been known for uniform step costs $\tau_i = \tau$ for some years (see e.g. [235]).

Recently these existing results have been extended with respect to the step costs. For example, the evaluation of F may involve implicit numerical integrators. Since they have to solve a nonlinear system of algebraic equations at each physical step F_i the complexity measures τ_i may vary widely. For evolutions with nonuniform step costs one has to apply a search algorithm to determine a time-minimal reversal schedule. Using dynamic programming and exploiting a certain monotonicity property one can perform the search in $O(c\,l^2)$ operations [494].

28.4   Parallel Reversal Schedules

On one hand serial reversal schedules allow an enormous reduction of the memory required to reverse a given evolutionary system F. On the other hand one has to pay for this improvement by a growth in the temporal complexity. If no increase of the time needed to reverse F is acceptable at all the use of a sufficient number of additional processors provides the possibility to reverse the evolutionary system F with drastically reduced spatial complexity and minimal temporal complexity. Then the functions $\bar{F}_i$ for $i = l-1,\ldots,0$ are executed successively without any interruption.

It seems natural to have the extra processors perform some of the repeated forward sweeps, i.e. actions b, concurrently. In contrast the reverse steps themselves cannot be parallelized, because they must happen in the prescribed reverse order. Hence the minimal runtime needed by a parallel reversal schedule is given by

$$t = \sum_{i=1}^{l} \tau_i + \sum_{i=1}^{l} \bar{\tau}_i + \max_{j \le l}\left(\hat{\tau}_j - \sum_{i=j}^{l} \tau_i\right) .$$

Here the maximum in the last term will usually be equal to $\hat{\tau}_l - \tau_l$ so that the time is taken up by an advancing sweep to the penultimate state $l-1$, the recording step to the final state l and a subsequent returning sweep from l down to 1. In particular this must be the case if we assume uniform step costs in that for the constants $\hat{\tau}, \bar{\tau} \in \mathbb{R}$ and $\hat{\omega}, \bar{\omega} \in \mathbb{N}$ we have $\tau_i = \tau$, $\hat{\tau}_i = \hat{\tau} = \hat{\omega}\tau$ and $\bar{\tau}_i = \bar{\tau} = \bar{\omega}\tau$ for all i. Without this simplifying assumption it seems quite hard to coordinate the various advances such that they are performed concurrently by a limited number of processors. Moreover, then it is possible to divide each time-minimal parallel reversal schedule into $t = (l-1)\tau + \hat{\tau} + l\bar{\tau}$ computational cycles. Now one may look for the maximal number of time steps $l = l(c,p)$ that can be reversed without any delay using p processors and space for c checkpoints.

After extended but fruitless attempts to derive $l(c,p)$ theoretically, we simply wrote an exhaustive search procedure for implicitly enumerating all possible reversal schedules given c and p with the aim of maximizing l. The most efficient implementation of this brute force approach yielded for the simplest case $\tau = \hat{\tau} = \bar{\tau}$ the maximal values l as shown in Table 28.1 on one node of the Origin 2000 in a few hours of computing time. The numbers

TABLE 28.1. Maximal l_ϱ for different values of c and p

$c \backslash p$	1	2	3	4	5	6	7	8
1	1	3	5	8	13	21	34	55
2	1	4	8	13	21	34	55	89
3	1	5	12	21	34	55	89	144
4	1	6	17	33	55	89	144	233*
5	1	7	23	47	84	144	233*	377*
6	1	8	30	67	?	?	377*	610*

marked with an asterisk were not computed but derived theoretically afterwards. The values of the combinations $c = 6$ and $p = 5$ or $p = 6$ are still
unknown. The first observation was that the values of l "above the diagonal," i.e., with $p > c$, were constant along counterdiagonals, i.e., dependent
only on the "resource number" $\varrho = p + c$. Therefore, as a first step we concentrate our effort on finding time-minimal parallel reversal schedules for
a given value of ϱ. Hence, we assume that a resource can be used in each
computational cycle either as a checkpoint or as processor.

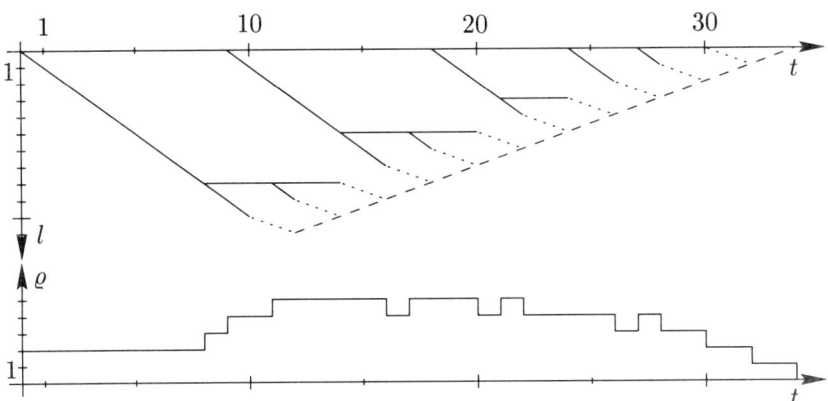

FIGURE 28.1. Parallel Reversal Schedule for $\hat{\tau} = \bar{\tau} = 2$

One corresponding optimal schedule is displayed in Figure 28.1. Horizontal lines represent checkpoints and slanted lines represent running processes. The profile on the lower margin illustrates the number of resources
used in any one of the computational cycles. Not surprisingly the resource
usage reaches a maximum at and near the vertex, i.e., at the end of the
forward and the beginning of the return sweep. During the warm-up and
cool-down phase some checkpoints and processors are idle, which happens
quite often in parallel scheduling.

Because all schedules with the same maximal $l = l_\varrho$ involve the same

total number of computational cycles, point-wise comparison of their resource profiles turns them into a partially ordered set. Modifications that
do not increase resource usage in any one of the computational cycles will
be labeled as "without loss of efficiency". We modified the exhaustive search
procedure to find amongst the schedules reaching the maximal l_ϱ a representative that was minimal with respect to the needed resources in each
computational cycle. This leads to the detection of recursive patterns. They
form the basis for proving the following assertion (see [493]):

Theorem: *Suppose the number of available resources $\varrho = p + c$ and the
temporal complexities $\hat{\tau}$ and $\bar{\tau}$ of the $\hat{F}_i$ and $\bar{F}_i$ are given. Then the maximal
length of an evolution that can be reverted in parallel without interruption
is given by*

$$l_\varrho = \begin{cases} \varrho & \text{if } \varrho < 2 + \hat{\tau}/\bar{\tau} \\ l_{\varrho-1} + \bar{\tau}\, l_{\varrho-2} - \hat{\tau} + 1 & \text{else} \end{cases} . \qquad (28.1)$$

This result was derived in two parts. First the upper bound l_ϱ on the
number of physical steps that can be reversed with a given number ϱ of
processors and checkpoints is established. For that several structural properties of the parallel reversal schedules are used. For example one can derive
checkpoint persistence and processor persistence. This can be interpreted
as saying that the horizontal and slanted lines running to the right in Figure 28.1 may bifurcate from each other but never merge or throw hooks
until they reach the unique slanted line running towards the left.

Then parallel reversal schedules that attain the upper bound l_ϱ are constructed theoretically. This construction process is by no means trivial,
especially since some of the subschedules involved are not optimal reversal schedules themselves. Nevertheless, constructing appropriate parallel
reversal schedules one also finds that the number of processors p_ϱ needed
to reverse l_ϱ physical steps can be bounded above for $\varrho \geq 2 + \hat{\tau}/\bar{\tau}$ by the
equation

$$p_\varrho \equiv \lceil (\varrho + 1)/2 \rceil + \lceil \lfloor (\hat{\tau} - 1)/\bar{\tau} \rfloor /2 \rceil \approx \varrho/2 .$$

At first one might think that demanding at least as many processors,
namely p_ϱ, as checkpoints $c = \varrho - p_\varrho$ is asking a little much. On the other
hand it was found that having more that $p \approx \varrho/2$ processors does not help
at all so that $p \approx c$ is indeed enough. Moreover, one may derive from the
linear difference equation (28.1) that asymptotically

$$l_\varrho \approx \left[1 + \sqrt{(1 + 4\bar{\tau})}/2 \right]^\varrho .$$

Therefore $l = l_\varrho$ grows exponentially as a function of $\varrho \approx 2p$ and conversely
$p \approx c$ grow logarithmically as a function of l. For example, we find that 8
processors and 8 checkpoint are sufficient to reverse an evolution over 10923
time steps when $\hat{\tau} = \bar{\tau} = 2$ and 16384 time steps when $\hat{\tau} = 1, \bar{\tau} = 2$. Hence,
the case $p \leq c$ seems only important for machines with 2 or 4 processors.

There are many other questions regarding parallel reversal schedules that warrant further investigation. In our analysis we have assumed that communication costs are not critical, i.e., do not effect the overall execution time. This assumption is not as unrealistic as it may seem because whenever a checkpoint is set by some advancing process, there is a gap of at least one computational cycle before another process starts from this checkpoint. The setting processor may use a nonblocking send to dispatch the checkpoint to some shared memory location or directly to the second processor. As long as this state transfer takes less time than is needed to execute a single forward step it does not alter the optimal schedule. Also, the whole communication pattern is known beforehand so that it can be organized a priori, possibly by setting up a suitable processor topology. This notion relies again on the assumption that all intermediate states have roughly the same size and that all physical steps take roughly equal time. Failing this all schedules designed for the uniform case are likely to get out of sync. This may also happen if some advancing sweep or communication is delayed due to currents system load. Naturally, a practical implementation would have to contain recovery strategies for these contingencies.

28.5 Conclusions

The potentially enormous memory requirement of program reversal by complete logging can be drastically reduced by keeping at most c intermediate states as checkpoints. Rather than accepting an increase in run-time we considered the use of p processors to reverse evolutionary systems with minimal (wall-clock) time. For our optimal parallel reversal schedules the treatable number of physical steps l grows exponentially as a function of the resource number $\varrho = c + p$. A corresponding software tool implementing these schedules is under development.

Part VII

Exploiting Structure and Sparsity

29

Elimination Techniques for Cheap Jacobians

Uwe Naumann

ABSTRACT The generation of optimized code for evaluating the Jacobian matrix of a vector function is known to result in a remarkable speedup of three and more compared to standard methods of Automatic Differentiation in most cases. So far, this optimization has been built on the elimination of vertices in the computational graph. We show that vertex elimination in general does not lead to optimal Jacobian code. We introduce two new elimination methods and demonstrate their superiority over the vertex elimination approach.

29.1 Accumulation of Jacobians - Why and How

Let $F' = F'(\mathbf{x}_0) = \left(\frac{\partial y_i}{\partial x_j}(\mathbf{x}_0)\right)_{j=1..n}^{i=1..m}$ denote the Jacobian matrix of a nonlinear vector function $F : \mathbb{R}^n \supseteq D \to \mathbb{R}^m : \mathbf{x} \mapsto \mathbf{y} = F(\mathbf{x})$ evaluated at a given argument $\mathbf{x}_0$. The run time of numerous numerical algorithms is dominated by the time it takes to accumulate F' or to evaluate products of the form $(\mathbb{R}^{m \times l_1} \ni) \dot{Y} = F' \dot{X}$ and $(\mathbb{R}^{l_2 \times n} \ni) \bar{X} = \bar{Y} F'$. We present various methods for accumulating F' efficiently, anticipating that these ideas also will be useful for computing higher order derivative tensors. Automatic Differentiation (AD) will be looked at from the point of view of graph theory and combinatorial optimization.

F is assumed to be given as a computer program which decomposes into a sequence of scalar elemental functions $(\mathbb{R} \ni) v_j = \varphi_j(v_i)_{i \prec j}$ where $j = 1..q$ and $p = q - m$. The direct dependence of v_j on v_i is denoted by $i \prec j$. The relation $i \prec^* j$ indicates that there exist $k_1..k_p$ such that $i \prec k_1 \prec \ldots \prec k_p \prec j$. Hence, $\{i | i \prec j\}$ is the index set of the arguments of φ_j, and we denote its cardinality by $|\{i | i \prec j\}|$. Within F we distinguish between three types of variables $V = X \cup Z \cup Y$, independent $(X \equiv \{v_{1-n}..v_0\})$, intermediate $(Z \equiv \{v_1..v_p\})$, and dependent $(Y \equiv \{v_{p+1}..v_q\})$. We set $x_i \equiv v_{i-n}$, $i = 1..n$, and $y_j \equiv v_{p+j}$, $j = 1..m$. The numbering $\mathcal{I} : V \to \{(1-n)..q\}$ of the variables of F is expected to be consistent, i.e. it must induce a topological order with respect to dependence as $i \prec^* j \Rightarrow \mathcal{I}(v_i) < \mathcal{I}(v_j)$.

Since the differentiation of F is based on the differentiability of its elemental functions, it will be assumed that the φ_j, $j = 1..q$, have jointly continuous partial derivatives $c_{ji} \equiv \frac{\partial}{\partial v_i}\varphi_j(v_k)_{k \prec j}$, $i \prec j$, on open neighborhoods $\mathcal{D}_j \subset \mathbb{R}^{n_j}$, $n_j \equiv |\{i | i \prec j\}|$, of their domain. The computational graph (or c-graph) $\mathbf{G} = (V, E)$ of F is a directed acyclic graph with $V = \{i | v_i \in F\}$ and $(i, j) \in E$ if $i \prec j$. We assume $\mathbf{G}$ to be linearized in the sense that all partial derivatives of the elemental functions are attached to their corresponding edges, i.e. (i, j) is labeled with c_{ji}.

The accumulation of F' can be regarded as the process of transforming $\mathbf{G}$ into a subgraph $\mathbf{G}'$ of the complete bi-partite graph $K_{n,m}$ [247]. Different application sequences of the chain rule to $\mathbf{G}$ yield up to round-off identical numerical results, whereas the number of arithmetic operations actually performed may vary drastically as shown in [398].

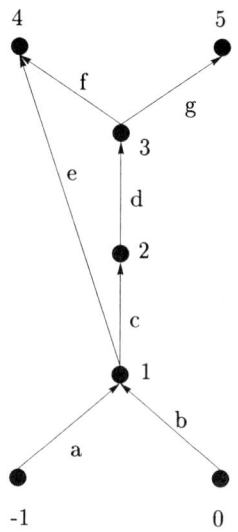

FIGURE 29.1. Computational graph

Prior to the introduction of the elementary elimination techniques on which the transformation $\mathbf{G} \to \mathbf{G}' \subseteq K_{n,m}$ is built, let us have a brief look at an example to illustrate the above. Consider $F : \mathbb{R}^2 \to \mathbb{R}^2$ given by the following system of nonlinear equations:

$$
\begin{aligned}
h_1 &= x_1 \cdot x_2 \\
h_2 &= \exp(\sin(h_1)) \\
y_1 &= h_1 \cdot h_2 \\
y_2 &= \cos(h_2) \ .
\end{aligned}
\tag{29.1}
$$

Its c-graph is shown in Figure 29.1 where edges (or local partial derivatives) have been given names a,b,..... From the chain rule it follows that an entry $F'(i, j)$ of the Jacobian can be computed by multiplying the edge labels

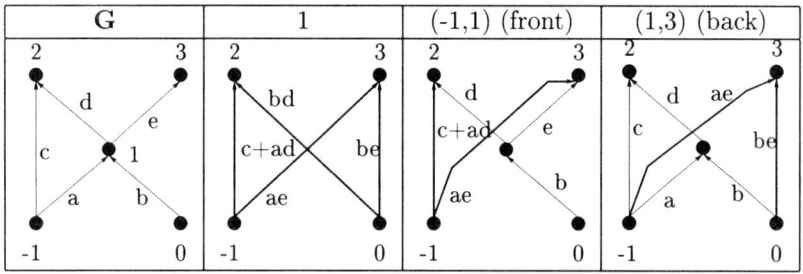

FIGURE 29.2. Elimination techniques in **G**

over all paths connecting the minimal vertex j with the maximal vertex i followed by summing these products [247]. Consequently, the Jacobian of (29.1) is

$$F' = \begin{pmatrix} ae + acdf & acdg \\ be + bcdf & bcdg \end{pmatrix} . \tag{29.2}$$

A naive approach to computing (29.2) requires 14 scalar multiplications and 2 additions. The pre-accumulation of $r = cd$, $s = rf$, and $t = rg$ yields

$$F' = \begin{pmatrix} a(e + s) & at \\ b(e + s) & bt \end{pmatrix} . \tag{29.3}$$

at a cost of only 7 multiplications and 2 additions. This chapter presents a theoretical basis for the constructing (29.3) starting with (29.2). Refer to [398] for a collection of heuristic methods for solving this computationally hard combinatorial optimization problem.

29.2 Elimination Techniques

Initially, we consider three types of eliminations in **G** to transform it into **G'**. We assume the following terminology. By the chain rule *updating the existing or generating new edge labels* means that the labels of successive edges (i, j) and (j, k) are multiplied to form the new label of (i, k), whereas labels of parallel edges having the same source and target are added. Parallel edges will always be merged by performing this addition.

Graphically, the **elimination of a vertex** j from **G** is equivalent to connecting all $i|i \prec j$ with all $k|j \prec k$ followed by updating the existing or generating new edge labels and, finally, the deletion of j. A vertex is deleted together with all edges incident to it.

The direct dependence of j on i can be eliminated in two ways. The **back-elimination** of (i, j) is equivalent to connecting all $k|k \prec i$ with j and updating the existing or generating new edge labels. (i, j) is **front-eliminated** by introducing new edges connecting i with all $l|j \prec l$. Again the existing edge labels have to be updated or new edge labels have to be generated. The edge (i, j) must be deleted in both cases.

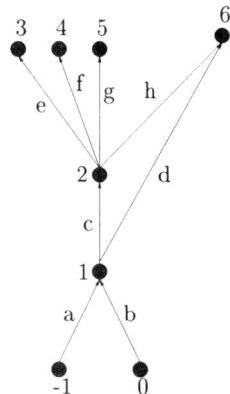

FIGURE 29.3. Lion

Counting on the abilities of modern floating-point units [308], we assume equal execution times of a *multiply-add-fused* $ab+c$, $a, b, c \in \mathbb{R}$, and a scalar multiplication. Furthermore, we expect memory accesses to take constant time, independent on the actual elimination order. Thus, the front- [back-] elimination of (i, j) takes $|\{l | j \prec l\}|$ $[|\{l | j \prec l\}|]$ scalar multiplications. The number of multiplications involved in the elimination of a vertex j is called the *Markowitz degree* of j and is equal to the product $|\{k | k \prec j\}| \cdot |\{l | j \prec l\}|$. All three elimination techniques are illustrated by Figure 29.2, with the local partial derivatives are represented by $a, b, \ldots$.

Our objective is to solve the *Optimal Jacobian Accumulation* (OJA) problem of minimizing the number of multiplications needed for the accumulation of F' using some method M, i.e. $\text{MULTS}_{F'}(M) \to \min$. In the existing literature this problem has been considered equivalent to the problem of finding an optimal vertex elimination sequence [82, 247]. While first numerical results based on the *Markowitz rule* have been presented in [247], several new local heuristics as well as dynamic programming [282] and simulated annealing [485] algorithms have been developed and implemented in [398]. They turned out to speed up the resulting Jacobian codes by factors of three and more. In addition to the development of heuristic methods for its approximate solution the OJA problem has been investigated further from a theoretical point of view leading to the results presented here.

Example 1 *The graph in Figure 29.3 represents an example where an optimal vertex elimination sequence is not the solution of the OJA problem.*

EXPLANATION: Motivated by its shape (turn it 90 degrees clockwise) the graph displayed in Figure 29.3 is called the *lion graph*. It is a c-graph **G** of a vector function $F : \mathbb{R}^2 \to \mathbb{R}^4$ given by

$$\begin{aligned}
v_1 &= x_1 \cdot x_2 \\
v_2 &= \varphi_2(v_1) \\
y_j &= \varphi_{j+2}(v_2) \quad \text{for } j = 1, 2, 3 \\
y_4 &= \varphi_6(v_2, v_1) .
\end{aligned} \tag{29.4}$$

G contains two intermediate variables. The two vertex elimination sequences require both 12 multiplications. Try to *back-eliminate* $(2,6)$ before 1 and 2 to verify that $\mathrm{MULTS}_{F'}((2,6), 1, 2) = 11$. Consider

$$F' = \begin{pmatrix} ace & bce \\ acf & bcf \\ acg & bcg \\ ach + ad & bch + bd \end{pmatrix},$$

which is one possible representation the Jacobian matrix of (29.4). Undoubtedly, $r = ac$, $s = bc$, and $t = ch + d$ should be pre-accumulated to get

$$F' = \begin{pmatrix} re & se \\ rf & sf \\ rg & sg \\ at & bt \end{pmatrix}$$

at the minimal cost of 11 multiplications. The fact that both ac and ch have to be computed explicitly in order to assure optimality implies that 1 should be eliminated before 2, and that 2 should be eliminated before 1. Obviously this is a contradiction. ∎

Example 1 gives rise to the fundamental question on how large the *vertex discrepancy* can possibly become. In [398] we showed that it does not exceed a small constant factor for c-graphs containing two intermediate vertices. The problem remains unsolved for $p > 2$.

Example 2 *The graph in Figure 29.4 represents an example where an optimal edge elimination sequence is not the solution of the OJA problem.*

EXPLANATION: Motivated by its shape we will refer to the graph displayed in Figure 29.4 as the *bat graph* (turn it upside down). The Jacobian of the corresponding vector function is given by

$$F' = \begin{pmatrix} ae + aif & be + bif & cgi & dgi \\ afj & bfj & cgj & dgj \\ afk & bfk & cgk & dgk \\ afl & bfl & ch + cgl & dh + dgl \end{pmatrix}. \tag{29.5}$$

Pre-accumulate

$$r = if + e; s = cg; t = dg; u = af; v = bf; w = gl + h$$

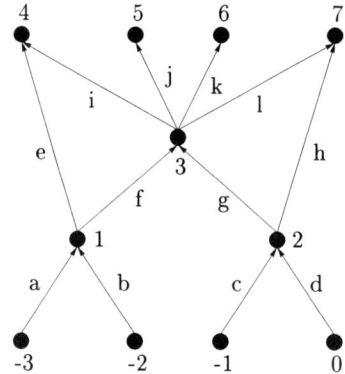

FIGURE 29.4. Bat

to get

$$F' = \begin{pmatrix} ar & br & si & ti \\ uj & vj & sj & tj \\ uk & vk & sk & tk \\ ul & vl & cw & dw \end{pmatrix} \tag{29.6}$$

at a cost of 22 multiplications. To see that this will never be achieved by an edge elimination sequence we should notice that

1. $F'(1,1) = a(e + if)$ and $F'(1,2) = b(e + if) \Rightarrow (3,4)$ should be back-eliminated before $(1,3)$ is back-eliminated;

2. af and bf are computed while fl is not $\Rightarrow (1,3)$ should be back-eliminated before $(3,7)$;

3. $F'(4,3) = c(h + gl)$ and $F'(4,4) = d(h + gl) \Rightarrow (3,7)$ should be back-eliminated before $(2,3)$ is back-eliminated;

4. cg and dg are computed while gi is not $\Rightarrow (2,3)$ should be back-eliminated before $(3,4)$.

Consequently, we get

$$(3,4) < (1,3) < (3,7) < (2,3) < (3,4) \tag{29.7}$$

where $(i,j) < (k,l)$ denotes the situation where (i,j) is eliminated before (k,l). Obviously, (29.7) yields a contradiction. ∎

29.3   Conclusion and Outlook

We may conclude that neither vertex nor edge elimination in c-graphs can be used to solve the OJA problem in general. One should rather look at the elimination of *faces* in **G**, which is equivalent to eliminating edges in the *line-graph* [44] associated with **G**. This is what we actually did in the explanation of Example 2. The implications of these results are being investigated in current research.

Apart from being an interesting "playground" for graph theory, combinatorial optimization, and complexity analysis, the above elimination techniques have a very relevant practical impact on AD and its application to real-world problems. A remarkable speedup of the resulting derivative code can be guaranteed in most cases (see [398]). The effort put into the optimization will well pay off provided that the code is meant to be run a large number of times.

As one of the next steps, we plan to develop simulated annealing algorithms based on homogeneous Markov chains [485] with several new heuristics for neighborhood definitions. The 500 node N-cube 2 available at the University of Hertfordshire, UK, will be used to implement a parallel version. It is planned to use several solutions of subproblems to generate *crossings* representing better solutions of the combined problem. However, there are many questions remaining open regarding this interesting problem.

The resulting elimination sequences will be used to generate optimized Jacobian code automatically. There are many so far unsolved problems associated with that. As a generalization of the trajectory problem in the reverse mode of AD [185], these questions are dealt with in ongoing research.

30

AD Tools and Prospects for Optimal AD in CFD Flux Jacobian Calculations

Mohamed Tadjouddine, Shaun A. Forth and John D. Pryce

ABSTRACT We consider the problem of linearising the short (approximately 100 lines of) code that defines the numerical fluxes of mass, energy and momentum across a cell face in a finite volume compressible flow calculation. Typical of such formulations is the numerical flux due to Roe, widely used in the numerical approximation of flow fields containing moderate to strong shocks. Roe's flux takes as input 10 variables describing the flow either side of a cell face and returns as output the 5 variables for the numerical flux. We present results concerning the efficiency of derivative calculations for Roe's flux using several currently available AD tools. We also present preliminary work on deriving near optimal differentiated code using the node elimination approach. We show that such techniques, within a source transformation approach, will yield substantial gains for application code such as the Roe flux.

30.1 Differentiation of Roe's Numerical Flux

In previous work [198] we have reported on the application of our own operator overloaded, forward AD library to calculate the linearisation of Roe's numerical flux [440]. We validated the linearisation by showing that a finite difference evaluation converged to the AD linearisation as the finite difference step size was reduced. Here we linearise the flux using currently available AD software (ADIFOR 2.0 [78], TAMC [211] and ADO1 [427]). Results are presented in Table 30.1 where time(∇F) denotes the CPU time for a Jacobian (and function) evaluation, time(∇F)/ time(F) represents the ratio of the Jacobian evaluation to the function evaluation in CPU time, and the final column gives the maximum absolute difference in the Jacobian (assuming that the ADIFOR results are correct). Note that the ADIFOR results are for all arrays and loops of fixed length 10 and with the *Performance* exception handling option.

We see that all the AD methods produced the same results to within machine relative precision whereas, as expected, the finite difference results are in error by around the square root of machine precision. The source

TABLE 30.1. CPU timings (in seconds) on SGI IRIX64 IP27

Method	time (∇F)	$\dfrac{\text{time } (\nabla F)}{\text{time } (F)}$	Deviation vs ADIFOR
Finite difference, 1-sided	0.12	10.64	4.34E-07
ADIFOR	0.15	13.37	—
TAMC (Forward)	0.13	11.94	4.66E-15
TAMC (Reverse)	0.11	10.28	5.77E-15
AD01 (Forward)	1.55	134.68	7.99E-15
AD01 (Reverse)	0.95	82.90	4.88E-15

transformation methods (ADIFOR and TAMC) are substantially faster than the library AD01 using operator overloading. The forward methods ADIFOR and TAMC (forward) are comparable in execution time and only slightly slower than the finite differences while giving exact results. The reverse modes of TAMC and AD01 have performed better than the forward ones with this SGI machine. However TAMC reverse performed worse with an older COMPAQ Alpha AXP 250-330 workstation. We attribute this to cache misses during the reverse pass on this smaller cache machine.

In the rest of this chapter, we consider AD techniques applied to a restricted class of Fortran codes important in CFD applications: subroutines with typically 5 to 30 inputs x_i and outputs y_i and some hundreds v_i of intermediate values, with no loops but allowing branches. The aim is to efficiently compute the Jacobians associated with such subroutines.

30.2   Elimination Techniques

An alternative to the conventional forward and reverse modes of AD is the elimination approach [247, 238, 398].

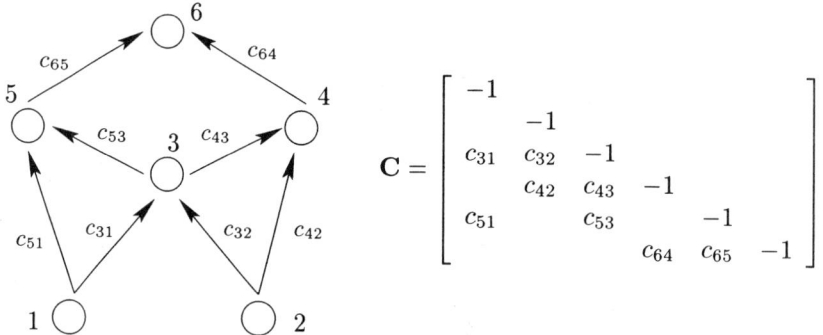

FIGURE 30.1. An example of CG (left) and its matrix representation (right)

To illustrate this technique, consider the graph sketched in Figure 30.1, which shows a Computational Graph (CG) for derivative calculation and its (sparse) matrix representation. The nodes 1 and 2 represent the inputs x_1 and x_2; nodes 3, 4, and 5 the intermediates; and node 6 the output y. The matrix represents the equations $\mathbf{C}\dot{v} = -(\dot{x}_1, \dot{x}_2, 0, \dots, 0)^T$, that is:

$$\dot{v}_1 = \dot{x}_1 \qquad\qquad \dot{v}_4 = c_{42} * \dot{v}_2 + c_{43} * \dot{v}_3$$
$$\dot{v}_2 = \dot{x}_2 \qquad\qquad \dot{v}_5 = c_{51} * \dot{v}_1 + c_{53} * \dot{v}_3$$
$$\dot{v}_3 = c_{31} * \dot{v}_1 + c_{32} * \dot{v}_2 \qquad \dot{v}_6 = y = c_{64} * \dot{v}_4 + c_{65} * \dot{v}_5$$

The coefficients $c_{i,j}, 1 \leq i, j \leq 6$ represent partial derivatives $\frac{\partial v_i}{\partial v_j}$. The Jacobian $\frac{\partial y}{\partial(x_1, x_2)}$ is determined by eliminating intermediate nodes or edges from the graph until it becomes bipartite. In terms of the matrix representation, node elimination is equivalent to successively choosing a diagonal pivot element from rows 3 to 5, eliminating all the coefficients under that pivot and leaving the Jacobian as elements c_{61} and c_{62}. Adopting the notations $\prec$ and $\prec^*$ of [238], we define the Markowitz and VLR costs at an intermediate node v_j respectively as follows:

$$\text{mark}(v_j) = |\{i : i \prec j\}||\{k : j \prec k\}|$$
$$\text{VLR}(v_j) = \text{mark}(v_j) - |\{i : i \prec^* j\}||\{k : j \prec^* k\}|. \qquad (30.1)$$

Ordering the elimination process using heuristics based on choosing the node with minimum Markowitz or VLR cost generally gives a further improvement [398].

30.2.1 Application of node elimination methods to Roe flux

We have written and used a Fortran 90 AD module using operator overloading to build up the CG of the Roe flux code. The graph consists of 221 nodes and 342 edges. Then, we applied the standard node elimination methods (Forward, Reverse, Markowitz, VLR). We also employed what we term *reverse bias* variants of Markowitz and VLR which use the last node with the minimum Markowitz/VLR cost as opposed to the first in the conventional implementations.

Figure 30.2 shows the elimination sequences taken by these 6 methods and Table 30.2 the required number of multiplications. Classical forward AD requires 3420 multiplications and reverse 1710. Hence the elimination techniques use up to some 40% fewer multiplications than reverse AD and 70% fewer than forward AD for this problem. We observe that Markowitz

TABLE 30.2. Number of multiplication of the different strategies

Forward	Reverse	Mark.	VLR	Mark.(rev. bias)	VLR (rev.bias)
1959	1335	1031	1075	998	1073

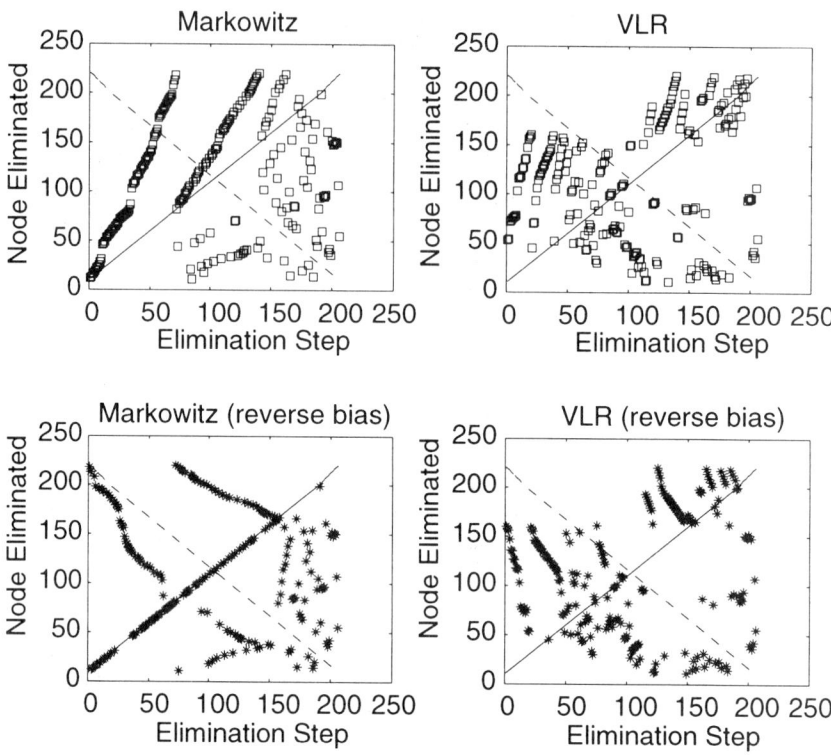

FIGURE 30.2. Elimination sequence for different strategies applied to the Roe Flux test case (forward/reverse given by solid/dashed line in each graph)

has taken fewer operations than the other methods and incorporating the reverse bias has improved Markowitz more than it did VLR. Furthermore the behaviour of the Markowitz (reverse bias) looks like a hybrid of forward and reverse eliminations.

Interestingly we see that VLR and VLR (reverse bias) perform no elimination at the top and bottom of the CG for approximately the first third of their elimination sequences. We explain this as follows. The VLR cost (see equation 30.1) involves a fixed value for each node which is given by the product of the number of input/output nodes it is eventually connected to. If we assume that a node v_j in the centre will be connected to all inputs and outputs, then in our case (10 inputs and 5 outputs) it will have a constant cost of $\mathrm{mark}(v_j) - 5 \times 10$. Nodes close to the outputs will only be connected to a few (e.g., 2) outputs but to all the 10 inputs and have a higher (e.g., $\mathrm{mark}(v_j) - 2 \times 10$) constant cost. Similarly nodes close to inputs will have a higher constant cost again (e.g., $\mathrm{mark}(v_j) - 2 \times 5$). Thus when selecting the node with minimum cost for elimination, VLR is heavily biased for many steps towards the nodes in the centre of the CG.

30.3   Development of a Source Transformation Tool

We are developing an AD tool to efficiently compute Jacobians of a restricted class of functions via source transformation. The input code is parsed to get the Abstract Syntax Tree (AST) using the freeware ANTLR translator [413]. Two passes through the AST allow us to build an Abstract Computational Graph (ACG) of the program from which we compute the Jacobian via an elimination approach.

30.3.1   Building the abstract computational graph

```
t = x1**2+1

if (t>x2) then

  y1 = abs(x1)
else
  y1 = (t+x2)/x2
endif

y = ln(y1)
```

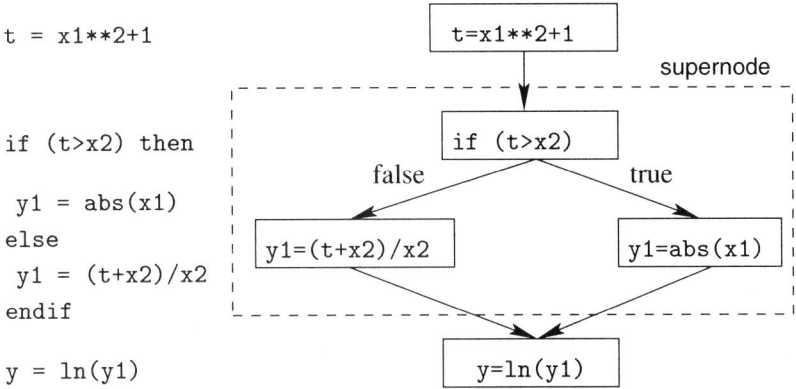

FIGURE 30.3. Fortran code (left) and Control Flow Graph (right)

Unlike the CG that represents one execution of the program, the ACG takes into account all execution paths. For the class of codes above with single assignment, the derived ACG is a DAG describing the chain of operations from the data inputs to the outputs. The ACG is viewed as a flowchart i.e. a control flow graph in which each basic block is expanded to a computational graph (see Figures 30.3, and 30.4). It is built by using the AST to rewrite the input code as a code list [238]. Then we analyse the code to top-down propagate the active variables and compute the local partial derivatives.

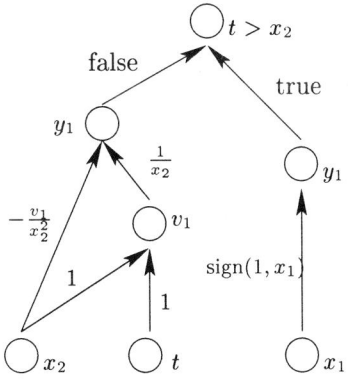

FIGURE 30.4. The subgraph related to the IF block (supernode)

30.3.2   Dealing with IF blocks

A novel feature of our work is that we intend to deal with IF blocks. They are viewed as supernodes, which we term *branching nodes* whose inputs and outputs are determined using a read/write analysis. The inputs are the union of variables imported (read and not written) into the IF block. The outputs are the union of variables written in the IF block and exported to (read from) another region of the program. This may be seen as an extension of Bischof and Haghighat's hierarchical AD [82] from the subroutine level to the basic block level. The branching node represents a DAG whose reduction gives rise to a bipartite graph that connects all possible inputs to all possible outputs according to the value of the test controlling the IF block (see Figures 30.4 and 30.5). From Figure 30.4, we eliminate nodes at the lowest level, so that we get an augmented bipartite graph as in Figure 30.5 (right) with one of 2 sets of local derivatives used depending on the branch taken.

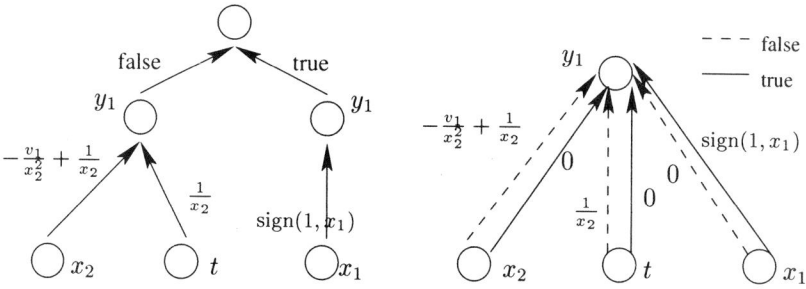

FIGURE 30.5. Elimination process to build up the bipartite graph from a supernode

30.3.3   Jacobian accumulation

Generation of the source text that computes the Jacobian proceeds as follows:

1. Apply the node elimination process from the ACG to build up the bipartite graph. The innermost branching nodes will be treated first, providing local Jacobians whose entries will then be used to recursively complete the overall elimination.

2. Generate derivative code consisting of the input code interspersed with derivative calculations obtained from the node elimination process positioned so as to attempt to maximise cache performance.

30.4   Conclusions

We have applied several different AD software systems ADIFOR, ADO1, and TAMC, to Roe's numerical flux. For such dense code with more inputs than outputs, reverse mode has been shown to be more efficient than forward provided the machine cache is sufficiently large. Source transformation is shown to be consistently superior to operator overloading. For a restricted class of codes, we have introduced the Abstract Computational Graph to implement the node elimination methods via source transformation. Preliminary results using operator overloading indicate that using elimination strategies such as Markowitz or VLR will give a substantial improvement over the conventional forward and reverse modes.

31

Reducing the Number of AD Passes for Computing a Sparse Jacobian Matrix

Shahadat Hossain and Trond Steihaug

ABSTRACT A reduction in the computational work is possible if we do not require that the nonzeros of a Jacobian matrix be determined directly. If a column or row partition is available, the proposed substitution technique can be used to reduce the number of groups in the partition further. In this chapter, we present a substitution method to determine the structure of sparse Jacobian matrices efficiently using forward, reverse, or a combination of forward and reverse modes of AD. Specifically, if it is true that the difference between the maximum number of nonzeros in a column or row and the number of groups in the corresponding partition is large, then the proposed method can save many AD passes. This assertion is supported by numerical examples.

31.1 Introduction

To determine a sparse Jacobian matrix, one can compute a partition of the columns and then use finite differences (FD) or automatic differentiation (AD) to obtain the nonzeros. In a direct method [23, 138, 152, 286], the nonzero entries corresponding to a group of columns can be read off the difference approximations or a forward pass directly, i.e., without any other arithmetic operations. In an indirect method such as a substitution method the unknowns are ordered such that the ordering leads to a triangular system [132, 138, 285]. The additional cost of substitution is often small compared with the savings in the number of AD passes. Substitution methods have been investigated mainly in the context of symmetric matrices [132, 227, 423]. In this chapter we extend the results in [285] and present numerical experiments. The techniques presented here assume that a column (row) partition or a complete direct cover [286] or bipartition [140] is available. Using the sparsity information we define systems of linear equations that can be solved by substitution. If the number of groups in the partition is larger than the maximum number of nonzeros in any row of the Jacobian matrix, then the proposed techniques reduce the number of AD passes by at least one.

In this chapter, §31.2 begins with a brief discussion of the CPR method [152]. We show how to reduce the number of AD passes by defining a new grouping of columns from a given column partition. Our main result is presented in this section. We illustrate the usefulness of the proposed techniques by a motivating example. In §31.3, we show that under certain sparsity restrictions, the number of AD passes can be reduced further by grouping together columns from more than two structurally orthogonal column groups into one. §31.4 presents the numerical experiments. In the final section, we review the methods presented in the chapter and discuss future research directions.

In this chapter if a capital letter is used to denote a matrix (e.g., A, B, X), then the corresponding lower case letter with subscript ij refers to the (i,j) entry (e.g., a_{ij}, b_{ij}, x_{ij}). The elements in an arbitrary row will be denoted by Greek letters.

31.2   Partitioning and Substitution Ordering

Assume that we have a priori knowledge of the sparsity structure of the Jacobian matrix $A \in R^{m \times n}$. We want to obtain a *seed matrix* $X \in R^{n \times p}$ such that nonzeros of A can be recovered by substitution with p as small as possible. Solve for nonzero a_{ij} in

$$X^T A^T = B^T, \tag{31.1}$$

where the columns of B are matrix-vector products of A computed from one AD pass or a difference in function evaluations $B \leftarrow AX$. A group of columns are said to be *structurally orthogonal* if they do not contain nonzeros in the same row position. In the CPR approach columns of A are partitioned into structurally orthogonal groups. The resulting partition is called a *consistent partition*. Corresponding to each group of structurally orthogonal columns C_k, $k = 1, 2, \ldots, p$, where C_k refers to a set of column indices, we define $x_{jk} = 1$ if column j of A is in C_k and $x_{jk} = 0$ otherwise. Then, (31.1) defines m diagonal linear systems in the nonzeros of each row of A.

If we do not require that the nonzeros be determined directly, it may be possible to reduce the computational work further. To illustrate, we consider the following matrix

$$A = \begin{bmatrix} a_{11} & 0 & a_{13} \\ a_{21} & a_{22} & 0 \\ 0 & a_{32} & a_{33} \end{bmatrix}, \text{ and choose } X = \begin{bmatrix} 1 & 0 \\ 1 & 1 \\ 0 & 1 \end{bmatrix}.$$

For example, the second row of A can be determined by solving for a_{21} and a_{22}

$$\begin{bmatrix} a_{21} & a_{22} & 0 \end{bmatrix} \begin{bmatrix} 1 & 0 \\ 1 & 1 \\ 0 & 1 \end{bmatrix} = \begin{bmatrix} b_{21} & b_{22} \end{bmatrix}.$$

Eliminating row 3 of X and transposing the system, we get

$$\begin{bmatrix} 1 & 1 \\ 0 & 1 \end{bmatrix} \begin{bmatrix} a_{21} \\ a_{22} \end{bmatrix} = \begin{bmatrix} b_{21} \\ b_{22} \end{bmatrix},$$

an upper triangular system. The nonzeros in two other rows of A can be solved for in an analogous way, requiring only two AD passes. On the other hand, the associated column intersection graph $G(A)$ is a complete graph, and hence the columns have only a trivial partition of 3 structurally orthogonal groups requiring 3 AD passes.

Assume that $A \in R^{m \times n}$ with at most ρ nonzeros per row. Let the columns be partitioned into $p \geq \rho + 1$ groups C_i, $i = 1, 2, \ldots, p$. Consider a new grouping of the columns,

$$\mathcal{C}_i = \{C_i, C_{i+1}\}, \quad i = 1, 2, \ldots, p - 1. \tag{31.2}$$

In the above, we have adopted the notation $\{C_i, \ldots, C_k\} = C_i \cup \cdots \cup C_k$. We claim that (31.2) defines a substitution ordering of the unknowns. Without loss of generality, let $p = \rho + 1$, and let $\alpha_1, \alpha_2, \ldots, \alpha_p$ denote the elements in row i of A that we want to determine. Then

$$\begin{bmatrix} 1 & 1 & 0 & \cdots & 0 & 0 \\ 0 & 1 & 1 & \cdots & 0 & 0 \\ 0 & 0 & 1 & \cdots & 0 & 0 \\ \vdots & \vdots & \vdots & \vdots & \vdots & 0 \\ 0 & 0 & 0 & \cdots & 1 & 0 \\ 0 & 0 & 0 & \cdots & 1 & 1 \end{bmatrix} \begin{bmatrix} \alpha_1 \\ \alpha_2 \\ \alpha_3 \\ \vdots \\ \alpha_{p-1} \\ \alpha_p \end{bmatrix} = \begin{bmatrix} \beta_1 \\ \beta_2 \\ \beta_3 \\ \vdots \\ \beta_{p-1} \end{bmatrix}, \tag{31.3}$$

where $\beta_1, \ldots, \beta_{p-1}$ is row i in B (31.1). Since there are at most $\rho < p$ nonzeros in any row of A, one of the variables α_k, $k \in \{1, 2, \ldots, p\}$ in (31.3) is zero, so that at least one of α_{k-1} and α_{k+1} can be determined directly, and the remaining nonzeros can be determined by substitution. The above discussion establishes the following result.

Lemma 1 *If $p > \rho$ then $\{C_1, C_2\}, \{C_2, C_3\}, \cdots, \{C_{p-1}, C_p\}$ defines a substitution ordering of the nonzero elements of A.*

An idea along the lines of Lemma 1 has been considered in [421] as "multicoloring". Lemma 1 shows that the substitution ordering saves one AD pass provided that $p > \rho$. This result also indicates that the problems where the Jacobian matrix has a relatively few nonzeros per row but the chromatic number of its intersection graph is large [470] are good candidates for a substitution scheme.

If each row of the compressed matrix has at least two consecutive zero elements, then the matrix formed by $\{C_1, C_2\}, \{C_2, C_3\}, \cdots, \{C_{p-1}, C_p\}$ will have at least one zero element in every row, and the process could be repeated. However, as will be shown in the next section, merging of the successive index sets is done only once.

We note that any arbitrary combination of $\rho + 1$ column groups saves one AD pass. If $p \bmod (\rho + 1) \neq 0$, then it is likely that some ρ or fewer column groups has to be determined directly. Thus, the total number of AD passes in the substitution method is at most $p - \lfloor \frac{p}{(\rho+1)} \rfloor$.

Let us assume that a consistent partition of the columns of A contains

$$p = k(\rho + 1)$$

groups, where $k \geq 1$. Then we save k forward passes.

We illustrate the above substitution method on an example. Consider a 15×6 matrix A with the following sparsity pattern [470].

$$
\begin{bmatrix}
\times & \times &   &   &   &   \\
       &   & \times & \times & \times & \times \\
\times &   & \times &   &   &   \\
\times &   &   & \times &   &   \\
       & \times & \times &   &   &   \\
       & \times &   & \times &   &   \\
\times &   &   &   & \times &   \\
\times &   &   &   &   & \times \\
       & \times &   &   & \times &   \\
       & \times &   &   &   & \times \\
       &   & \times &   & \times &   \\
       &   & \times &   &   & \times \\
       &   &   & \times & \times &   \\
       &   &   &   &   & \times
\end{bmatrix}
\tag{31.4}
$$

The column intersection graph $G(A)$ is a graph on 6 vertices, and the chromatic number is 6, hence any consistent column partition must have at least 6 column groups. Since $\rho = 2$ for this example we can define the two parts

$$\{C_1, C_2\}, \ \{C_2, C_3\} \text{ and } \{C_4, C_5\}, \ \{C_5, C_6\}, \tag{31.5}$$

where C_j contains column j of A for $j = 1, 2, \ldots, 6$. Using our substitution method, we need only 4 passes (or 4 extra function evaluations).

31.3   Exploiting Row (Column) Sparsity

In the discussion of §31.2, we tacitly assume that there are many rows in the Jacobian matrix with ρ nonzeros and that in the new grouping of columns (31.2) each time a structurally orthogonal group is added to a new group C_i, the number of nonzero elements in a row in C_i is also increased by one. Including more than two structurally orthogonal groups in each of the new groups will reduce the number of new groups and therefore the number of AD passes. Whether the nonzeros in a row of the Jacobian matrix can be recovered by substitution depends on the zero-nonzero structure of the corresponding row of the compressed Jacobian matrix. The compressed Jacobian matrix has p columns, where column j consists of the nonzero elements of the columns in C_j of the Jacobian matrix. Let $p = 8$, $\rho = 5$, and define

$$\mathcal{C}_i = \{C_i, C_{i+1}, C_{i+2}\} \quad i = 1, 2, \ldots, p - 2.$$

Further, assume that $\alpha_3 = \alpha_4 = 0$. Then

$$
\begin{bmatrix}
1 & 1 & 0 & 0 & 0 & 0 \\
0 & 1 & 0 & 0 & 0 & 0 \\
0 & 0 & 1 & 0 & 0 & 0 \\
0 & 0 & 1 & 1 & 0 & 0 \\
0 & 0 & 1 & 1 & 1 & 0 \\
0 & 0 & 0 & 1 & 1 & 1
\end{bmatrix}
\begin{bmatrix}
\alpha_1 \\
\alpha_2 \\
\alpha_5 \\
\alpha_6 \\
\alpha_7 \\
\alpha_8
\end{bmatrix}
=
\begin{bmatrix}
\beta_1 \\
\beta_2 \\
\beta_3 \\
\beta_4 \\
\beta_5 \\
\beta_6
\end{bmatrix},
$$

from which the unknowns can be solved for using forward and back substitution. The observation here is that if at least two consecutive elements are known to be zero, then the above system would split and lead to triangular systems.

Consider the matrix (31.4). Each row has at least 2 consecutive zero elements hence the matrix can be computed with 4 passes. In general, if each row of the compressed Jacobian matrix (defined by the consistent partition of columns) has at least d consecutive elements that are known to be zero and $d \leq p - \rho$, then we can define a substitution scheme with $p - d$ passes.

Lemma 2 *If $p > \rho$ and each row of the compressed matrix has at least d consecutive zero entries, $d \leq p - \rho$, then the $p - d$ groups $\{C_1, \ldots, C_{d+1}\}$, $\{C_2, \ldots, C_{d+2}\}, \cdots, \{C_{p-d}, \ldots, C_p\}$ defines a substitution ordering of the nonzero elements.*

All our discussions concerning the new techniques are equally applicable to consistent row partitions and the reverse mode of AD. That is, given a consistent row partition, we define a new grouping of rows that leads to a substitution ordering of the nonzeros, where the nonzeros are obtained from the reverse AD passes.

The above techniques can be extended to the complete direct cover [286] method or bipartition [140]. Given a complete direct cover of the rows and columns of a Jacobian matrix, we can determine the number of consecutive zeros in the compressed Jacobian matrices corresponding to row and column groups and choose the one that gives the largest reduction in the number of AD passes. If the nonzeros in column groups are determined by substitution then the nonzeros in row groups are determined directly and vice versa.

31.4   Computational Results

In this section, we describe the computational experiments for the substitution schemes of §§31.3 and 31.4. Our test problems are drawn from the Harwell-Boeing [167] test matrix collection. For each of the test problems, we compute a row partition, a column partition, and a complete direct cover. We use DSM [133] to compute one directional partitioning and the algorithm in [286] to compute the direct cover. From the sparsity and partition information of the Jacobian matrix, we construct the compressed matrix pattern. For each row of the pattern matrix we calculate the maximum number of consecutive zeros. The minimum of the maximum number of consecutive zeros over all rows (d) is the number of AD passes saved from the substitution ordering. The results on test problems are summarized in Tables 31.1 and 31.2. The following results are presented:

- ρ: Maximum number of nonzeros in any row of A
- mngp: Lower bound on the number of groups in any consistent partition, computed in DSM
- p: Number of groups in the DSM computed consistent partition
- d: Number of consecutive zeros in the compressed matrix obtained from the corresponding consistent partition

TABLE 31.1. Results for SMTAPE collection

| | Direct(DSM) | | | | | | Substitution | |
| | A | | | A^T | | | A | A^T |
Name	ρ	mngp	p	ρ	mngp	p	d	d
abb313	6	10	10*	26	26	26	1	0
ash219	2	3	4*	9	9	9	1	0
ash331	2	6	6*	12	12	12	2	0
ash608	2	5	6*	12	12	12	2	0
ash958	2	6	6*	13	13	13	2	0
will199	6	7	7*	9	9	9	1	0
str	34	34	34	34	34	35*	0	1
str200	34	34	34	26	26	31*	0	2
str400	33	33	33	34	36	36*	0	1
str600	33	33	33	34	36	36*	0	1
fs541-1	11	11	13*	541	541	541	1	0
fs541-2	11	11	13*	541	541	541	1	0
lundA	12	12	12	12	12	13*	0	1
lundB	12	12	12	12	12	13*	0	1
Total	200	217	223	1315	1319	1327	11	7

In Table 31.1 we compute consistent partitions for both A and A^T. If we consider the problems where $p > \rho$ (marked with asterisk *) and consider the column computation, then DSM requires 65 matrix-vector products (forward pass of AD), and the substitution technique based on the partition from DSM will reduce the number of matrix-vector products to 54, a reduction of 17%. Even when DSM produces optimal partitioning, e.g., ash331, we can reduce the number of AD passes (to 4 from 6).

Our second test suite consists of problems from the CHEMIMP (chemical engineering plant models) collection. We compute a consistent column partition and a complete direct cover for the test problems and report the number of AD passes saved in the corresponding substitution scheme.

TABLE 31.2. Results for CHEMIMP collection

		Direct(DSM)	Substitution	Complete Direct Cover	
Name	ρ	p	d	cgrp(d)	rgrp(d)
impcola	8	8	0	0(0)	6(1)
impcolb	7	11*	2	10(2)	0(0)
impcolc	8	8	0	4(1)	1(0)
impcold	10	11*	1	4(1)	1(0)
impcole	12	21*	1	21(4)	0(0)
Total	45	59	4	39(8)	8(1)

In Table 31.2, cgrp and rgrp are the number of column and row groups respectively in the direct cover. The number of AD passes that can be saved using substitution are enclosed in parenthesis. Taking the sum of the matrix-vector products or AD passes over the problems where $p > \rho$ (marked with asterisk *):

- DSM requires 43 matrix-vector products

- DSM with substitution requires 39 matrix-vector products

- Direct Cover requires 36 AD passes

- Direct Cover with substitution requires 29 AD passes

We get a 19% reduction on the number of AD passes with the Direct Cover technique.

We note that the maximum number of nonzeros in any row or column (ρ) is a lower bound on the number of AD passes needed to determine the Jacobian matrix using either column or row computation. The technique of Newsam and Ramsdell [209, 403] requires only ρ AD passes, and the nonzeros are recovered by solving m small linear systems. With the Vandermonde seed matrices [209], an efficient solver is obtained. However, the numerical reliability of the computed values is a major concern due to ill-conditioning. Other choices, e.g., Chebyshev polynomials for the seed matrices [403], with acceptable condition numbers are possible with an increased computational cost. Our numerical results show that the proposed substitution technique is close to this minimum.

31.5   Concluding Remarks

Our approach to determining the structure of a sparse Jacobian matrix efficiently is based on a given consistent column (or row) partition or a complete direct cover. If the partition is not minimal, the proposed substitution technique saves at least one AD pass. On several test problems, however, we save more than one AD passes.

The number of consecutive zeros in the compressed Jacobian matrix determines how many AD passes we can save. An important research question is whether the number of consecutive zeros in the compressed matrix can be increased by a reordering of the groups or by an interchange of columns (of the Jacobian matrix) between groups.

Acknowledgments: This research was supported in part by the Research Council of Norway and Natural Sciences and Engineering Research Council of Canada under RGPIN.

32

Verifying Jacobian Sparsity

Andreas Griewank and Christo Mitev

ABSTRACT In a recent paper [244] we describe an automatic procedure for successively reducing the set of possible nonzero elements in a Jacobian matrix until eventually the exact sparsity pattern is obtained. The dependence information needed in this probing process consists of "Boolean products" Jacobian-vector products and vector-Jacobian products, which can be evaluated by propagating Boolean variables forward or backward through the function evaluation procedure. Starting from a user specified (or by default initialized) prior probability distribution, the procedure suggests a sequence of probing vectors. The resulting information is then used according to Bayes' law to update the probabilities that certain elements are nonzero.

We analyze the special situation where the sparsity pattern is provided as a guess and merely needs to be verified with a small number of probes. For banded $n \times n$ matrices with h superdiagonals it is shown that the greedy heuristics developed in [244] requires in verification mode only $O(h + \log_2 n)$ probes. Our numerical results indicate that this is still true if nothing is known a priori about the sparsity pattern.

32.1   Introduction

The first AD tool that allowed the automatic detection and exploitation of Jacobian sparsity was the SparsLinC [83] facility of ADIFOR. It propagates the derivative objects of all intermediate quantities in a sparse vector format. This approach is very flexible but incurs a significant run time overhead at each new evaluation point. Recently ADOL-C [242] and TAMC [214] have acquired the capability to propagate sparsity patterns as bit-vectors. This approach avoids the manipulation of index sets but requires roughly $n/32$ times as much storage as the original function itself.

It has been found in [23, 138, 209] and [286] that the compression of sparse Jacobians by suitable seed matrices can be very effective for exploiting sparsity patterns, provided they are correctly known beforehand. Therefore we investigate ways of determining the pattern automatically but without the overhead of a dynamically sparse implementation.

We assume that some vector function $F \equiv (F_i)_{i=1\ldots m} : \mathbb{R}^n \mapsto \mathbb{R}^m$ is continuously differentiable on some neighbourhood $\mathcal{D}$ of a given vector argument $x \equiv (x_j)_{j=1\ldots n} \in \mathbb{R}^n$. Then we may define the *sparsity pattern* of the Jacobian as the Boolean matrix

$$S \equiv (s_{ij})_{j=1\ldots n}^{i=1\ldots m} \in \{0,1\}^{m \times n}$$

with

$$s_{ij} = 0 \quad \Longleftrightarrow \quad \partial F_i(x)/\partial x_j \equiv 0 \quad \text{at all} \quad x \in D \quad .$$

Given any vector $v \in \{0,1\}^n$ we must be able to evaluate the Boolean vector

$$r \equiv (r_i)_{i=1\ldots m} \equiv S \, v \in \{0,1\}^m$$

such that

$$r_i = 1 \quad \Longleftrightarrow \quad s_{ij} = 1 = v_j \quad \text{for at least one } j \leq n \quad . \tag{32.1}$$

We will refer to the pair (v, r) as a "Boolean probe." A related but different concept of probing is used in domain decomposition methods to precondition matrices with a large number of very small entries (see [461] and the original proposals [152, 403]).

Using the reverse mode one may obtain for a given vector $\bar{v} \in \{0,1\}^m$ the result $\bar{r} = S^T \bar{v} \in \{0,1\}^n$. This kind of adjoint dependence information is extremely useful when the Jacobian has dense rows, in which case one would need n probing vectors v to completely identify the sparsity pattern S on the basis of direct products $r = Sv$ alone.

Either way we have much better chances of "nailing down" zeros rather than identifying nonzeros. To see this let us define the index set

$$\mathcal{J} \equiv \mathcal{J}(v) \equiv \{j : v_j = 1\} \quad \text{and} \quad \bar{\mathcal{J}} \equiv \bar{\mathcal{J}}(v) \equiv \{i : r_i = 0\} \tag{32.2}$$

where $r = S v$. Then it follows from the converse of (32.1) that s_{ij} must vanish for all $(i, j) \in \bar{\mathcal{J}} \times \mathcal{J}$. Exactly the same block of zeros is found by the direct product $r = Sv$ and the transpose product $\bar{r} = S^T \bar{v}$ provided the test vectors are chosen as logical negation such that $\bar{v} = 1 - r$ and $v = 1 - \bar{r}$. The last observation implies that direct and transpose products are equivalent as far as the following *verification task* is concerned.

Verification Task:
> Given a sparsity pattern $S \in \{0,1\}^{m \times n}$, find a small set of K minors $\bar{\mathcal{J}}_k \times \mathcal{J}_k$ with $\bar{\mathcal{J}}_k \subset [1\ldots m]$ and $\mathcal{J}_k \subset [1\ldots n]$ that cover all zeros of S in that

$$\{(i, j) : s_{ij} = 0\} = \bigcup_{k=1\ldots K} \bar{\mathcal{J}}_k \times \mathcal{J}_k \quad .$$

Simply verifying a (correctly) given sparsity pattern is in some sense our easiest task. It can be done column- or row-wise so that $K = \min(n, m)$.

However, the task of trying to find a cover of the absolutely minimal size K_{opt} is listed as *biclique cover* in Chapter 10 of the book [278] on approximate algorithms for NP-hard problems. Under the rather safe assumption $P \neq NP$, it is furthermore shown there that for some sufficiently small $\varepsilon > 0$, no polynomial algorithm can always find a cover of size $\mathcal{O}(K_{opt}(n + m)^\varepsilon)$. Hence we will be forced to design heuristic procedures for finding reasonably small covers at an acceptable overhead cost.

32.2 Theoretical Results

In this chapter we concentrate on the verification task even though the ultimate goal is certainly to detect arbitrary sparsity patterns without any a priori information. The latter, more general task has been attacked with considerable success in the paper [244]. There, a matrix of probabilities

$$p_{ij} \equiv P(s_{ij} = 1) \in [0, 1]$$

is updated after each probe using a Bayesian formula. By suitable initialization and appropriate updating all probabilities generated satisfy the consistency condition

$$p_{ij} = 0 \Rightarrow s_{ij} = 0 \quad \text{and} \quad p_{ij} = 1 \Rightarrow s_{ij} = 1 \quad .$$

For any such matrix $P = (p_{ij})_{j=1\ldots n}^{i=1\ldots m}$ the remaining uncertainty can be quantified by the function

$$U(P) \equiv \sum_{p_{ij} > 0} (1 - p_{ij}) \quad ,$$

where i and j range over $1 \ldots m$ and $1 \ldots n$, respectively. It was shown in [244] that $U(P)$ can be interpreted as the expected number of zeros that remain as yet unidentified.

Bayesian Updating: After the taking of a direct probe $r = Sv(\mathcal{J})$ the probabilities are updated according to the formula

$$p_{ij} \equiv \begin{cases} p_{ij} & \text{if } j \notin \mathcal{J} \\ 0 & \text{if } j \in \mathcal{J} \text{ and } r_i = 0 \\ p_{ij}/(1 - w_i) & \text{if } j \in \mathcal{J} \text{ and } r_i = 1 \end{cases}, \quad (32.3)$$

where

$$w_i = \prod_{j \in \mathcal{J}} (1 - p_{ij}) = 1 - O\left(\sum_{j \in \mathcal{J}} p_{ij}\right) . \quad (32.4)$$

It was also shown in [244] that the resulting reduction in U for a candidate index set $\mathcal{J}$ has the expected value

$$\Delta U(P, \mathcal{J}) \equiv \sum_{i=1}^m l_i w_i = \sum_{i=1}^m l_i - O\left(\sum_{p_{ij} < 1} p_{ij}\right)$$

where

$$l_i \equiv \left| j \in \mathcal{J} : p_{ij} > 0 \right| \cdot \mathrm{sign}(w_i) \quad .$$

This expression is used as a merit function for judging the quality of prospective probing sets $\mathcal{J}$.

Verification Algorithm: Suppose we initialize with some small $\varepsilon > 0$

$$p_{ij} \equiv \begin{cases} 1 & \text{if} \quad s_{ij} = 1 \\ \varepsilon & \text{if} \quad s_{ij} = 0 \end{cases} \quad . \tag{32.5}$$

This initialization corresponds to the verification mode, where all nonzeros of the sparsity pattern are known for sure and only the small probability ε is attached to the possibility that one of the other entries is also nonzero. As a consequence of (32.3) and (32.4) we see that throughout any resulting sequence of probes $\mathcal{J}_k$ with $\bar{\mathcal{J}}_k$ as defined in (32.2)

$$p_{ij} \equiv \begin{cases} 1 & \text{if} \quad s_{ij} = 1 \\ 0 & \text{if} \quad \{i, j\} \in \bar{\mathcal{J}}_k \times \mathcal{J}_k \quad \text{for some k} \\ O(\varepsilon) & \text{otherwise.} \end{cases}$$

Consequently the expected reduction is always of the form

$$\Delta U(P, \mathcal{J}) \equiv \sum_{i=1}^{m} l_i - O(\varepsilon) \quad . \tag{32.6}$$

Neglecting the $O(\varepsilon)$ term we find that $\Delta U(P, \mathcal{J})$ simply counts the number of zeros $s_{ij} = 0$ that could be newly identified by a probe $v = v(\mathcal{J})$. Hence any method that selects a sequence of $\mathcal{J}$ maximizing $\Delta U(P, \mathcal{J})$ for the current $P \equiv S + O(\varepsilon)$ may be viewed as a greedy algorithm for finding a cover $\bigcup \bar{\mathcal{J}}_k \times \mathcal{J}_k$ of all zeros in S. Equivalently we could select the corresponding sets $\bar{\mathcal{J}}$ since ΔU becomes invariant with respect to transposition as $\varepsilon \to 0$.

Naturally, the chosen sequence of $\mathcal{J}_k$'s should be the same for all sufficiently small values ε so that we may view ε as $+0$ with $1 - \varepsilon = 1$ but $\varepsilon > 0$. Numerically this situation can be achieved by selection ε somewhat smaller than the relative machine precision. With this convention we have the following verification algorithm.

Verification Algorithm:

  0) Initialize P according to (32.5).
  1) Select a $\mathcal{J} \subseteq [1 \dots n]$ minimizing $\Delta U(P, \mathcal{J})$.
  2) Perform the probe $r = S v$ with $v = v(\mathcal{J})$.
  3) Update P according to (32.3) using $w_i = 0$.
  4) Continue with **1)** until $U(P) = 0$.

Triangular Case: For a dense upper triangular sparsity pattern the situation is quite simple. Given any $v = v(\mathcal{J})$ we must have $r_i = 1$ for all

$i \leq l \equiv \max(\mathcal{J})$. Hence zeros can only be verified in the last $n - l$ rows, and we might as well define $\mathcal{J}$ as the contiguous range $[1..l]$ unless there are some other reasons to select a subset thereof. In any case a set $\mathcal{J}$ with $\max(\mathcal{J}) = l$ is the only possibility to verify the subdiagonal zero in the position $(l + 1) \times l$. Hence verifying a triangular matrix of dimension n requires exactly $n - 1$ probes. The algorithm described above would first choose $\mathcal{J} \equiv [1 \ldots l]$ with l an integer closest to $n/2$. This choice verifies the maximal number of zeros possible, namely $(n - l)l$, which equals $n^2/4$ when n is even and $(n^2 - 1)/4$ when n is odd. After the update of P the original triangular verification task has been partitioned into two subtasks of the same kind but roughly half its size. More specifically we find for $\mathcal{J} \subseteq [1 \ldots n]$

$$\Delta U(P, \mathcal{J}) = \Delta U(P, \mathcal{J}_1) + \Delta U(P, \mathcal{J}_2) \quad , \tag{32.7}$$

with $\mathcal{J}_1 = \mathcal{J} \cap [1 \ldots l]$ and $\mathcal{J}_2 = \mathcal{J} \setminus \mathcal{J}_1$. Hence the decomposition process can continue, and one can easily check that our algorithm verifies a triangular pattern using just $n - 1$ probes.

Diagonal Case: Let us consider the situation $s_{ij} = 1 - \text{sign}(|i - j|)$, where for simplicity $n = 2^p$ with some integer p. Then

Proposition 1 *On diagonal sparsity patterns the algorithm defined above terminates after exactly $2p = 2 \log_2 n$ probes.*

Proof. With $l \equiv |\mathcal{J}|$ denoting the number of elements in $\mathcal{J}$ we find that $l_i = l$ when $i \notin \mathcal{J}$ and $l_i = 0$ if $i \in \mathcal{J}$. Hence ΔU has the value $(n - l)l$, whose maximum is attained exactly when $l = n/2$. Since our procedure is invariant with respect to simultaneous row and column permutations, we may assume without loss of generality that the first $\mathcal{J}$ is chosen as the continuous range $[1 \ldots n/2]$. As a result the bottom left quarter of the diagonal sparsity pattern is identified as zero. The new value $\Delta U(P, \mathcal{J})$ is for any $\mathcal{J}$ bounded by $(n - l)l$ with $l = |\mathcal{J}|$. Moreover, $\Delta U(P, \mathcal{J})$ is smaller than this upper bound if $\mathcal{J}$ contains any element of the first range $[1 \ldots n/2]$ because then all l_i with $i \in [n/2 + 1 \ldots n]$ equal at most $l - 1$. Hence the maximum $\Delta U(P, \mathcal{J}) = \frac{1}{4}n^2$ is attained exactly by the complement $\mathcal{J} = [n/2 + 1 \ldots n]$ of the first $\mathcal{J}$. After the second probe the remaining nonzero entries in P form a block diagonal structure. Subsequently, the sets $\mathcal{J}$ and the corresponding values ΔU can be decomposed into a first part $\mathcal{J}_1 \equiv \mathcal{J} \cap [1 \ldots n/2]$ and the rest $\mathcal{J}_2 \equiv \mathcal{J} \setminus \mathcal{J}_1$ so that equation (32.7) holds. Finding optimal choices for $\mathcal{J}_1$ and $\mathcal{J}_2$ corresponds exactly to the original problem with n replaced by $n/2$, so that we must have $|\mathcal{J}_1| = n/4 = |\mathcal{J}_2|$ and thus again $|\mathcal{J}| = n/2$. This halving continues p times with two probes required at each level, which completes the proof. $\Box$

For general n the diagonal verification task can be solved using $2\lceil \log_2 n \rceil$ probes by enlarging the dimension to the next power of 2. At least theoretically the number of probes can be reduced to about half.

Blocking Invariance: A simple but useful observation is that the verification task for a partitioned matrix $S = (S_{ij})_{i=1...m, j=1...n}$ with each block $S_{ij} \in \{0,1\}^{m_i \times n_j}$ either zero or dense can be solved using the same number of probing steps as needed for the much smaller *quotient* system $(s_{ij})_{i=1...m, j=1...n} \in \{0,1\}^{m \times n}$ with $s_{ij} = 0$ if $S_{ij} \equiv 0$ and $s_{ij} = 1$ otherwise. The probing vectors for the quotient system can be transformed into those of the full system by expanding their jth scalar component to a subvector of length n_j with the original entry being replicated n_j times. In particular, the number of probes required to verify a block diagonal pattern is given by $2\lceil \log_2 \tilde{n} \rceil$, where $\tilde{n} = \tilde{m}$ is the number of block rows and columns. Even more useful is this blocking invariance principle when it comes to verifying general banded matrices by a sequence of probes derivable from following special situation.

Tridiagonal Case: For tridiagonal matrices of dimension $n = 1 + 2^p$ one may use almost the same probing vectors as in the diagonal case of dimension 2^p. The case $p = 4$ and thus n=17 is displayed in Figure 32.1. The structures of the probing vectors labeled by the letters $a - h$ and the resulting vectors r are sketched as columns to the right of the square sparsity pattern where the blocks of zeros verified in each probe are labeled accordingly. As one can see all probing vectors except for the first two a and b are symmetric about the centre, a property that plays a crucial role in the proof of the following result.

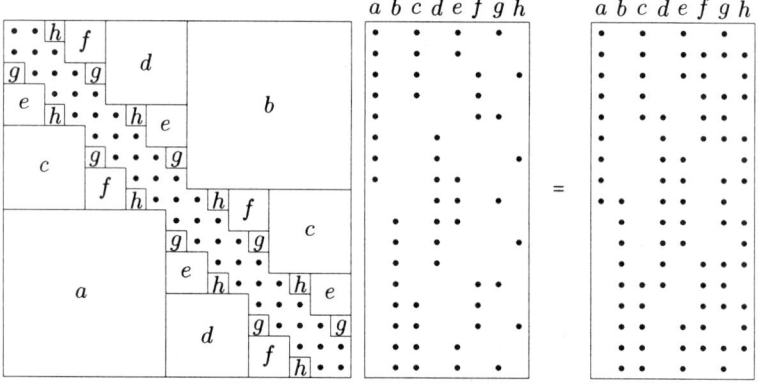

FIGURE 32.1. Verification of tridiagonal sparsity pattern

Proposition 2 *On tridiagonal sparsity patterns the Verification Algorithm terminates after exactly $2p = 2\log_2(n-1)$ probes.*

Proof. Assuming again without loss of generality that the first set $\mathcal{J}$ takes the form $[1...l]$, we find that $\Delta U(P, \mathcal{J}) = (n - l - 1)l$ attains its minimum $\frac{1}{4}(n-1)^2$ at $l = (n-1)/2$. By the same argument as in the proof of

Proposition 1 we find that the second index set takes the complementary form $[\frac{1}{2}(n+3)\ldots n]$ so that the zeros in the square boxes labeled with a and b on the left of Fig. 32.1 are verified. Abbreviating $\bar{n}=(n+1)/2$ we can subsequently decompose the expected reduction $\Delta U(P,\mathcal{J})$ using (32.6) into

$$\Delta U_1 = l_- + \sum_{i=0}^{\bar{n}-1} l_i \quad \text{and} \quad \Delta U_2 = l_+ + \sum_{i=\bar{n}+1}^{n} l_i \ ,$$

where $l_- = |\bar{n} > j \in \mathcal{J}|$ if $\bar{n} \notin \mathcal{J} \not\ni (\bar{n}-1)$ and $l_+ = |\bar{n} < j \in \mathcal{J}|$ if $\bar{n} \notin \mathcal{J} \not\ni (\bar{n}+1)$. These integers must vanish when the conditions on the right are not met. Partitioning accordingly $\mathcal{J}_1 = \mathcal{J} \cap [1\ldots\bar{n}]$ and $\mathcal{J}_2 = \mathcal{J} \cap [\bar{n}\ldots n]$, we obtain the representation

$$\Delta U(P,\mathcal{J}) = \Delta U_1(P,\mathcal{J}_1) + \Delta U_2(P,\mathcal{J}_2) + l_{\bar{n}} - l_- - l_+ \ .$$

Now it can be easily checked that $l_{\bar{n}} \leq l_- + l_+$ with equality holding if $\bar{n} \in \mathcal{J}$ or $(\bar{n}-1) \in \mathcal{J} \ni (\bar{n}+1)$ or $(\bar{n}-1) \notin \mathcal{J} \not\ni (\bar{n}+1)$. The latter condition holds in particular if $\mathcal{J}$ is symmetric in the sense that $(\bar{n}-j) \in \mathcal{J} \Leftrightarrow (\bar{n}+j) \in \mathcal{J}$. Since $\Delta U_1(P,\mathcal{J}_1)$ and $\Delta U_2(P,\mathcal{J}_2)$ have exactly the same form as the original $\Delta U(P,\mathcal{J})$ except that the dimensional exponent p has been reduced by one, they obtain the maximal values $\frac{1}{2}(\bar{n}-1)^2$ with $\mathcal{J}_1$ and $\mathcal{J}_2$ containing exactly $(\bar{n}-1)/2$ elements, which we may assume to be consecutive after a suitable permutations. In particular we may assume that the maximizer takes the form $\mathcal{J}_1^- = [1\ldots(\bar{n}-1)/2]$, $\mathcal{J}_1^+ = [(\bar{n}+3)/2\ldots\bar{n}]$, $\mathcal{J}_2^- = [n\ldots(\bar{n}-1)3/2]$ and $\mathcal{J}_2^+ = [n-(\bar{n}-1)/2\ldots n]$. Then $\mathcal{J}^- \equiv \mathcal{J}_1^- \cup \mathcal{J}_2^+$ and $\mathcal{J}^+ \equiv \mathcal{J}_1^+ \cup \mathcal{J}_2^-$ are symmetric index sets at which $\Delta U(P,\mathcal{J})$ attains its global maximum. This argument can be repeated to conclude the proof as two more probing vectors are added at each level of induction.   □

The Banded Case: Here we only give a possible probing sequence without proving that the current version of our algorithm actually realizes it. Any banded sparsity pattern with h superdiagonals is contained in a block tridiagonal pattern with blocks of size h, provided n is a multiple of h. We will make the even stronger assumption that $n = h(1+2^p)$, which of course can always be achieved by a suitable enlargement of n to a value with such a representation. A situation with $h = 3$, $p = 2$ and thus $n = 3 \cdot 5 = 15$ is displayed in Figure 32.2. Using the blocking invariance principle discussed above one may verify all zeros outside the 13 blocks by the $2p = 4$ probes labeled a, b, c and d. To verify the internal zeros in each triangular block requires $h - 1 = 2$ probes. On closer inspection one finds that the two triangle in the same block column can be verified using just h Cartesian probing vectors. Moreover, these probing vectors can be combined with those for every third block column, so that we obtain an upper bound of $3h + 2\lceil\log_2(n/h)\rceil$ probes for verifying a banded sparsity pattern with h superdiagonals. For sizable h a reduction in the leading term $3h$ might be possible, but the key observation is that the number of probes is still of order $2\lceil\log_2 n\rceil$ just as in the diagonal case.

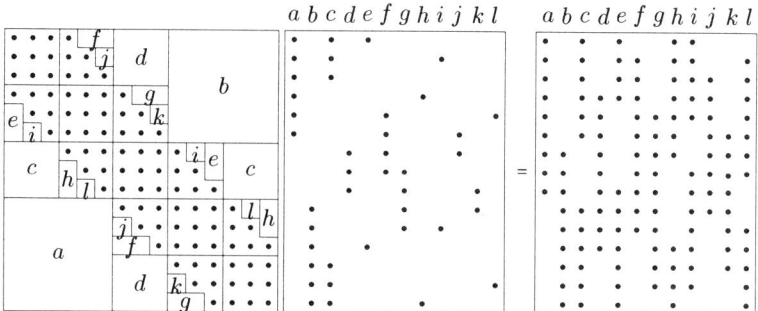

FIGURE 32.2. Two Stage Verification of Banded Pattern

32.3   Experimental Results

The specification of our algorithm requires the minimization of $\Delta U(P, \mathcal{J})$ at each stage, which is a rather difficult task similar to that of finding a maximal bipartite clique. Our current implementation uses a greedy strategy that successively enlarges or reduces the index set $\mathcal{J}$ by one element at a time until no further increase in ΔU in possible in this way. As one can see in Figure 32.3 the results are nevertheless quite close to our theoretical expectations as diagonal, tridiagonal, and 5-point stencil sparsity pattern are verified using a number of probes that is clearly a small multiple of $\log_2 n$. The same observations applies also to square matrices with 10 nonzeros per row that are randomly distributed.

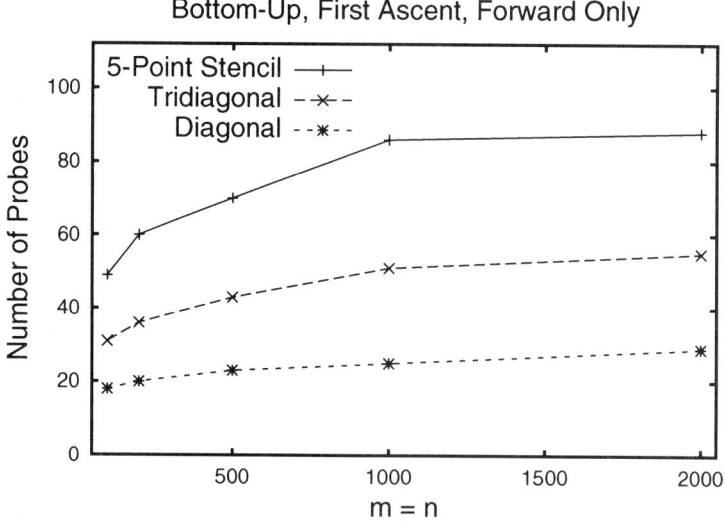

FIGURE 32.3. Heuristic Verification on Diagonal, Tridiagonal and 5-Point Stencil

32.4   Summary and Conclusion:

For verifying all zeros in a given banded sparsity pattern, our greedy type algorithm requires only $O(\log n)$ probing steps. In [244] it was found that without prior knowledge about the sparsity pattern a similar logarithmic growth can be achieved not only on banded matrices, but also other sparse problems where the number of nonzeros per row or possibly per column is bounded. In the latter case the reverse mode becomes important, which has played no role in the verification scenario considered here.

Part VIII

Space-Time Tradeoffs in the Reverse Mode

33

Recomputations in Reverse Mode AD

Ralf Giering and Thomas Kaminski

ABSTRACT The main challenge of the reverse (or adjoint) mode of automatic differentiation (AD) is providing the accurate values of required variables to the derivative code. We discuss different strategies to tackle this challenge. The ability to generate efficient adjoint code is crucial for handling large scale applications. For challenging applications, efficient adjoint code must provide at least a fraction of the values of required variables through recomputations, but it is essential to avoid unnecessary recomputations. This is achieved by the Efficient Recomputation Algorithm implemented in the Tangent linear and Adjoint Model Compiler and in Transformation of Algorithms in Fortran, which are source-to-source translation AD tools for Fortran programs. We describe the algorithm and discuss possible improvements.

33.1 Introduction

For a differentiable function represented by a numerical algorithm, automatic differentiation (AD) [232] is the process of constructing a second algorithm that represents the function's derivative. This second algorithm is an implementation of the chain rule, i.e. of a product of the Jacobian matrices representing the derivatives of the individual steps in the algorithm that defines the function. If the number of the function's input variables exceeds the number of output variables, it is favorable to have this derivative algorithm operate in the reverse mode of AD (adjoint code), i.e. to evaluate this matrix product in the order reverse to that of the function algorithm. The entries of these Jacobian matrices, however, contain values of variables of the function code, which are called required variables. Another type of required variables are those whose values determine the control flow in the function code [180]. Providing the accurate values of required variables to the derivative code in an efficient way is by no means trivial but the key challenge for constructing efficient adjoint code. Given that for many applications the function code alone consumes a large fraction of the available computer resources already, efficiency of the adjoint code is indispensable.

One way to provide the values of required variables is to store them in memory or on disk during an initial function evaluation and then access

```
1     y = 2*x
2     w = cos(y)
3     y = sin(y)
4     z = y*w
```

FIGURE 33.1. Example of a block with overwriting of y. All variables are active

them while evaluating the derivative. For large scale applications check pointing algorithms [235] arrange multiple use of the same storing space. Since this multiple use requires multiple forward integrations of the function code, the additional CPU requirements still make it infeasible to provide all required values exclusively by storing. An alternative consists of recomputing the values of the required variables within the derivative code. Again, it is not difficult to imagine relevant examples for which the cost of these recomputations exceeds the available CPU time. Hence, many large scale applications can only be handled by applying an efficient combination of storing and recomputation [213].

The Tangent linear and Adjoint Model Compiler (TAMC, [211]) is a source-to-source translation AD tool for Fortran routines, which supports a combination of storing and recomputation. By default TAMC generates code for recomputation based on the Efficient Recomputation Algorithm (ERA), which we present in this chapter. In addition, storing and reading of required variables is handled automatically by a few library functions, which arrange the necessary bookkeeping. This feature is triggered for variables indicated by the user through directives in the function code.

In §33.2 we use a simple example to introduce the basic strategies of storing and recomputation. In §33.3 we discuss the analysis steps and techniques upon which ERA is based and present the algorithm itself. §33.4 contains our conclusions.

33.2   Strategies of Providing Required Variables

Using a simple example of a sequence of statements, this section introduces the general strategies for generating corresponding adjoint code. Although a *sequence of statements* or *block*[1] is not an explicit statement in some programming languages (e.g., Fortran), in terms of reverse mode AD it is one of the most important structures. A block is a group of statements that is executed consecutively, i.e. there is no control flow changing statement inside a block. The statements comprising a block are elementary statements such as assignments, subroutine calls, conditional statements, or loops, which can contain blocks themselves. Hence, blocks define a multi-level hierarchy of partitions of the function code, which on the coarsest level partitions the

---

[1] The block we define here is different from a basic block defined in code analysis.

top level routine to be differentiated and on the lowest level reaches the elementary assignments.

```
S1    s1 = y
1     y = 2*x                          1     y = 2*x
S2    s2 = w
2     w = cos(y)                       2     w = cos(y)
S3    s3 = y                           S3    s3 = y
3     y = sin(y)                       3     y = sin(y)
S4    s4 = z
4     z = y*w                          4     z = y*w


R4    z = s4
A4    adw += y*adz                     A4    adw += y*adz
      ady += w*adz                           ady += w*adz
      adz = 0                                adz = 0


R3    y = s3                           R3    y = s3
A3    ady = cos(y)*ady                 A3    ady = cos(y)*ady


R2    w = s2
A2    ady += -sin(y)*adw               A2    ady += -sin(y)*adw
      adw = 0                                adw = 0


R1    y = s1
A1    adx += 2*ady                     A1    adx += 2*ady
      ady = 0                                ady = 0
```

FIGURE 33.2. Adjoint of the block of Figure 33.1 generated by a *store-all-mod-ified-variables* strategy (SA, left panel) and a *minimal-store* strategy (SM, right panel). a += b is a shorthand notation for a = a+b.

The block shown in Figure 33.1 contains four assignments, one of which (statement 3) is overwriting the variable y, which was defined previously (statement 1). The first algorithm employs the *store-all-modified-variables* (SA) strategy implemented in the AD tool Odyssée [441] for blocks defined by subroutine bodies and yields the code depicted in the left hand panel of Figure 33.2. The adjoint statements (denoted by a leading A) reference both adjoint variables (marked by the prefix ad) and required variables. The generation of the pure adjoint statements, i.e. without recomputations, is discussed in more detail by Giering and Kaminski [212]. In the joint mode [238] during a preceding forward sweep, the value of every variable is saved to an auxiliary variable (S1, ..., S4) before a new value is assigned. During the following reverse sweep through the adjoints of the individual statements, the value previously saved in front of a statement is restored in front of the corresponding adjoint statement (R4, ..., R1). By this algorithm

```
E4.1 y = 2*x
E4.2 w = cos(y)
E4.3 y = sin(y)
A4    adw += y*adz
      ady += w*adz
      adz = 0
```

```
E4.1 y = 2*x
E4.2 w = cos(y)
E4.3 y = sin(y)
A4    adw += y*adz
      ady += w*adz
      adz = 0
```

```
E3.1 y = 2*x
E3.2 w = cos(y)
A3    ady = cos(y)*ady
```

```
E3.1 y = 2*x


A3    ady = cos(y)*ady
```

```
E2.1 y = 2*x
A2    ady += -sin(y)*adw
      adw = 0
```

```
A2    ady += -sin(y)*adw
      adw = 0
```

```
A1    adx += 2*ady
      ady = 0
```

```
A1    adx += 2*ady
      ady = 0
```

FIGURE 33.3. Adjoint of the block of Figure 33.1 generated by the *recompute-all* strategy (RA, left panel) and the *minimum-recomputation* strategy (RM, right panel).

it is ensured that all required variables have their correct values.

In the previous example several saves and restores are unnecessary. Applying a *minimum-store* strategy (SM), we insert only stores and restores of those variables that are referenced by the adjoint statements. In our example the restores R1, R2, and R4 as well as the corresponding stores are not needed, because the restored value is not required. Such an algorithm is described by Faure and Naumann [185].

Alternatively, instead of saving required variables, code for recomputing their values can be inserted. Applying the straightforward *recompute-all* strategy (RA) to a block precedes every adjoint statement by the fraction of the block that precedes the corresponding statement. For our code example this yields the adjoint code shown in Figure 33.3.

Obviously, some of the recomputations are unnecessary and can be avoided by a more sophisticated strategy. In the adjoint code shown in the lefthand panel of Figure 33.3 recomputations E3.2 and E.2.1 are not needed, because variable w is not referenced thereafter, and variable y already has its required value. This *minimum-recomputation* strategy (RM) yields the code shown in the right hand panel of Figure 33.3. ERA follows this strategy.

The strategies SA and RA are easy to implement, while their respective counterparts SM and RM require a sophisticated data flow analysis. However, this additional analysis improves the efficiency of the code. Comparing both more sophisticated strategies RM and SM, RM is more efficient in terms of memory, while SM is more efficient in terms of the number of arithmetic operations. For a given problem, machine characteristics, namely the access time to memory and the speed of arithmetic operations, determine the ratio of the execution times of RM and SM.

33.3   Efficient Recomputation Algorithm

This section presents ERA, which is based on the multi-level hierarchy of partitions of the function code into blocks introduced in §33.2. ERA is based on a data flow analysis, which analyzes the memory accesses of statements. In general there are two kinds of accesses: read and write. In higher programming languages, memory is accessed by means of scalar variables or array variables. Especially for array variables, their memory access pattern can be very complex. In general every reference of a variable accesses a set of memory locations.

We consider here neither the representation of the sets of memory locations nor the implementation of the operations (union, intersection) on them. Nor are the representation of incomplete knowledge and its influence on the operations in order to gain "must" and "may" information discussed. For more details on these topics see e.g. [150]. We assume in the following that there is a representation of memory accesses including all necessary operations available and that the knowledge is complete. For readability we will denote the set of memory locations that variables access simply as *set of variables*.

Figure 33.4 shows the heart of ERA, the function AD_EFFREC represented in the notation ICAN [391]. The input of AD_EFFREC is a block of statements S = $\langle S_1, ..., S_N \rangle$ and all code F that precedes S. The output is the adjoint A of the block. ERA starts by calling AD_EFFREC with the coarsest partition of the top level routine of the function code and F=F_0, which is the code to reconstruct the values of the independent variables[2]. Then AD_EFFREC works top down by recursively calling itself, with the elements of the partition as arguments. When the level of the simple assignment is reached, AD_EFFREC constructs the corresponding adjoint. Going bottom up again, AD_EFFREC then level by level composes the adjoint code and propagates all information that is necessary for doing so. The algorithm uses several functions explained briefly.

SLICE(stmt, vars) takes as input a statement or sequence of statements (stmt) and a set of variables (vars). It computes the subset of variables that can be computed by the input set of statements and returns the smallest subsequence of statements that still generates this subset of variables (by removing all 'unnecessary statements').

---

[2]Since there usually is no code given to recompute the independent variables, they have to be stored at the very beginning of the adjoint top-level routine.

function AD_EFFREC(S, F) : A
if S is not block **then**
  A = AD_ELEMENTARY(S,F)
else
  required := $\emptyset$; killed := $\emptyset$; A := $\langle\,\rangle$
  for $i = N$ to 1 step -1 **do**
    if required $\neq \emptyset$ **then**
      B_i := SLICE(S_i, required)
      required := (required - GEN(B_i)) $\cup$ USE(B_i)
      killed $\cup$ = KILL(B_i); A := $B_i \odot$ A
    end if
    if IS_ACTIVE(S_i) **then**
      A_i := AD_EFFREC(S_i, F $\odot \bigoplus_{k=1}^{i-1} S_k$)
      invalid := killed $\cap$ USE(A_i)
      valid := USE(A_i) - invalid
      if invalid $\neq \emptyset$ **then**
        lost := $\emptyset$
        repeat
          invalid $\cup$ = lost
          valid $-$ = lost
          E_i := SLICE(F$\odot \bigoplus_{k=1}^{i-1} S_k$, invalid)
          lost := KILL(E_i) $\cap$ valid
        until lost = $\emptyset$
        killed $\cup$ = KILL(E_i); A := A $\odot$ E_i
      end if
      required $\cup$ = valid; killed $\cup$ = KILL(A_i); A := A $\odot$ A_i
    end if
  end for
end if
return A

FIGURE 33.4. Function AD_EFFREC, which forms the heart of ERA. Sequences of statements are denoted by upper case letters; $\odot$ denotes concatenation of two sequences, and $\bigoplus_{k=1}^{i}$ denotes a concatenation of i of these sequences. $\langle\,\rangle$ is the empty sequence. Functions are denoted by upper case words, explained in the text. Sets of variables are denoted by lower case words; the standard notation for the operators is used.

IS_ACTIVE(stmt) returns 'true' if the statement is active, i.e. computes active variables. This attribute is determined in a foregoing data flow analysis.

AD_ELEMENTARY(stmt, leading_stmt) generates the adjoint of an elementary statement. For more complex statements containing blocks themselves, AD_ELEMENTARY will call AD_EFFREC, in which case the adjoint statement might contain recomputations. In some cases leading

$$\mathrm{B}_i \odot \boxed{\mathrm{B}_{i+1} \ \ldots \ \mathrm{B}_{N-1} \odot \mathrm{A}_N \ \ldots \ \mathrm{E}_{i+1} \odot \mathrm{A}_{i+1}} \odot \mathrm{E}_i \odot \mathrm{A}_i$$

FIGURE 33.5. Principle of enlargement of A, the sequence of statements forming the adjoint of a block S. For a given statement S_i of the block, A_i is the adjoint, B_i a slice of S_i, and E_i a slice of all statements in the entire function before S_i (see text). B_N and E_N are empty because the empty set is initially required is.

statements (a slice of leading_stmt) have to be inserted to recompute lost values.

KILL(stmt) returns the set of variables overwritten by the input sequence of statements (stmt). Adjoint variables do not need to be taken into account here.

GEN(stmt) returns the set of variables, of which every variable is defined by a statement of the input sequence of statements (stmt) but not overwritten by any of the following statements of the sequence.

USE(stmt) returns the set of variables that are used before possibly being overwritten by the input sequence of statements (stmt). Adjoint variables do not need to be taken into account here.

When called with a block S that contains only an elementary statement, AD_EFFREC calls the function AD_ELEMENTARY which constructs the corresponding adjoint code according to the rules given in [212]. Otherwise AD_EFFREC walks backward through the block statement by statement and updates three quantities: the sequence of generated statements (A), the set of variables (required) required by the sequence, and the set of variables (killed) overwritten by it. For the current statement, A is enlarged in two steps as depicted by Figure 33.5. Firstly, for the current statement S_i a slice of it (B_i) is generated that computes accumulated required variables and prepended to A. Secondly, if the statement S_i is active (see [212]), the adjoint statement (A_i) is generated. The adjoint statement might require variables ($\mathrm{USE}(A_i) \subset \mathrm{USE}(S_i)$) which must be provided by statements to be included. However, some of these required variables (invalid) might have been overwritten by recomputations of previously generated adjoint statements (A_k, $k > i$) or by previously included slices of recomputations (B_k, $k > i$), in which case they cannot yet be provided by the first slicing. For this set of variables additional recomputations are generated by slicing the statements of the sequence up to but not including the current statement of interest ($\bigoplus_{k=1}^{i-1} S_k$) concatenated to all statements (F) preceding the block. Note that $\mathrm{USE}(E_i) = \emptyset$, since $\mathrm{USE}(F) = \emptyset$. This slicing is done in a repeat loop in order to find the minimum of recomputations that do not overwrite required variables. The resulting statements are appended to A, and the sets of variables killed and required by A are updated accordingly.

In summary, A is built by walking backwards through the sequence of statements. Recomputations are added on the left and, if necessary, on the right, and adjoint statements are added on the right. The sets of killed and required variables are stored and thus are accessible by the functions

USE(A) and KILL(A).

ERA is not always able to generate the code according to the RM strategy of §33.2. For example, in Figure 33.3 ERA will unnecessarily include statement E2.1. A more sophisticated algorithm requires a more detailed analysis of the validity of values that already have been recomputed. TAMC applies an extension of ERA which handles this case. Further discussion is beyond the scope of this chapter.

33.4   Conclusions

We demonstrated that to generate efficient adjoint code, which is indispensable for large scale applications, a sophisticated analysis of the function code is needed. We have described the Efficient Recomputation Algorithm (ERA) which, based on a multilevel hierarchy of partitions of the function code into blocks, works recursively from the top level routine to the elementary statements on the lowest level (top down) and then walks its way up (bottom up) again generating the adjoint code. It requires a data flow and data dependence analysis of the code and the ability to do program slicing for arbitrary code segments.

The current version of the Tangent linear and Adjoint Model Compiler (TAMC) uses an extended version of ERA, which in some cases is capable of avoiding recomputations of values that are still valid. Other possible extensions are not yet included in TAMC:

 a) The minimum recomputation strategy could be modified to work on loops. Those can be looked upon as sequences of statements of which each statement is the kernel of the loop (loop unrolling).

 b) TAMC currently does not take into account the array access patterns and thus may generate unnecessary recomputations, if array sections are required which do not overlap with further array sections that are overwritten.

 c) Currently the scope of ERA is limited to Fortran routines (subroutines and functions), i.e. the starting point of recomputations begins with the input variables of routines, which are stored and restored if necessary. The extended algorithm would require slicing of routines to generate clones of them that compute only a subset of their original output.

For most large scale problems, the most efficient adjoint code applies a combination of storing and recomputation. TAMC supports storing and restoring of a subset of the required variables that is indicated by the user through directives in the function code. For those required variables, storing and restoring is handled automatically through library routines. The new AD tool Transformation of Algorithms in Fortran (TAF), by default already chooses a suboptimal combination of storing and recomputation automatically, while the user has the opportunity to improve this combination by inserting directives in the function code. To determine the optimal combination would require the ability to quantify the cost of recomputations and storing/restoring for a particular platform and compiler.

Although TAMC and TAF are restricted to Fortran routines, ERA can be implemented to differentiate functions represented in other numerical programming languages.

Acknowledgments: Kaminski was supported by the Bundesministerium für Bildung und Forschung (BMBF) under contract number 01LA9898/9.

34
Minimizing the Tape Size

Christèle Faure and Uwe Naumann

ABSTRACT For a real world program, most of the execution time of the adjoint code can be spend for the storage/retrieval of the tape recording operations for subsequent play-back. This execution time is proportional to the size of the tape. When writing adjoint code by hand, this is well known, and complex strategies are applied to limit the tape size. This chapter is a first step towards the automatic control of the size of the tape.

34.1 Motivations

The storage/retrieve of the tape of a real world program can easily occupy one half of the total execution time of the corresponding adjoint code as shown in [181]. This component of the total execution time is mainly proportional to the maximum size of the tape, but it also depends on the adjoining strategy chosen as well as on the scheduling: scalar or buffer moves [180]. In this chapter we only deal with the size of the tape, studying ways to diminish it using a static analysis.

To evaluate our method, we have chosen a large industrial code (70000 lines, 500 sub-programs, 1000 parameters). The code THYC-3D is a thermal-hydraulic in bundles code developed at EDF-DER [311]. The standard version of Odyssée generates adjoint codes in the *joined mode* [238]. For the purpose of this chapter, Odyssée has been modified to generate the recording part of the adjoint code using the *split program reversal mode* [238]. This way only the original statements plus the tape storage treatment are generated. Two recording codes (see §34.3) have been automatically generated by Odyssée [186]. One uses the standard tape management (the value of every modified variable is recorded before modification), and the second uses the static analysis described in this chapter to diminish the size of the tape. Both codes can be run in enumeration or storage mode. In enumeration mode, counters are incremented whenever a value has to be recorded. In storage mode, the values are actually recorded.

Using the tape analysis described here, the tape size is decreased by a factor of 5, from $213,920 \cdot 10^6$ scalar values (or $1,711,360$ MBytes if every value is a double) to only $40,486 \cdot 10^6$ scalar values (or $323,888$ MBytes). The gain in execution time could not be evaluated because the standard recording part could not even be executed on the available computer. The

```
                                                       if (z .lt. y(j))
                                                       then A
z = k*y(i)*x(2)              z = x(i+j)                else B
```

(a) A   (b) B   (c) C

```
xccl(2) += k*y(i)*zccl   xccl(i+j) += zccl   if (z .lt. y(j))
zccl     = 0.            zccl       = 0.     then adjoint(A)
                                             else adjoint(B)
```

(d) adjoint (A)   (e) adjoint (B)   (f) adjoint (C)

FIGURE 34.1. Elementary examples of required values

gain obtained automatically is of the same order as the gain obtained by hand optimization [38].

34.2   The Taping Problem

Let us call **required**, a value used within a statement s which is necessary for the computation of the adjoint statements $\bar{s}$. Why is a value required?

Figure 34.1 shows three code examples A, B, and C that are supposed to be differentiated with respect to x. Certain original values are required for the computation of the corresponding adjoint statements $\bar{A}$, $\bar{B}$, and $\bar{C}$. Each example illustrates one of the three reasons why a value used within a statement can also be used within its adjoint statement.

Assumption 1 *An original value can be required for the computation of the elementary partial derivatives. It then represents a* **derivative information**.

Figures 34.1(a) and 34.1(d) show the original code A and its adjoint code $\bar{A}$, respectively. In this example, the value of $k * y(i)$ is derivative information required for the computation of $\bar{A}$.

Assumption 2 *An original value can be required for the computation of the indices of an adjoint variable. This type of required value is* **index information**.

The value $i + j$ is index information required for the computation of the adjoint $\bar{B}$ of B (Figures 34.1(b) and 34.1(e)).

Assumption 3 *An original value can be required to recapitulate the branch conditions, loop entry and exit criteria. We then call it* **control information**.

Figures 34.1(c) and 34.1(f) show the original code C and its adjoint code $\overline{C}$. The value of z.lt.y(j) is control information required for the computation of $\overline{C}$.

The intersection of the three classes of required values is not empty in general. For some specific operation such as abs, the required values can belong to different classes depending on the local differentiation strategy chosen.

The set of values of variables involved in the computation of required values will be called the **tape**. Amongst these values, some will be modified by the original execution, and their intermediate values will be lost at the end of the forward run. However, some others others are never modified, and their values are still available at the end of the forward run. Only the modified values must be handled. They must be recomputed during the returning phase or recorded during the recording phase and restored during the returning phase. In this chapter, we only analyse the solution of storing the modified required values. We do not consider the possibility of recomputing these values [214]. To determine the tape from the source code, a static analysis has to be performed as shown in the next section.

34.3   Tape Analysis

Let $F : \mathbb{R}^n \to \mathbb{R}^m$ be a program evaluating a corresponding vector function. We assume that F decomposes into a sequences of scalar assignments of the results of its elementary functions to (program) variables. Arrays are considered to consist of several variables distinguished by their indices. The aim of the **Tape Analysis (TA)** is to determine for each statement in F those variables whose values have to be recorded for the generation of a feasible and efficient adjoint code $\overline{F} = \{\overrightarrow{F}; \overleftarrow{F}\}$. The recording part $\overrightarrow{F}$ evaluates the original variables, and the returning part $\overleftarrow{F}$ computes the adjoint variables.

In this chapter we concentrate on the **TA** of a single subroutine. We consider three kinds of statements: **assignments**, **control statements**, and **sub-program calls**. A finite sequence of statements is called a **block**.

Notation 1 s^- (B^-) and s^+ (B^+) will be used as notation for "before" and "after" the execution of a statement s (block B).

While derivative and index information can be analyzed together, the control information must be considered separately. Consequently, the tape is implemented as two stacks: **TAPE$_V$** to record both the derivative and index information and **TAPE$_C$** to record the control information. More specifically, **TAPE$_V$** records all values required for the determination of the arguments of the Jacobian of F. **TAPE$_C$** contains the values required to revert the control flow of F.

W.l.o.g. we assume that all assignments are scalar in the sense that they can be written as $v = \mathtt{rhs}$ such that the expression on the right hand side represents a scalar function $f : \mathbb{R}^{n_u} \supseteq D_\mathbf{u} \to \mathbb{R} : \mathbf{u} \mapsto v = f(\mathbf{u})$. Regarding control statements we will distinguish between loops and branches. A loop has the form $\mathtt{WHILE}\ (c)\ \mathtt{DO}\ B\ \mathtt{OD}$ with a Boolean control value c and a block B. Any other loop can be converted into this form. A branch has the form $\mathtt{IF}\ (c)\ \mathtt{THEN}\ B_1\ \mathtt{ELSE}\ B_2\ \mathtt{FI}$. A subprogram call (w.l.o.g. user-defined functions can be converted into sub-programs) has the form $\mathtt{CALL}\ \sigma$, where σ is the name of the subprogram.

An initially $\mathtt{False}$ Boolean **To-Be-Recorded** status (TBR-status) is attached to all variables. It is set to $\mathtt{True}$ whenever the current value of the corresponding variable is required during the following reverse sweep. Consequently, the value of this variable will be recorded before it is possibly overwritten. Storing the value of a variable results in its TBR status being set to $\mathtt{False}$. Storing entire arrays will be avoided. However, since an element-wise analysis generally is impossible, the TBR status of an array will not be reset once it was set to $\mathtt{True}$.

TA of an assignment $v = \mathtt{rhs}$, where the value of some right-hand side $\mathtt{rhs}$ is assigned to a variable v, is performed in two stages: First we decide whether the value on the left-hand side has to be recorded before the execution of the statement. Then we determine the possible changes in the TBR status of the variables involved in the right-hand side, using a recursive algorithm, which will not presented in detail.

TA is based on the notion of activity of variables and terms. A bi-directional static analysis can be performed to determine whether a given variable is active. In fact, a variable v is called active at a given point in the program if its current value depends on at least one of the independent variables and if there is at least one dependent variable depending on the current value of v. The variable v is called passive at this point in the program otherwise. A term $w = f(\mathbf{v})$, $\mathbf{v} = (v_i)_{i=1..k}$, is called active if at least one of its arguments v_i is active. Otherwise, w is called passive.

Provided the right hand side φ of an assignment statement is sufficiently smooth it can be decomposed into the form

$$\varphi(b, a, p, q) = C\, b + \varphi_1(a, p) + \varphi_0(p, q) \ ,$$

where all components of the sub-vectors b and a are active variables and those of p and q are passive. The C represents a matrix of suitable dimensions so that the leading term $C\, b$ is linear. We will refer to the components of a and p as nonlinear active and passive arguments, respectively.

TA applied to an assignment statement has the following effect:

1. It performs an analysis of the right-hand side $\mathtt{rhs}$ with the goal to detect possible changes in the TBR status of variables representing the arguments of $\mathtt{rhs}$.

2. The value of the variable v on the left-hand side is recorded, and the

TBR status of v is set to False if the current TBR status of v is True.

For $F = \{A; s; B\}$ and provided that the current value of v has to be recorded, the adjoint code of F is $\overline{F} = \{\overrightarrow{F}; \overleftarrow{F}\}$, where

$$\overrightarrow{F} = \{\overrightarrow{A};\ \textbf{PUSH}(v, \textbf{TAPE}_V);\ s\ ;\ \overrightarrow{B}\}$$
$$\overleftarrow{F} = \{\overleftarrow{B};\ v = \textbf{POP}(\textbf{TAPE}_V);\ \overleftarrow{s};\ \overleftarrow{A}\}.$$

The analysis of rhs must precede the decision on whether the current value of v has to be recorded, since an initially False TBR status of v might possibly become True if v occurs on the right-hand side of s.

Rule 1 *The TBR status of an argument w of rhs has to be set to True if*

1. *w is a nonlinear active argument;*

2. *w is a nonlinear passive argument;*

3. *w is the index of some active array z, e.g., $v = z(w)$;*

4. *w is the index of some passive array z such that $z(w)$ occurs in an active term, e.g., $v = z(w) \cdot a$, where a is some active variable;*

TA of blocks of assignments is performed by applying it successively to all statements thus propagating the TBR status of all variables. **TA** applied to a sub-program call is equivalent to the analysis of the sub-program itself.

A general program certainly contains branches and loops. The application of **TA** to these complex statements is defined as follows.

Rule 2 *Let $F = \{A; s; C\}$ such that*

$$s = \{\texttt{IF } (c) \texttt{ THEN } B_1 \texttt{ ELSE } B_2 \texttt{ FI}\}$$

and $c = f(v_1, \ldots, v_k)$ for some variables v_i, $i = 1..k$. The current value of c has to be recorded whenever any of v_i, $i = 1..k$ is written inside B_1, B_2, or C If so, the resulting adjoint code is $\overline{F} = \{\overrightarrow{F}; \overleftarrow{F}\}$, where

$$\overrightarrow{F} = \{\overrightarrow{A};\ \textbf{PUSH}(c, \textbf{TAPE}_C);\ \texttt{IF } (c) \texttt{ THEN } \overrightarrow{B}_1 \texttt{ ELSE } \overrightarrow{B}_2 \texttt{ FI};\ \overrightarrow{C}\}$$
$$\overleftarrow{F} = \{\overleftarrow{C};\ c = \textbf{POP}(\textbf{TAPE}_C);\ \texttt{IF } (c) \texttt{ THEN } \overleftarrow{B}_1 \texttt{ ELSE } \overleftarrow{B}_2 \texttt{ FI};\ \overleftarrow{A}\}.$$

An analogous rule holds for loops.

In the following let $\mathbf{tbr}(v, s^-)$ denote the TBR status of a variable v before the execution of a statement s. With Notation 1 $\mathbf{tbr}(v, s^+)$, $\mathbf{tbr}(v, A^-)$, and $\mathbf{tbr}(v, A^+)$ are defined correspondingly. The TBR status propagation through loops and branches is based on the following two propositions. The proofs are straightforward and are not presented here.

Proposition 1 *Consider the statement* $s = \{\text{IF } (c) \text{ THEN } B_1 \text{ ELSE } B_2 \text{ FI}\}$ *and let v be some variable. For a known value of* $\mathbf{tbr}(v, s^-)$ *one states:*

$$\mathbf{tbr}(v, B_1^-) = \mathbf{tbr}(v, s^-)$$
$$\mathbf{tbr}(v, B_2^-) = \mathbf{tbr}(v, s^-) \, ,$$

and the following value is deduced:

$$\mathbf{tbr}(v, s^+) = \mathbf{tbr}(v, B_1^+) \vee \mathbf{tbr}(v, B_2^+) \, .$$

Proposition 2 *Consider the statement* $s = \{\text{WHILE } (c) \text{ DO } B \text{ OD}\}$ *and let v be some variable. Knowing the value of* $\mathbf{tbr}(v, s^-)$ *one deduces:*

$$\mathbf{tbr}(v, s^+) = \mathbf{tbr}(v, s^-) \vee \mathbf{tbr}(v, B^+)^\star,$$

where $\mathbf{tbr}(v, B^+)^\star$ *is deduced after at most two analyses of B. The first analysis of B is initialized with* $\mathbf{tbr}(v, B^-) = \mathbf{tbr}(v, s^-)$. *No fixed point iteration is necessary to analyse this loop.*

34.4   Outlook

The gain of $4/5$ in terms of tape size obtained by the static analysis on the tape is really encouraging and could be even increased by some vector-component analysis. One other improvement could be obtained from some context static analysis, where a context includes all elementary values necessary to restart a computation when using the subprogram (joined program reversal mode [238]) or loop checkpointing techniques. The union of these contexts all along the recording phase is also to be diminished [182] to improve the checkpointing efficiency.

Various strategies can be studied to deal with the tape. Each value can be recorded or recomputed. In this chapter, we only consider the whole storage strategy but the proposed TBR analysis can be interpreted in a different way. The recomputation strategy is analyzed in [214].

The elementary values recorded to compute the required values are also to be discussed. In this chapter the elementary values are memory locations. One can also think of partial derivatives or common sub-expressions to be elementary information. The TBR status is then to be associated to sub-expressions, which makes the analysis more difficult.

Referring to [399], the tape analysis could also be regarded in the context of cross-country elimination methods in computational graphs. In this case the strictly sequential access patterns of the tape in the reverse mode of automatic differentiation cannot be guaranteed. This leads to further complications and interesting problems yet to be solved.

35

Adjoining Independent Computations

Laurent Hascoët, Stefka Fidanova and Christophe Held

ABSTRACT The *reverse* or *adjoint* mode of automatic differentiation is a software engineering technique that permits efficient computation of gradients. However, this technique requires a lot of temporary memory. In this chapter, we present a refinement that reduces memory consumption in the case of parallel loops, and we give a proof of its correctness based on properties of the *data-dependence graph* of adjoint programs and parallel loops. This technique is particularly suitable for assembly loops that dominate in mesh-based computations. Application is done on the kernel of a realistic Navier-Stokes solver.

35.1 Introduction

Adjoints are a popular numerical analysis tool to compute efficient and accurate gradients [357, 420], for optimisation [382], data assimilation [472], etc. Given an original program that computes a function f, the *reverse* mode of Automatic Differentiation (AD) generates an adjoint program that computes the gradient of f. In its basic version, sketched in Figure 35.1, the adjoint program is composed of two successive phases: the forward sweep, that reproduces the original program, with extra instructions that push intermediate values (the '*tape*') on a stack, and the reverse sweep, that (left-)multiplies the vector of adjoint variables with the local Jacobian of each elementary instruction, in the order *reverse* to that of the original program. These Jacobian matrices use values that are popped from the *tape*.

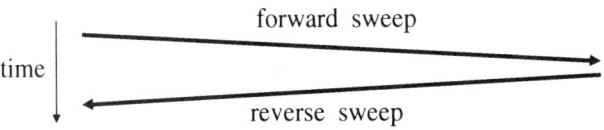

FIGURE 35.1. Forward and reverse sweeps of reverse AD

The tape mechanism is the main drawback of this adjoint program: all intermediate values required by the reverse sweep must be stored *before* any of them is used. In this chapter, we prove that parallelisable or reduction loops can be adjoined in a special way, with each loop iteration immediately followed by its adjoint iteration, therefore vastly reducing the total need of storage. Such loops are frequent and can be detected automatically. See for example *assembly loops* in mesh-based computations.

Our technique is a transformation of the standard adjoint program. Its correctness relies on an isomorphism between the data-dependence graphs of the forward and reverse sweeps, which is proven in §35.2. Using this isomorphism, §35.3 shows how any parallelisable loop can be transformed. §35.4 gives some experimental results, and we conclude with a comparison with two related approaches.

35.2   Isomorphism between Data-Dependence Graphs

Classically, a program transformation that only changes the execution order is correct if and only if the new order is compatible with the Data-Dependence Graph (DDG). Actually, the DDG is the sub-order of the original execution order that must be respected by any reordering.

Let us define the nodes of our original program's DDG as the elementary instructions (e.g., assignments). For each original elementary instruction, we define exactly one adjoint DDG node that contains all instructions achieving the product with the local Jacobian. By definition, we have a bijection between original and adjoint nodes.

The arrows of a DDG reflect *data dependencies* that relate uses of variables. Dependencies may be write-to-read, read-to-overwrite, or write-to-overwrite. We shall set one unique arrow in the DDG from node N_1 to node N_2 iff there is at least one dependence from some variable use in N_1 to a use of the same variable in N_2. In this chapter, we refine the usual notion of dependencies by making a special case for *increments* (x += ...). Successive increments to a variable can be done in any order, because of commutativity-associativity of the sum. Therefore this variable must not generate a dependence between these increments. However, this requires that increments are *atomic*. Two increments on x done exactly in parallel lead to race conditions. There are various ways to ensure atomicity. For example when the program is executed sequentially, atomicity is guaranteed.

The DDG is different from the *computational graph* [238]. One node of the computational graph represents one run-time instance of a variable, i.e. a value, whereas one node of the DDG represents one instruction in the program, and summarizes all its executions at run-time.

For example, Figure 35.2 shows on the left a small DDG and on the right its adjoint DDG. Only DDG arrows due to variables x and $\bar{x}$ (adjoint of x) are shown. Nodes and adjoint nodes face each other: notice therefore that the DDG on the right is executed bottom-up! There is no dependence between the two increments on $\bar{x}$. Notice also the two dotted arrows reaching the uses of variables **a** and **b** on the right: these are *tape dependencies*, which denote uses of intermediate values in the reverse sweep. Each tape dependence requires one value stored into the tape.

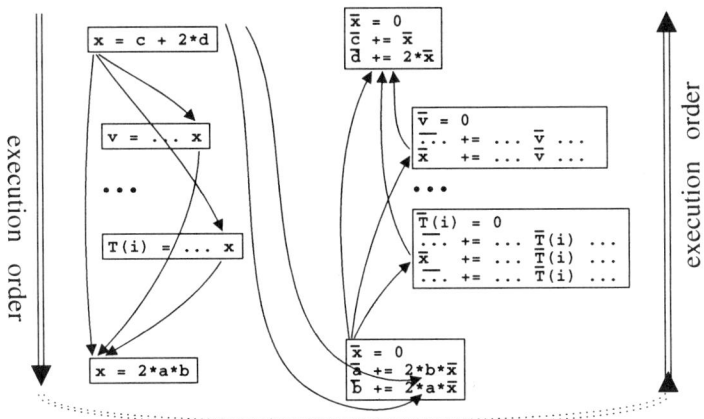

FIGURE 35.2. Adjoining a Data-Dependence Graph

With the above definitions, we prove the following isomorphism:
There exists an arrow from node N_1 to N_2 in the original DDG if and only if there exists an arrow from node adjoint(N_2) to adjoint(N_1) of the adjoint DDG.

Proof sketch: (cf. [270]) Consider any variable **x** that has an adjoint variable $\bar{x}$. DDG arrows due to **x** can be deduced from a partition of the DDG nodes in four categories: (**n**) (no use of **x**), (**i**) (**x** incremented only), (**r**) (**x** only read), and (**w**) (other cases). Variable **x** generates a dependence from a node to another in all cases except when one node is (**n**), or from (**r**) to (**r**), or from (**i**) to (**i**). Adjoining operates on these categories. Precisely: (**i**)↦(**r**), (**r**)↦(**i**), (**w**)↦(**w**), and (**n**)↦(**n**). Exploring all cases proves that **x** generates an arrow from node N_1 to N_2 iff $\bar{x}$ generates an arrow from adjoint(N_2) to adjoint(N_1). □

35.3 Adjoining a Parallelisable Loop

Consider a terminal parallelisable loop. "Terminal" means that the loop's results are not used until the reverse sweep starts. Therefore, all data dependencies leaving the loop go directly to the program's output. "Parallelisable" means that the loop iterations are *independent*, i.e. there are no *loop-carried* data dependencies. Figure 35.3 shows the standard adjoint. On top is the original loop, with tape storage, and under it is the reverse loop, executing the adjoint of the original instructions in the reverse order. Each vertical grey shape represents a particular loop iteration. Thin arrows are the DDG arrows *plus* tape dependencies. Since the original loop is paral-

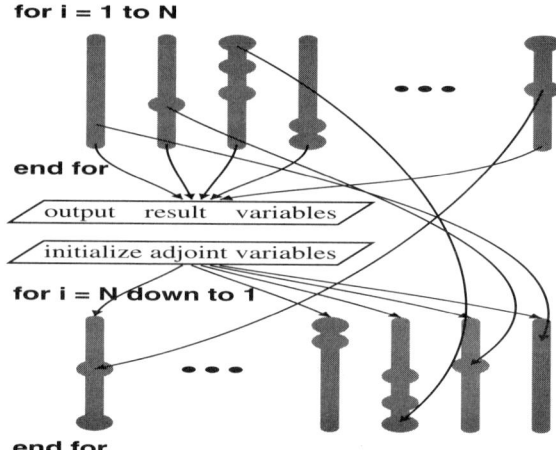

FIGURE 35.3. The adjoint for a terminal parallel loop, and its dependencies

lel, there are no loop carried dependencies. Because of DDG isomorphism, there are no loop-carried dependencies on the second loop either. Furthermore, there are no dependencies from the first loop to the initial adjoint values, nor from the original results to the second loop. We can thus inverse the reverse loop's iterations, then merge the two loops, yielding figure 35.4. Since the tape dependencies become local to each iteration, the total

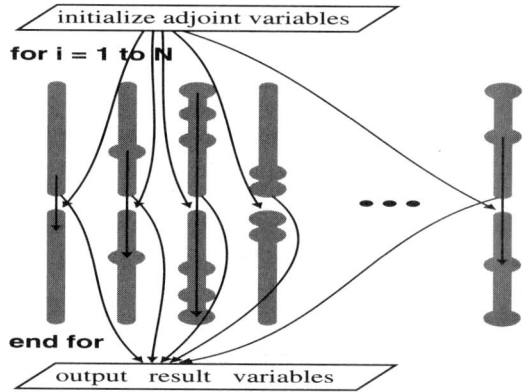

FIGURE 35.4. Transformed adjoint equivalent to Figure 35.3

memory required for tape storage is divided by **N**.

This can be generalised to loops containing "additive reductions" i.e. global sums, because the adjoint of a global sum simply requires the "spread" of a unique adjoint value to each iteration.

Finally, we lift the restriction to *terminal* loops. Figure 35.5 illustrates our method, called *dry-running,* a sort of *checkpointing* [235]. To make a program *fragment* terminal, first execute it without tape storage (thin arrow on figure). Then, when the reverse sweep reaches the end of the fragment, execute it again *with* tape storage, and continue immediately with its reverse sweep. Of course this requires that we save and restore enough data for recomputation (black and white dots on figure).

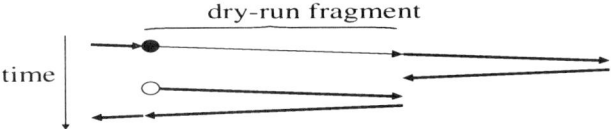

dry-run fragment

time

FIGURE 35.5. Dry-running a selected fragment of a program

35.4   Experimental Results

We took the kernel of an industrual-strength Navier-Stokes solver (7 parallel loops) on an unstructured triangular mesh ($\simeq$ 1000 elements). The standard adjoint was built by Odyssée [186], then we transformed parallel loops by hand. Dry-running was necessary because not all 7 loops were "terminal". The gradients found are identical. Table 35.1 shows the savings

TABLE 35.1. Compare memory usage for a Navier-Stokes solver

	reference	transformed
memory for original computation	171,000	171,000
memory for adjoint variables	297,000	297,000
memory for dry-running	0	47,000
memory for tape storage	2,400,000	800
Total	2,868,000	515,800

by a factor of 3000 on the tape corresponds to what could be expected, i.e. the product of the number of parallel loops by the number of parallel iterations. On the other hand, the dry-running mechanism used 47 kilobytes of memory. Rather surprisingly, the execution time of the transformed version is only slightly longer (3%). The increase due to dry-running was probably partly compensated by the economy in memory allocation time and a better data locality.

35.5   Related Work and Conclusions

The above technique can be combined profitably with others, such as fine-grain reduction of the tape, as described in [185]. Notice that [211] also provides a special strategy for adjoining parallel loops. However, further comparison is difficult because TAMC is based on recomputation rather than tape storage. Let us compare with two very similar techniques that reorder iterations of a loop and its adjoint. In the general case, the loop is not terminal, and thus we need dry-running, as sketched on Figure 35.6. In [292], the technique amounts to dry-running each iteration separately

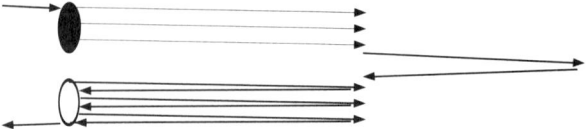

FIGURE 35.6. Our transformation, using dry-running

(Figure 35.7). Each iteration stores enough data for its own recomputation, which is generally more expensive than above. The *tape* savings are the same. On the other hand, the loop need not be parallel. Similarly, J.C.

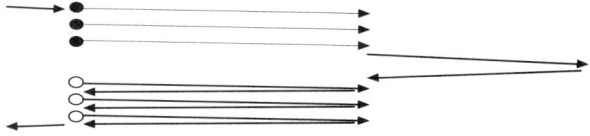

FIGURE 35.7. Dry-running on each iteration

Gilbert and F. Eyssette computed the explicit Jacobian matrix for each iteration (Figure 35.8). There is no more *tape* to be stored, nor checkpoints, but the Jacobians may use considerable memory. Again, the loop need not be parallel. A general strategy emerges from the above. If the loop is

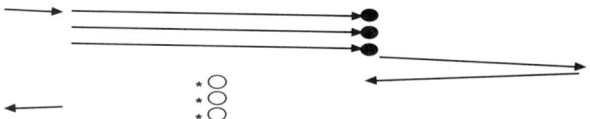

FIGURE 35.8. Local Jacobian technique

parallel, we recommend application of our method, especially when the loop is terminal. Otherwise, dry-running each iteration (Figure 35.7) is a good backup technique. The local Jacobian technique could make sense if loop body is computationally very expensive relative to the number of entries in the Jacobian.

This technique is a very efficient optimisation, already used by people who write adjoint programs by hand [451]. So far, this was done with no guarantee of correctness. Under precise hypotheses, that can be verified mechanically by any parallelisation tool, we proved that this transformation is correct. Thus it can be done automatically by source code analysis and transformation. This makes the link with notions from parallelisation theory such as data dependencies, and this improves our understanding of automatic differentiation techniques.

36

Complexity Analysis of Automatic Differentiation in the Hyperion Software

José Grimm

ABSTRACT
One important feature of the hyperion software is the rational approximation problem: given the m first terms of the power series expansion of a stable transfer function, find a stable approximation of it under the form P/q, of McMillan degree n. This leads to minimising $\psi(Q)$, for some Q. In this chapter, we show different ways of computing ψ and its derivatives, and we indicate the complexity of these computations.

36.1 The Rational Approximation Problem

The aim of the hyperion software, developed at INRIA, is the following: given a physical device, some measurements of it, and a mathematical model, find the values of some parameters of the device. See for instance [28], where we give an example that shows all steps needed to go from the measurements to the interpretation of the results.

The main assumption we make is that the device is governed by time-invariant, linear, stable, differential equations, and hence has a stable rational transfer function. One of the problems that hyperion tries to solve is the identification of this transfer function: given $F(z) = \sum F_k/z^{k+1}$, find a stable rational function P/q that is as near as possible to F. In order to simplify things, we consider here only the L^2 norm. This is a classical problem, but is in general solved only in the scalar case.

Hyperion can equally identify discrete or continuous time transfer functions (internally, continuous time systems are implicitly transformed into discrete time systems). The stability condition for discrete time systems is that the coefficients of the power series expansion of F at infinity are such that $\sum ||F_k||^2$ is finite. A rational function P/q, where P and q are co-prime, is stable if the roots of q are in the unit disk (of modulus less than one). In other words, $P/q = \sum X_k/z^{k+1}$, where $\sum ||X_k||^2$ is finite.

What we minimise is $\sum ||X_k - F_k||^2$, where the sum is over all k from 0 to infinity. Only a finite number m of quantities F_k can be computed; in the previous formula we let $F_k = 0$ for $k > m$. The degree n of the function

P/q is imposed and much smaller than m (if $m \leq n$, the problem has a trivial solution $q = z^m$).

Three integer parameters will be important in what follows. In the case of a multi-input, multi-output device, quantities F and P are matrices. For simplicity, we may assume that they are $p \times p$ square matrices. In the single-input, single-output case, quantities F and P are scalars, but we can consider them as 1×1 matrices (we shall refer to this case as the scalar one). In any case, q is a polynomial.

The second parameter is the McMillan degree n of P/q (in the scalar case, this is just the maximum of the degrees of P and q). This is the number of differential equations in the model, hence is given a priori. Sometimes, only a very general model is known, and the idea is to find a good fit between F and P/q for n as small as possible (this is called model reduction). We do not explain here what the McMillan degree is. The important point is that it is always possible to write P as a matrix whose entries are polynomials of degree less than n, and q is a polynomial of degree n, and P/q holds $2np$ free parameters (thus, there are many relations between the entries of P and q).

Using the inner-unstable (or Douglas-Shapiro-Shields) factorisation, we can write $P/q = Q^{-1}C$, eliminate C, and we are reduced to the following problem: find Q such that $\psi(Q)$ is as small as possible. The matrix Q, of size $p \times p$, must be inner of McMillan degree n. That is, there exists a polynomial matrix D and a polynomial q such that

$$Q = \frac{D}{\tilde{q}}, \quad Q^{-1} = \frac{\tilde{D}}{q}, \quad \det Q = \frac{q}{\tilde{q}}, \tag{36.1}$$

where the notation $\tilde{X}$ means $z^n X(1/z)^t$ (replace z by $1/z$, multiply by z^n, and transpose). For instance $\tilde{q}$ is just q, with the order of coefficients reversed. Note that Q contains np free parameters (half of what we had previously).

If we denote by G the quantity $\sum F_k z^k$, the function ψ we wish to minimise is defined by

$$\psi = ||V||^2 \qquad G\tilde{D} = Vq + R, \tag{36.2}$$

where V and R are the quotient and remainder in the Euclidean division of the product $G\tilde{D}$ by q.

The last parameter is m, the number of coefficients of F that are known. This number is typically between 100 and 1000 (if F_k is deduced by some measurements), but it can be much larger (10^5 in some cases, if F is completely known, and model reduction is wanted).

We wish to find Q such that $\psi(Q)$ is minimal. In the non-scalar case, we use the Schur algorithm (see for instance [202]) to write Q as a function of y, which is a vector of size np. In the scalar case, this is trivial. The objective of this chapter is to compute the complexity of ψ, ψ' and ψ'', the function and its derivatives, as a function of the parameters n, m and p.

36.2   Study of the Scalar Case

In the scalar case, $p = 1$. This is a trivial case, because we can always write $D = q$ in equations (36.1) and (36.2). This means that we have to minimise $\psi(q)$, where $q = q_0 + q_1 z + \ldots + q_{n-1} z^{n-1} + z^n$, a function of n variables, that depends on the m quantities F_k (which are constant).

In this section, we try to analyse the complexity of computing ψ, ψ' and ψ'', as functions of n and m. For simplicity, we count only the number of multiplications. The a priori time complexity of ψ is $m(2n + 1)$, and the space complexity is $m + 2n$. A naive implementation of ψ' and ψ'' gives a time complexity of $\alpha_1 mn^2$ and $\alpha_2 mn^3$ for some numbers α_1 and α_2. Such an implementation has existed since 1990 in the SCILAB software. The question we are interested in is "Can we do better, for instance using automatic differentiation?"

The first idea consists of differentiating ψ in reverse mode, and the result in direct mode. This gives a complexity of $3mn$ and $(8n + 1)nm$, for the first and second derivatives, respectively (method 1 in Figure 36.1).

If we differentiate (36.2), we get

$$Gz^{n-k} - Vz^k = q\frac{\partial V}{\partial q_k} + \frac{\partial R}{\partial q_k}, \quad \frac{\partial \psi}{\partial q_k} = 2\left\langle V \middle| \frac{\partial V}{\partial q_k} \right\rangle, \tag{36.3}$$

where $\langle a|b \rangle = \sum a_k b_k$ if $a = \sum a_k z^k$ and $b = \sum b_k z^k$. The interest of these formulas is that the first derivative of ψ can be obtained easily from the quotient in the division by q of Gz^n and Vz^n. This gives an algorithm for computing ψ' in time $(3m-1)n$ and ψ'' in time $m(n(n+1)/2+6n-2)-3n$. This was the first implementation of the algorithm in hyperion.

It is possible to rewrite equations (36.2) as

$$G = V_1 q + R_1, \quad R_1 \tilde{q} = V_2 q + R_2, \quad \psi = ||G||^2 - ||R_1||^2 + ||V_2||^2. \tag{36.4}$$

This means that we compute ψ in time $mn + 2n + n^2$, ψ' in time $mn + 3n^2$, and ψ'' in time $n(2mn + 10n^2 + 2n)$. Since n is much smaller than m, this is in general better, and is currently implemented (method 2 in Figure 36.1).

For equation (36.4), we obtain the equivalent of (36.3). Hence, we compute ψ' with a complexity of $n(mn + 5n - 4)$ and ψ'' with a complexity of $(m - 2)n + 2n^2(n + 1) + n^3(n + 1)$. Since we have a term with n^4, this is not so good (method 3 in Figure 36.1).

There is another way to compute ψ, and hence its derivatives. For fixed q, the scalar product of F and P/q is a linear function of P, and the coefficients of this linear function can be precomputed. Details can be found in [253]. Hence, we can find an algorithm that computes ψ' with a cost of $3mn + n^2$ and ψ'' with a cost of $5mn + n^2 + n^2(n + 1)$ (method 4 in Figure 36.1).

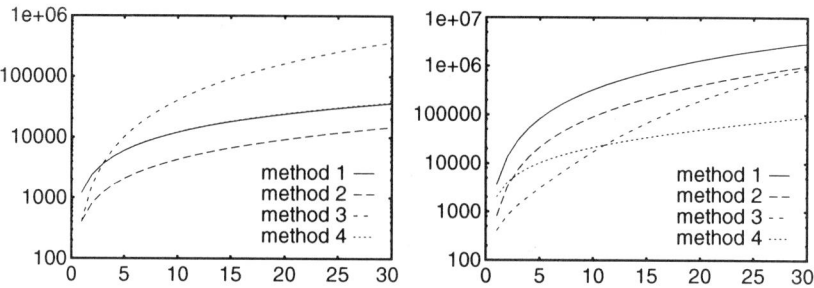

FIGURE 36.1. Cost of first and second derivatives, in the scalar case. We show the number of operations as a function of n, for $m = 400$, for the four methods indicated in the text.

36.3 The Non-Scalar Case

In the non-scalar case, the matrix Q defined by (36.1) is not trivial (remember that there are many relations between the entries of D and q). We shall explain here a method based on Schur parameters. If ω is a scalar, and y and u are vectors of size p, the formulas

$$b = (z - \omega)(1 - \omega), \quad \alpha = (b - \tilde{b}\|y\|^2), \quad \beta = \tilde{b} - b$$
$$q_+ = \alpha q + \beta y^* Du$$
$$X = (Duy^* D - y^* DuD)/q$$
$$D_+ = \alpha D + \beta[Duu^* + yy^* D - yu^* q] - \beta X$$

allow us to compute D_+ and q_+ as functions of D and q.

The quantities D and q that appear in (36.2) are the quantities D_n and q_n defined by the following equations:

$$(X_1, Y_1) = f(D_0, q_0, y_0, v_0)$$
$$(D_1, q_1) = g(X_1, Y_1, D_0, q_0, y_0, v_0)$$
$$(X_2, Y_2) = f(D_1, q_1, y_1, v_1)$$
$$(D_2, q_2) = g(X_2, Y_2, D_1, q_1, y_1, v_1)$$
$$\cdots$$
$$(X_n, Y_n) = f(D_{n-1}, q_{n-1}, y_{n-1}, v_{n-1})$$
$$(D_n, q_n) = g(X_n, Y_n, D_{n-1}, q_{n-1}, y_{n-1}, v_{n-1}),$$

where y_i is a vector of size p, v_i stands for (u_i, ω_i), X_i and Y_i are some auxiliary variables, f and g are some functions, and D_0, q_0 are initial values.

The quantity X that appears above is the quotient of a matrix of polynomials by a polynomial. The result is a matrix of polynomials (division is exact). In fact, it is possible in theory to compute X without division, but this is of interest only if $p = 2$ (the case $p = 3$ could be managed). It is important to notice that this term is the most costly.

36.3.1 Direct mode

We have studied the cost of computing Q and its derivatives in direct mode, optimising as much as possible for speed. This means that common subexpressions are only computed once. The ratio of the cost of one element of the second derivative by the cost of the function is $14/5$ for n and p large. This ratio is 4 in general. It is much smaller here because roughly half of the computation time is due to the division, and the ratio is $1/2$ for the other terms (because the formulas are linear in y and $||y||^2$). The fact that D_i is independent of y_j for $j > i$ also reduces the complexity. For n and p large, the complexity of computing Q is $15p^2n^2/2$, the cost of the first derivative is $3n^4p^3/2$, and the cost of the second derivative is $7p^4n^5/5$, and the space complexity is $p^4n^4/2$.

Once we have computed Q and its derivatives, we have to compute ψ. The memory cost is small, and the time is $p^4n^3(\alpha + \beta)/2$, where $\alpha = m(4n+3)$ and $\beta = 4pn^2$. We get other values for α and β if we implement (36.2) instead of (36.4).

The gain between these implementations is r_m. If m is large, the limit is $r = 1 + p$ for the function, $1 + p/2$ for the derivative and $1 + p/4$ for the second derivative. For instance, if $p = 2$, $m = 200$ and $n = 20$, these numbers are 2.6, 1.7 and 1.4.

Computing $\psi(Q)$ and its derivatives takes much more time than computing $Q(y)$ and its derivatives. On the other hand, the code for $Q(y)$ and its derivatives is much more complicated.

36.3.2 Reverse mode

The algorithm described above is very complicated, and not yet implemented. It uses a lot of memory, essentially because we store every derivative of X. This is perhaps not the best way to do it, so that we have still some work to do. In fact, we are convinced that the best way to use the sparsity of the derivatives of Q with respect to y_i is to implement the reverse mode of differentiation for it. Hence, we have written a small differentiator that generates directly the code of ψ and its derivatives from the formulas, via pattern matching.

The differentiator understands operations of the form $X = X + f(A) * f(B)$, where A and B are scalars, polynomials, matrices of polynomials, or matrices of scalars, $f(A)$ is A, its transpose, its conjugate or $\bar{A}$. It also understands division by polynomials. To each basic operation is associated a pattern. This gives 200 lines of Lisp code (the largest part is the division, 50 lines). The code generator is formed of 500 lines of Lisp code, and generates 6000 lines of C++ code on this example, for the function and its derivatives.

The memory cost is $5p^2n^2/2$ (essentially, this is three times the space needed by all D_k). The complexity of computing $\psi(Q)$ and its derivatives

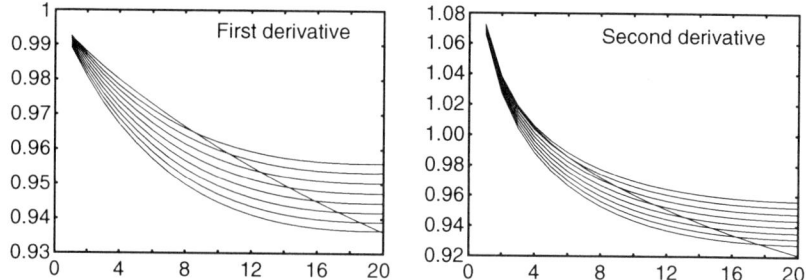

FIGURE 36.2. Relative costs of Schur parameters (reverse mode). On the left, we have $C(\psi')/2C(\psi)$, on the right, we have $C(\psi'')/6C(\psi)$. Here $C(\psi')$ is the cost of ψ', assuming that ψ has been computed, and $C(\psi'')$ is the cost of one row of the Hessian. The total cost of the Hessian is $npC(\psi'')$. On each curve p is fixed (between 2 and 10, note the special case $p = 2$), and n varies between 1 and 20.

via (36.4) is $p^3n^2 + p^2mn$, $2p^3n^2 + 2p^2mn$ and $6p^4n^3 + 6mn^2p^3$ (plus terms of smaller degree).

If we write $D_+ = \alpha D + \beta(X/q + Y)$, the cost of Q is essentially given by the complexity of X/q and Y. The high order terms of the cost of the quantities X/q and Y are p^2n^3 and p^3n^2, respectively. It is hence not easy to say which term is the largest. Moreover, we cannot always assume that p and n are large. For instance, in p^3n^2, the coefficient of n^2 is really $p^3 + 2p^2$, so that, for $p = 2$ we are wrong by a factor of 2, and in p^2n^3, the coefficient of p^2 is $3\sum(i+1)^2 = n(n+1)(n+1/2)$. Here, for $n = 10$, which is a typical value, the error is ten percent. As in the previous case, the cost of Q is much smaller than the cost of $\psi(Q)$, so that we can estimate the total cost to be $p^2n(m + pn)$. In general, we can assume that pn is smaller than m and simplify this to mnp^2.

As a conclusion, implementing the formulas in reverse mode seems in this case the best possibility, since both space and time complexity are smaller than for the direct mode. However, some variants of this problem need the inversion of a matrix that depends non trivially on the roots of q, and in this case, the situation might be quite different.

37

Expression Templates and Forward Mode Automatic Differentiation

Pierre Aubert and Nicolas Di Césaré

ABSTRACT This work deals with an implementation of automatic differentiation of C++ computer programs in forward mode using operator overloading and expression templates. In conjunction with a careful reuse of data, this technique also improves performance of programs involving linear algebra computations mixed or not with automatic differentiation. We give a broad view of implementation and explain some important concepts regarding code optimization. We conclude with some benchmarks applied to our optimal control software.

37.1 Introduction

This work deals with an implementation of automatic differentiation (AD) for C++ computer programs in forward mode using operator overloading and expression templates. This technique is a way to "teach" the C++ compiler how to handle variables. It is complementary to the classical overloading technique used for example in ADOL-C [242] or FADBAD [41] in its use of expression templates (ET). It is a drop-in replacement to the forward mode of FADBAD; the interface is compatible, but performance is better. The requirements are the same as for FADBAD. Pointers and dynamically allocated data are allowed, but memory aliasing must be checked by the programmer.

This library also addresses the blocking of data in memory. It is well known that on current processors, computations are faster than memory accesses. In order to get high performance, data accesses must be carefully scheduled in memory, first in registers and then in L1 cache. For instance, if we replace every `double` declaration with automatic differentiated floats, say `ad::F<double>`[1], we lose a lot in performance even if the linear algebra library is optimized, because we have replaced a `double` by an `ad::F<double>`, which might no longer fit in cache memory. The library described here overloads the assignment operator and performs computa-

---

[1] `ad::` stands for the namespace, and `F` is for forward.

tions by blocks to get high performance.

37.2   Expression Templates

ET [487] are a clean but complex way of dealing with return values for
classes in C++. Consider the following simple example:

```
ad::F<double> a,b,c,d;
d = a*b+c;
```

A classical overloading technique provides two operators * and + which
compute d and its derivatives with respect to some parameters from the
partial derivatives held in a, b and c. Usually, the compiler creates a temporary variable of type ad::F which holds the product, then another temporary variable which holds the sum, and then copies the result into d.

```
t1 = a*b;
t2 = t1+c;
d  = t2;
```

In contrast, expression templates build an expression which has a *static*
type, i.e. the precise type is known at compile time, in contrast to *dynamic*
types, for example, a tree built at run time:

```
meta::Expr<meta::Binary<
            meta::Binary<ad::F<double>,        d =          +
                         ad::F<double>,                     / \
                         meta::Mult>,                      *   c
            ad::F<double>,                                / \
            meta::Plus> >                                a   b
```

The parse tree is encoded as a C++ template. Then it is wrapped in a
generic class called meta::Expr. No computation is done at this stage,
and a smart C++ compiler removes all instances of classes meta::Expr
or meta::Binary. We then have to provide in the class defining a generic
assignment operator ad::F

```
template <class T_expr>
inline this_t&
operator=( meta::Expr<T_expr> const& rhs );
```

This operator is generic. It can cope with arbitrary expressions. The computations are done in this operator, and no temporary variable is created
in this process. Dynamic polymorphism is *not* used. This creates a large
improvement in performance for small objects. The cost of lookup in the
virtual function table becomes small with respect to the time spent in computations when the number of floats holding the derivatives is smaller than
ten. Thus expression templates are particularly efficient for a small number
of variables, which is the case in forward mode.

Static polymorphism [32] also can be used in combination with ET. It also gives good performance and has been implemented in [19]. It may be simpler to understand from a programmer's point of view, but compilers often crash on this kind of construction.

37.3   Memory Blocking

It is well known that level 3 BLAS operations must be blocked in memory to improve data locality. An optimized version of the BLAS library such as ATLAS [500] deals very efficiently with this problem, but it does not handle complex types such as forward differentiation or interval types. Following [153, 500], we have created a generic algorithm which blocks data at the appropriate memory level. It is recursive and can handle more that one level of blocking if needed. It also works well for standard BLAS subroutines. BLAS Level 1 and 2 are efficiently handled by ET as shown in BLITZ++ [486]. BLAS Level 3 is dealt with by combinations of ET and partitioning of data.

The algorithm [17] is generic in the sense that it will optimize $d = a \times b + c$ and also $d = c + a \times b$ without a special treatment. It works well with typical BLAS calls and with expressions such as $d = a \times b + c \times d$ or $d' = a' \times b + a \times b' + c' \times d + d \times c'$.

37.3.1   Using traits classes to carry information

A very useful technique in C++ is called traits [393]. We show how to use it to carry information, here the BLAS level associated with an expression:

```
double alpha,beta,
Matrix<> A,B,C,D;
C = alpha+A+B;              // BLAS level 2 call
D = alpha*A*B+beta*C;       // BLAS level 3 call
```

All computations are deferred via expression templates in the assignment operator that we provide in class Matrix

```
template <class T_expr>
inline this_t
operator= ( meta::Expr<T_expr> const& rhs ) {
   typedef typename mBLAS<T_expr>::blas_level level;
   blas::compute<level>::go(*this,rhs);
   return *this; }
```

In this operator, the trait class mBLAS statically computes the BLAS level associated with an expression, for example:

```
template <class S,class A>
struct mBLAS<Matrix<S,A> > {
```

```
    typedef mBLAS0 blas_level; }
```

and we specialize the generic case for a binary additive expression:

```
template <class E1,class E2>
struct mBLAS<meta::Binary<E1,E2,Plus> > {
    typedef typename E1::blas_level blas_level1;
    typedef typename E2::blas_level blas_level2;
    typedef typename PromoteBigger<blas_level1,blas_level2>
            ::blas_level blas_level; }
```

and for multiplication:

```
template <class E1,class E2>
struct mBLAS<meta::Binary<E1,E2,meta::Mult> {
    typedef mBLAS3 blas_level;   }
```

Many special cases are needed to treat all cases. In particular a `float` times a matrix does not yield to a call to a BLAS level 3 subroutine.

The main advantage of this technique is that it does not require dedicated matrix or vector classes. You need to specialize `struct mBLAS` for your own matrix vector classes. Algorithms already written are derived from this information and perform well.

37.3.2   Loop blocking

Let A, B, C, D be matrices and α a constant. Let a code compute $D = A \times B + \alpha C$ via a call to a BLAS level 3 subroutine. In forward mode, we replace matrices of floats by matrices of automatic differentiated floats. Thus each cell in a matrix has a different physical length than the simple float. These differences are not well handled by the BLAS library. We have rewritten the optimization strategy used in the design of ATLAS to cope with different sizes of floats and with different types of calls generated by differentiation of BLAS subroutines. ET have been used to deal with every case, and we do not have to explicitly code all cases. The main process is to compute how many matrices are involved in the assignment operator and then to compute the size of blocks such that they fit in L1 cache. An unroll and jam strategy improves performance at the register level.

37.4   Results

The code we consider comes from a shape optimal problem [20]. It spends most of its CPU time computing matrix-vector and matrix-matrix products. The first kind of products are from GMRES and conjugate gradient solvers. The second kind comes from the geometric transformation process between the CAO model and 3D meshes, which generate large dense rectangular matrices. The current version of the code [19] provides an optimized

BLAS library which works as well as ATLAS for floating points computations and also handles overloaded types. Genericity ensures almost no replications of code in order to handle `float`, `double` and associated overloaded types. On a Pentium III processor, the operation matrix times vector with overloaded variables is around 30 MFLOPS and goes to 270 MFLOPS after optimizations on a computer where peak performances are around 300 MFLOPS. On the whole code, performance improvement is about 5 − 6 between the blocked and the non blocked versions. Performance between hand-coded and overloaded gradient is under 5%, and accuracy is around 10^{-12} in double precision.

Our library performs additional optimizations including loop unrolling via meta-programming, loop re-ordering also with ET, and some loop fusion. Constant strides are optimized via some copying of data when needed and the use of iterators otherwise. The copy strategy is statically selected if needed.

Comparisons with other libraries are difficult. Neither FADBAD nor ADOL-C compile with an ANSI C++ compiler. We have slightly modified them to compile small examples, but we have been unable to use them in our large software. Comparisons are in [19] and show a large increase in performance when the number of variables in the gradient is small, i.e. less than ten. On the linear algebra part, the increase in performance is better on complex processors such as the PA8600 from HP than on an Intel Pentium III.

37.5   Conclusion

A technique called expression templates has been incorporated in a class of forward differentiation floats. Although this technique is not new, it seems that compiler technology is mature enough so that they can now be used almost everywhere. The main drawbacks are increases in compile time and an memory consumption at compile time only. This technique and careful memory access enable high performance in this library. The intrinsic genericity of the above method can be applied in other fields such as tensor computations or finite element computations.

A description and software version 2.06 source are available, under a LGPL license, from

http://pierre.aubert.free.fr/software/software.php3

Part IX

Use of Second and Higher Derivatives

38

Application of AD to a Family of Periodic Functions

Harley Flanders

ABSTRACT   We study the initial value problem (IVP) for the system $dx/dt = -y^m$, $dy/dt = x^n$, where m and n are positive odd integers, $x(0) = 1$, $y(0) = 0$. Its solutions are periodic, and we find relations between the periods. Automatic differentiation is used to enable interesting mathematical explorations.

38.1   The ODE System

This is an investigation of properties of, and relations between the solutions of a specific family of nonlinear, autonomous ODE. Each system generalizes the harmonic system. The trajectories are periodic, and there are interesting area relations for their closed orbits. The complexity of the system for high m and n provides a test for Automatic Differentiation (AD) methods of numerical solution. We use AD methods for estimating solutions of ODE and for numerical integration. We introduce a method for estimating intervals of convergence based on the Cauchy root test.

Let j and k be positive integers and consider the initial value problem

$$\frac{dx}{dt} = -y^{2j-1} \qquad \frac{dy}{dt} = x^{2k-1} \qquad (38.1)$$
$$x(0) = c > 0 \qquad y(0) = 0.$$

Its solution satisfies the "first integral"

$$\frac{x^{2k}}{k} + \frac{y^{2j}}{j} = \frac{c^{2k}}{k}, \qquad (38.2)$$

a closed, convex curve, symmetric about the origin and the two axes, something like a flattened ellipse. The solution is periodic; we denote its half-period by $\pi_{j,k,c}$. Thus $\pi_{1,1,1} = \pi$. As we shall see, it is always sufficient to consider the normalized case $c = 1$. Hence we define $\pi_{j,k} = \pi_{j,k,1}$.

To prove periodicity, we must show that the trajectory goes around the whole curve. Its velocity vector $\mathbf{v} = (dx/dt,\ dy/dt)$ satisfies

$$|\mathbf{v}|^2 = \left(\frac{dx}{dt}\right)^2 + \left(\frac{dy}{dt}\right)^2 = x^{2(2k-1)} + y^{2(2j-1)}.$$

This function is positive on the compact trajectory, hence bounded away from 0. That is, $|\mathbf{v}|^2 \geq B^2 > 0$, which shows that the point $(x(t), y(t))$ moves completely around the trajectory in finite time.

We next investigate the dependence of $\pi_{j,k,c}$ on c.

Theorem 1 $\pi_{j,k,c} = c^{(j+k-2jk)/j} \pi_{j,k}$.

Proof. With x and y determined by the system above, define functions $u(t)$ and $v(t)$ by

$$x(t) = cu(\alpha t) \qquad y(t) = \beta v(\alpha t),$$

where α and β are determined by $c\alpha = \beta^{2j-1}$ and $\beta\alpha = c^{2k-1}$. We find that $u(t)$ and $v(t)$ satisfy the system

$$\frac{du}{dt} = -v^{2j-1} \quad \frac{dv}{dt} = u^{2k-1} \quad u(0) = 1 \quad v(0) = 0,$$

whose half-period is $\pi_{j,k}$. This forces $\alpha = c^{(2jk-j-k)/j}$, and the theorem follows from the relation $\pi_{j,k} = \alpha \pi_{j,k,c}$. ∎

We have computed the numbers $\pi_{j,k}$ numerically for $1 \leq j, k \leq 63$, and have subjected them to extensive testing for relations. We have drawn several conclusions, then found proofs, all discussed later. Our computations use the package ODE developed by the author and available at ftp.osprey.unf.edu/flanders/ode.

38.2   AD Solution of the IVP

We use AD to compute Taylor polynomial approximations to solutions of the IVP (38.1). If we know the expansions of $x(t)$ and $y(t)$ to degree k, then AD computes the degree k terms of dx/dt and dy/dt, which are the degree $k + 1$ terms of $x(t)$ and $y(t)$, up to constants.

Starting with the initial data, Taylor coefficients of degree zero, we compute the Taylor polynomials of the solution to as high a degree as we please. In our numerical work, we use degree 200. Once we compute the expansions of degree 200 at $t = 0$, we use a modified Cauchy root test to determine a "radius of trust" (§38.2.2) considerably smaller than the radius of convergence, then start over again, to continue the solution as far along the t-axis as we please.

To minimize the number of multiplications involved with the powers in the right sides of (38.1), we use the usual (binary) algorithm that reduces computing a power to a sequence of squarings and multiplications.

38.2.1   Estimating half-periods

The function $y(t)$ is initially zero and increasing. By testing successive intervals of length 0.5 starting with $[0.5, 1]$, we find the first interval in which $y(t)$ changes sign from $+$ to $-$. Then a sequence of successive linear interpolations and IVP solutions pins down the zero of $y(t)$ in that interval; it is the desired half-period. (The zeros of y are simple; where $y = 0$, the first integral (38.2) implies $x \neq 0$, so $\dot{y} \neq 0$.) We have done extensive checking by graphing and tabulating the solutions.

38.2.2 Radius of trust

A standard property [230, pp. 386-388] of convergent power series is that they are dominated by geometric series. Suppose

$$x(t) \approx \sum_0^M x_j t^j$$

is a partial sum of a convergent power series. Geometric convergence means $|x_j t^j| \leq AG^j$, where $0 \leq G < 1$ is a fixed positive number we shall call the *geometric radius*. To make the geometric convergence rapid, we usually take $G \leq 0.5$.

To apply this practically, we should assume that the coefficients are noisy for small j, that is, there may be some transient behavior that makes some of the initial x_j larger than the subsequent pattern. Hence, we consider these inequalities only for $M/2 \leq j \leq M$. Assuming M is fairly large, we may assume $A^{1/j} \approx 1$. Our condition for convergence becomes

$$|x_j|^{1/j} \cdot |t| \leq G \quad \text{for} \quad M/2 \leq j \leq M.$$

That is

$$|t| \leq R = \frac{G}{\max\left(|x_j|^{1/j}\right)} \quad \text{taken over} \quad M/2 \leq j \leq M.$$

We take R as our *radius of trust*. If we assume that *all* the coefficients x_j behave in a regular way for all $j \geq M/2$, then we have a bound for the remainder when $|t| \leq R$:

$$\left| x(t) - \sum_0^M x_j t^j \right| = \left| \sum_{M+1}^\infty x_j t^j \right| \leq \sum_{M+1}^\infty |x_j|\, |t|^j$$

$$\leq \sum_{M+1}^\infty \left(\frac{G}{R}\right)^j |t|^j \leq \sum_{M+1}^\infty G^j = \frac{G^{M+1}}{1-G}.$$

This estimate is sufficiently small when M is large. If $\mathbf{x}(t)$ is a vector, then $|\mathbf{x}|$ is the sup norm. This estimate of the radius of trust, based on the Cauchy root test, seems to be far more robust and accurate than any of the tests that have been proposed based on generalizations of the ratio test.

In analytic continuation of a solution, we have to make an adjustment if R is extremely small, near MachEps.

38.3   Numerical Results

Our ODE program is compiled with Borland Pascal 7.01a. Its real numbers are IEEE 10-byte floating point reals, which is the native type of the Intel 80x87 family of numerical coprocessors. The 10 bytes are split into an 8 byte signed mantissa and a two byte exponent. The interval of the positive reals is about $[3.4 \times 10^{-4932}, 1.1 \times 10^{4932}]$, with 19 to 20 significant decimal digits accuracy.

We computed the half-periods for $1 \leq j$, $k \leq 63$, a total of 3969 solving and shooting problems in 7:57 hours on a Pentium II 400. The raw data is available on the author's FTP site `ftp.osprey.unf.edu/flanders/periods`. A few samples are given in Table 38.1, most with results truncated to 6 significant figures.

TABLE 38.1. $\pi_{j,k}$ for low values of j and k.

j	k: 1	2	3	4	5	6	7	8
1	3.14159	3.70814	4.20654	4.65437	5.06383	5.44315	5.79803	6.13260
2	3.11816	3.70814	4.28563	4.84493	5.38670	5.91267	6.42457	6.92393
3	2.91665	3.38294	3.85524	4.32397	4.78668	5.24288	5.69269	6.13645
4	2.76750	3.14154	3.52683	3.91384	4.29949	4.68263	5.06287	5.44010
5	2.66006	2.96935	3.29113	3.61669	3.94292	4.26853	4.59292	4.91582
6	2.58000	2.84251	3.11743	3.39691	3.67801	3.95943	4.24052	4.52095
7	2.51823	2.74571	2.98507	3.22922	3.47545	3.72249	3.96969	4.21671
8	2.46912	2.66954	2.88115	3.09756	3.31623	3.53599	3.75620	3.97650

38.4   The Area Formula

We return to the original system, with normalized initial data:

$$\frac{dx}{dt} = -y^{2j-1} \quad \frac{dy}{dt} = x^{2k-1} \quad x(0) = 1 \quad y(0) = 0,$$

with half-period $\pi_{j,k}$ and trajectory

$$\frac{x^{2k}}{k} + \frac{y^{2j}}{j} = \frac{1}{k}. \tag{38.3}$$

This convex curve bounds the closed, convex domain

$$D_{j,k} = \left\{ (x, y) \ \Big| \ \frac{x^{2k}}{k} + \frac{y^{2j}}{j} \leq \frac{1}{k} \right\}.$$

Theorem 2 *(Area Formula) The area of $D_{j,k}$ satisfies*

$$|D_{j,k}| = \frac{2j}{j+k} \pi_{j,k} \ , \ \text{that is,} \quad \pi_{j,k} = \frac{j+k}{2j} |D_{j,k}|.$$

In particular, $|D_{k,k}| = \pi_{k,k}$.

Proof. By Green's theorem,

$$\oint_{\partial D_{j,k}} \frac{x\,dy}{k} - \frac{y\,dx}{j} = \iint_{D_{j,k}} \left(\frac{1}{j} + \frac{1}{k}\right) dx\,dy = \frac{j+k}{jk}|D_{j,k}|.$$

By the differential equations,

$$\oint_{\partial D_{j,k}} \frac{x\,dy}{k} - \frac{y\,dx}{j} = \int_0^{2\pi_{j,k}} \left(\frac{x^{2k}}{k} + \frac{y^{2j}}{j}\right) dt = \int_0^{2\pi_{j,k}} \frac{1}{k}\,dt = \frac{2\pi_{j,k}}{k}.$$

∎

Corollary 1 $\lim_{k\to\infty} \pi_{j,k} = \infty.$

Proof. The convex domain $D_{j,k}$ contains the rhombus with vertices $(\pm 1,\ 0)$ and $(0, \pm(j/k)^{1/(2j)})$. It follows that $|D_{j,k}| > 2(j/k)^{1/(2j)}$, hence by the theorem,

$$\pi_{j,k} > \frac{j+k}{2j} \cdot 2(j/k)^{1/(2j)} > \frac{j+k}{j} \frac{j^{1/(2j)}}{k^{1/(2j)}} = j^{1/(2j)}\left(\frac{1}{k^{1/(2j)}} + \frac{k}{j \cdot k^{1/(2j)}}\right).$$

As $k \to \infty$, the first summand $\to 0$ while the second summand $\to \infty$ because $j \geq 1$. ∎

Lemma 3 *The sequence of domains* $\{D_{k,k}\}$ *is strictly increasing.*

Proof. Suppose $j < k$ and $(x,\ y) \in \partial D_{j,\ j}$. Then $x^{2j} + y^{2j} = 1$, so $|x| \leq 1$ and $|y| \leq 1$. Therefore $x^{2k} + y^{2k} \leq x^{2j} + y^{2j} = 1$, with equality only for $(0,\ \pm 1)$ and $(\pm 1,\ 0)$. It follows that $(x,\ y) \in D_{k,k}$ so $D_{j,j} \subseteq D_{k,k}$. The inclusion is strict because, for instance, each corner $(x,\ x)$ of $D_{k,k}$ does not belong to $D_{j,j}$. ∎

Theorem 3 *The sequence* $\{\pi_{k,\ k}\}_k$ *is strictly increasing, with limit 4.*

Proof. Strictly increasing follows from Lemma 3 and the area formula. It is clear that the regions D_k converge to the square whose vertices are $(\pm 1, \pm 1)$ and whose area is 4. ∎

Remark 1 The situation with the main diagonal $\pi_{j,\ j}$ is interesting. The corresponding functions $y_j(t)$ converge to a limit function $y(t)$ that is piecewise linear, with vertices $(0,\ 0)$, $(1,\ 1)$, $(3,\ 1)$, $(5,\ -1)$, $(7,\ -1)$, $(8,\ 0)$, extended both directions by the period 8. The corresponding limit function $x(t) = y(t + 2)$.

Remark 2 More generally, a similar conclusion is probably true for any diagonal $\{\pi_{a+j,\ j}\}_j$ or $\{\pi_{j,\ b+\ j}\}_j$ parallel to the main diagonal.

Conjecture 1 *For each j, the sequence* $\{\pi_{j,\ k}\}_k$ *is strictly increasing.*

38.5   Relation to the Γ Function

Let us recall the basic formulas for Eulerian integrals [5]:

$$B(x, y) = \int_0^1 t^{x-1}(1-t)^{y-1}\,dt = \frac{\Gamma(x)\,\Gamma(y)}{\Gamma(x+y)}, \quad \Gamma(x) = \int_0^\infty t^{x-1}e^{-t}\,dt,$$

$$(38.4)$$

for $x, y > 0$, and the recursion formula $\Gamma(x+1) = x\,\Gamma(x)$.

Theorem 4

$$\pi_{j,k} = \frac{1}{j}\left(\frac{j}{k}\right)^{1/(2j)}\frac{\Gamma\left(\frac{1}{2j}\right)\Gamma\left(\frac{1}{2k}\right)}{\Gamma\left(\frac{1}{2j}+\frac{1}{2k}\right)}.$$

Proof. The domain $D_{j,k}$ is symmetric in both coordinate axes, consequently

$$|D_{j,k}| = \iint_{D_{j,k}} dx\,dy = 4\iint_{\substack{D_{j,k}\\ x,y\geq 0}} dx\,dy = 4\int_0^1 y(x)\,dx,$$

where $y(x)$ is obtained by solving (38.3) for y. Hence

$$|D_{j,k}| = 4\left(\frac{j}{k}\right)^{1/(2j)}\int_0^1 \left(1-x^{2k}\right)^{1/(2j)}dx.$$

The substitution $t = x^{2k}$, with $dx = (1/2k)\,t^{(1/2k)-1}\,dt$, yields

$$|D_{j,k}| = \frac{2}{k}\left(\frac{j}{k}\right)^{1/(2j)}B\left(\frac{1}{2j}+1,\ \frac{1}{2k}\right).$$

By (38.4), the Beta function on the right can be expressed in terms of Gamma functions. After simplifications, we arrive at

$$|D_{j,k}| = \frac{2}{j+k}\left(\frac{j}{k}\right)^{1/(2j)}\frac{\Gamma\left(\frac{1}{2j}\right)\Gamma\left(\frac{1}{2k}\right)}{\Gamma\left(\frac{1}{2j}+\frac{1}{2k}\right)}.$$

The result follows from Theorem 2. ∎

38.6   Relations Between Half-Periods

Theorem 5 *For all $j, k \geq 1$, we have $\pi_{j,k} = (j/k)^{(j+k-2jk)/(2jk)}\pi_{k,j}$.*

Proof. Consider the system $\dot{x} = -y^{2j-1}$, $\dot{y} = x^{2k-1}$, $x(0) = 1$, $y(0) = 0$, with half-period $\pi_{j,k}$. Define functions $u = -y$ and $v = x$. They satisfy $\dot{u} = -v^{2k-1}$, $\dot{v} = u^{2j-1}$, $u(0) = 0$, $v(0) = 1$. We know the half-period for this system in two ways. First, it is obviously the same as the period of the x, y system, that is, $\pi_{j,\,k}$. Its half-period also is $\pi_{k,\,j,\,c}$, where c is determined by the trajectory

$$\frac{u^{2j}}{j} + \frac{v^{2k}}{k} = \frac{x^{2k}}{k} + \frac{y^{2j}}{j} = \frac{1}{k}.$$

Now $c > 0$ is the value of u when $v = 0$, that is, $c = (j/k)^{1/(2j)}$, so we have

$$\pi_{j,\,k} = \pi_{k;\,j,\,c} = c^{(j+k-2jk)/k}\,\pi_{k,\,j} = (j/k)^{(1/2j)[(j+k-2jk)/k]}\,\pi_{k,\,j}$$
$$= (j/k)^{(j+k-2jk)/(2jk)}\,\pi_{k,\,j}. \ \blacksquare$$

When $j = 1$, we have the special case $\pi_{1,\,k} = k^{(k-1)/(2k)}\,\pi_{k,\,1}$. More is true of these particular half-periods. On the basis of extensive computer experiments, we conjectured the following theorem, and then checked it to our computational accuracy, 16–18 significant decimal digits, for all $k \le 63$. The idea for a proof based on the Gamma function was suggested to the author independently by Jonathan M. Borwein and Richard A. Askey.

Theorem 6 $\pi_{1,\,k} = \frac{k^{1/2}}{2^{(k-1)/k}}\,\pi_{k,\,k}.$

Proof. By Theorem 4,

$$\pi_{1,\,k} = \left(\frac{1}{k}\right)^{1/2} \frac{\Gamma\left(\frac{1}{2}\right)\Gamma\left(\frac{1}{2k}\right)}{\Gamma\left(\frac{1}{2} + \frac{1}{2k}\right)} \quad \text{and} \quad \pi_{k,\,k} = \frac{1}{k}\frac{\Gamma\left(\frac{1}{2k}\right)^2}{\Gamma\left(\frac{1}{k}\right)},$$

so the required formula is equivalent to

$$\left(\frac{1}{k}\right)^{1/2} \frac{\Gamma\left(\frac{1}{2}\right)\Gamma\left(\frac{1}{2k}\right)}{\Gamma\left(\frac{1}{2} + \frac{1}{2k}\right)} = \frac{k^{1/2}}{2^{(k-1)/k}} \cdot \frac{1}{k}\frac{\Gamma\left(\frac{1}{2k}\right)^2}{\Gamma\left(\frac{1}{k}\right)}.$$

This relation simplifies to the case $x = 1/(2k)$ of the "Duplication Formula" of Gamma functions, combined with the known value $\Gamma(1/2) = \pi^{1/2}$. $\blacksquare$

For example, for $k = 4$ and $k = 7$, Theorem 6 asserts that $4.65437 = 2^{1/4} \cdot 3.91384$ and $5.79803 = (\sqrt{7}/2^{6/7}) \cdot 3.96970$, which check to 4 decimal places. Note that Theorem 5 could have been proved this way too.

In seeking other relations we have tested $\pi_{m,\,1} = (p/q)^{r/s}\pi$ for $m = 3$, 5, 7 and $\pi_{m,3} = (p/q)^{r/s}\pi_{m,m}$ for $m = 5$, 7 for $1 \le q < p \le 200$ and $1 \le r$, $s \le 200$ with negative results.

The following conjectures are based on examination of a lot of data.

Conjecture 2 *Each half-period $\pi_{j,\,k}$ is a transcendental number.*

Conjecture 3 *The set of diagonal half-periods $\{\pi_{k,\,k}\}$ is algebraically independent over the rationals Q.*

Conjecture 4 *For each k, the sequence $\{\pi_{j,\,k}\}_j$ is eventually strictly decreasing, with limit 0.*

"Eventually" may be quite far along. The sequence $\{\pi_{j,\,k}\}_j$ for $3 \leq k \leq 63$ increases as j goes from 1 to 2, then decreases. The sequence $\{\pi_{j,\,126}\}_j$ increases as j goes from 1 to 3, then decreases.

38.7 Integration

It is not well-known that AD supplies a fast and accurate tool for approximate integration. Briefly, to integrate a function on an interval, use AD to compute a Taylor expansion of the function about the center of the interval; then simply write down the Taylor expansion of its indefinite integral. This will be adequate for an accurate integral in some subinterval centered on the midpoint; if there are remaining intervals on each side, do these the same way, etc. Tests indicate that this is more accurate than and as fast as any of the standard methods of numerical integration.

To return to our study of half-periods: The area formula gives another way to compute the $\pi_{j,\,k}$:

$$\pi_{j,\,k} = \frac{2(j+k)}{j} \left(\frac{j}{k}\right)^{1/2j} \int_0^1 \left(1 - x^{2k}\right)^{1/2j} dx.$$

Because the integrand is singular at $x = 1$, it is better to integrate in the x-direction part of the way, do the remainder in the y-direction, then subtract the area counted twice. We choose $x = (1/2)^{1/2k}$ as the dividing value, with the corresponding $y = (j/2k)^{1/2j}$. The result is

$$\pi_{j,\,k} = A_{j,\,k} \cdot \left(\int_0^{(1/2)^{1/2k}} \left(1 - x^{2k}\right)^{1/2j} dx \right.$$
$$\left. + \int_0^{(1/2)^{1/2j}} \left(1 - y^{2j}\right)^{1/2k} dy - B_{j,\,k} \right),$$

where

$$A_{j,\,k} = \frac{2(j+k)}{j} \left(\frac{j}{k}\right)^{1/2j} \quad \text{and} \quad B_{j,\,k} = \left(\frac{1}{2}\right)^{(j+k)/2jk}.$$

We have done this calculation by AD for the same range as earlier. The results agree with those previously computed to 16-18 significant digits.

39

FAD Method to Compute Second Order Derivatives

Yuri G. Evtushenko, E. S. Zasuhina and V. I. Zubov

ABSTRACT We develop a unified methodology for computing second order derivatives of functions obtained in complex multistep processes and derive formulas for Hessians arising in discretization of optimal control problems. Where a process is described by continuous equations, we start with
a discretization scheme for the state equations and derive exact gradient
and Hessian expressions. We introduce adjoint systems for auxiliary vectors
and matrices used for computing the derivatives. A unique discretization
scheme is automatically generated for vector and matrix adjoint equations.
The structure of the adjoint systems for some approximation schemes is
found. The formulas for second derivatives are applied to examples.

39.1   Introduction

The numerical solution of optimization problems often is obtained using
first and second derivatives of composite functions. Hence, accuracy and
run time of derivative calculation essentially influence the effectiveness of
optimization algorithms. A unified methodology for computing exact gradients of functions obtained in complex multistep processes was suggested
in [176]. This approach based on the generalization of a FAD method has
been applied to boundary control problem of Burgers' equation [177], to a
problem of melting metals [7], and to an inverse problem of Burgers' equation. Here, we extend this approach to calculate the second order derivatives of composite functions. The problem of computation of high order
derivatives was considered also in [262].

39.2   Basic Expressions

Suppose the mappings $W : R^n \times R^r \to R^1$, $\Phi : R^n \times R^r \to R^n$ are twice
continuously differentiable. Let $z \in R^n$ and $u \in R^r$ satisfy the nonlinear
system of n scalar algebraic equations:

$$\Phi(z, u) = 0_n, \text{ where } 0_n \text{ is an } n\text{-dimensional null vector.} \quad (39.1)$$

We will use following notation. For $f : R^n \times R^r \to R^m$, let $f_z^\top(z, u)$ denote the $n \times m$ - matrix whose ij-th element is $\partial f^j / \partial x^i$. Let $f_{z^\top}(z, u)$ denote $(f_z^\top(z, u))^\top$. We assume that the matrix $\Phi_z^\top(z, u)$ is nonsingular. According to the implicit function theorem, system(39.1) defines a twice continuously differentiable function $z = z(u)$, $\Phi(z, u) = 0_n$. As a rule, z and u are referred to as dependent and independent variables respectively. Thus, the composite function $\Omega(u) = W(z(u), u)$ is twice continuously differentiable. As it was shown in [176], the way to calculate the gradient of function $\Omega(u) = W(z(u), u)$ suggested by the implicit function theorem is much more labour-consuming than another method consisting of solving an adjoint system and differentiating the Lagrange function.

This chapter extends this approach to the calculation of second order derivatives of function $\Omega(u) = W(z(u), u)$. Expressions for calculating the exact gradient of $\Omega(u)$ can be derived in the form

$$L_p(z, u, p) = \Phi(z, u) = 0_n, \tag{39.2}$$

$$L_z(z, u, p) = W_z(z, u) + \Phi_z^\top(z, u)p = 0_n, \tag{39.3}$$

$$L_u(z, u, p) = d\Omega(u)/du = W_u(z(u), u) + \Phi_u^\top(z(u), u)p, \tag{39.4}$$

where $L(z, u, p) = W(z, u) + p^\top \Phi(z, u)$ is a Lagrange function with the Lagrange multiplier $p \in R^n$.

We select the l-th component of the vector u and define by the same technique the gradient of the new function $\widehat{\Omega}(u) = \widehat{W}(z(u), u, p(u)) = d\Omega/du^l$ under functional relations (39.2)-(39.3). This gradient is the l-th row of the matrix $d^2\Omega/du^2$. We introduce the new Lagrange function

$$\widehat{L}(z, u, p, \mu, \nu) = \widehat{W}(z, u, p) + \mu^\top \Phi(z, u) + \nu^\top (W_z(z, u) + \Phi_z^\top p),$$

with $\mu \in R^n$, and $\nu \in R^n$. We calculate z, p, μ, ν, and $\nabla\widehat{\Omega}$ as

$$\widehat{L}_\mu(z, u, p, \mu, \nu) = \Phi(z, u) = 0_n, \tag{39.5}$$

$$\widehat{L}_\nu(z, u, p, \mu, \nu) = W_z(z, u) + \Phi_z^\top(z, u)p = 0_n, \tag{39.6}$$

$$\widehat{L}_p(z, u, p, \mu, \nu) = \widehat{W}_p(z, u, p) + (\Phi_z^\top(z, u))^\top \nu = 0_n, \tag{39.7}$$

$$\widehat{L}_z(z, u, p, \mu, \nu) = \widehat{W}_z(z, u, p) + \Phi_z^\top(z, u)\mu$$
$$+ (W_z(z, u) + \Phi_z^\top(z, u)p)_z^\top \nu = 0_n, \tag{39.8}$$

$$\widehat{L}_u(z, u, p, \mu, \nu) = \nabla\widehat{\Omega}(u) = \widehat{W}_u(z, u, p) + \Phi_u^\top(z, u)\mu$$
$$+ (W_z(z, u) + \Phi_z^\top(z, u)p)_u^\top \nu. \tag{39.9}$$

Systems (39.5)-(39.6) defining $z(u)$ and $p(u)$ coincide with systems (39.2)-(39.3). Therefore, for the second derivative calculation, we solve $2r$ linear systems in a matrix that has already been used during the computation of the gradient, where r is the dimension of independent variable or control.

System (39.8) is identical to the adjoint system (39.6) except for the right hand side, and the system (39.7) involves the transposed matrix of the adjoint system. Joining formulas (39.5)-(39.9) for all components of the gradient of $\Omega(u)$ and rewriting them in matrix form, we obtain

$$\tilde{L}^j_{\mu_i} = \Phi(z, u)\delta_{ij} = 0_n, i = 1, 2, \ldots, r, \tag{39.10}$$

$$\tilde{L}^j_{\nu_i} = L_z\delta_{ij} = 0_n, i = 1, 2, \ldots, r, \tag{39.11}$$

$$\tilde{L}_{p^\top} = L_{up^\top} + \Lambda\Phi^\top_z = 0_{rn}, \tag{39.12}$$

$$\tilde{L}_{z^\top} = L_{uz^\top} + \Gamma\Phi_{z^\top} + \Lambda L_{zz^\top} = 0_{rn}, \tag{39.13}$$

$$\tilde{L}_{u^\top} = d^2\Omega/du^2 = L_{uu^\top} + \Gamma L_{pu^\top} + \Lambda L_{zu^\top}, \tag{39.14}$$

where $\quad \tilde{L}(z, u, p, \Gamma, \Lambda) = L_u(z, u, p) + \Gamma L_p + \Lambda L_z,$

$\Gamma \in R^{r\times n}$, $\Lambda \in R^{r\times n}$, $\Gamma^\top = (\gamma_1 \ldots \gamma_r)$, $\Lambda^\top = (\lambda_1 \ldots \lambda_r)$, γ_i, λ_i are the Lagrange multipliers corresponding to $L_{u^i}, i = 1, 2, \ldots, r$.

In multistep problems, the variables z and u are usually partitioned into k variables of lower dimensionality:

$$z^\top = [z_1^\top, z_2^\top, \ldots, z_k^\top], \quad u^\top = [u_1^\top, u_2^\top, \ldots, u_k^\top],$$
$$z_i \in R^s, \quad u_i \in R^m, \quad 1 \le i \le k.$$

The system (39.1) is split into k relations:

$$z_i = F(i, Z_i, U_i), \quad 1 \le i \le k, \quad n = s \cdot k, \quad r = m \cdot k, \tag{39.15}$$

where Z_i and U_i are given sets of variables z_j and u_j, respectively. As in [176], we introduce following sets of indices

$$\tilde{Q}_i = \{j \in D : z_i \in Z_j\}, \quad \tilde{K}_i = \{j \in D : u_i \in U_j\},$$

where $D = \{1, \ldots, k\}$. In agreement with partitioning the variables z and u into k variables of dimension s and m, respectively, divide the sets of rows and columns of all matrices from (39.2)-(39.4) and (39.7)-(39.9) into k block-rows and k block-columns. After such separation these matrices may be considered as matrices of block elements. The matrix $\tilde{F}_z^\top$ (denoted by B), defines the structure of all systems (39.3), (39.7)-(39.8). This matrix B consists of k^2 blocks B_{ij}

$$B_{ij} \in R^{s\times s}, \quad i \in D, \quad j \in D.$$

In the i-th block-row of matrix B

$$\begin{cases} B_{ij} = F_{z_i}^\top(j, Z_j, U_j), & j \in \tilde{Q}_i, \\ B_{ij} = 0_{s\times s}, & j \notin \tilde{Q}_i. \end{cases}$$

Splitting systems (39.2)-(39.4) and (39.7)-(39.9) like z and u, we may rewrite expressions for calculating first and second order derivatives of the composite function $\Omega(u)$ as

$$L_{p_i}(z, u, p) = F(i, Z_i, U_i) - z_i = 0_s, \quad i \in D, \tag{39.16}$$

$$L_{z_i}(z, u, p) = W_{z_i}(z, u) + (i\text{-th block-row of } B\ )p - p_i = 0_s, \tag{39.17}$$

$$L_{u_i}(z, u, p) = \frac{d\Omega}{du_i} = W_{u_i}(z, u) + (i\text{-th block-row of } \tilde{F}_u^\top\ )p, \tag{39.18}$$

$$\widehat{L}_{p_i}(z, u, p, \mu, \nu) = \widehat{W}_{p_i}(z, u, p) + (i\text{-th block-column of } B)\nu - \nu_i = 0_s, \tag{39.19}$$

$$\widehat{L}_{z_i}(z, u, p, \mu, \nu) = \widehat{W}_{z_i}(z, u, p) + (i\text{-th block-row of } B\ )\mu$$
$$- \mu_i + (L_z)_{z_i}^\top \nu = 0_s, \tag{39.20}$$

$$\widehat{L}_{u_i}(z, u, p, \mu, \nu) = \widehat{\Omega}_{u_i} = \widehat{W}_{u_i}(z, u, p)$$
$$+ (i\text{-th block-row of } \tilde{F}_u^\top\ )\mu + (L_z)_{u_i}^\top \nu, \tag{39.21}$$

$$\text{where} \quad p^\top = [p_1^\top, p_2^\top, \ldots, p_k^\top], \quad p_i \in R^s, \quad 1 \le i \le k.$$
$$\mu^\top = [\mu_1^\top, \mu_2^\top, \ldots, \mu_k^\top], \quad \nu^\top = [\nu_1^\top, \nu_2^\top, \ldots, \nu_k^\top]$$
$$\mu_i \in R^s, \ \nu_i \in R^s, \ 1 \le i \le k,$$
$$\widehat{W}(z, u, p) = L_{u^l}(z, u, p)\ ,$$

and u^l is the l-th component of u.

39.3   The Optimal Control Problem

Let a process be governed by a system of ordinary differential equations:

$$dz/dt = f(t, z, u, \xi), \quad T_1 \le t \le T_2, \quad z(T_1, z_1) = z_1, \tag{39.22}$$

where $z \in R^s$ is the state vector, and the control u is an arbitrary piecewise continuous function of t with values in U, a given compact subset of the space R^m. The vector of design parameters is $\xi \in V \subset R^q$. As a rule, T_1, T_2, and z_1 are fixed. If T_1, T_2, and z_1 must be optimized, then they are included into the vector ξ.

The problem is to find a control function $u(t) \in U$ and a vector of design parameters $\xi \in V$ that minimize the cost function $W(z(T_2, z_1), \xi)$ subject to mixed constraints on the state, the control, and the vector of design parameters.

$$g(t, z(t), u(t), \xi) = 0, \quad q(t, z(t), u(t), \xi) \le 0, \quad T_1 \le t \le T_2. \tag{39.23}$$

The application of various finite difference approximations to the problem leads us to a variety of multistep processes and systems (39.17)-(39.21). The structure of these adjoint systems is determined by matrix B. We study matrices B obtained by various schemes of discrete approximations. To pass from a continuous problem to its discrete analogue, the segment

$[T_1, T_2]$ is divided into $(N-1)$ parts $(t_1 = T_1, t_2, \ldots, t_n = T_2$ are the points of division, $\tau_i = t_{i+1} - t_i, \quad i = 1, \ldots, N-1)$, the functions $z(t)$ and $u(t)$ are substituted by piecewise constant ones, and the constraints (39.23) are considered at each point of the finite-difference grid.

The Euler discretization of (39.22) is given by the formulas

$$z_i = z_{i-1} + \tau_{i-1} f(t_{i-1}, z_{i-1}, u_{i-1}, \xi) = F(t_{i-1}, z_{i-1}, u_{i-1}, \xi). \qquad (39.24)$$

Obviously,

$$\tilde{Q}_i = \{i+1\}, \quad \tilde{K}_i = \{i+1\}, \quad i = 1, \ldots, N-1, \tilde{Q}_N = \emptyset, \quad \tilde{K}_N = \emptyset.$$

Therefore, in each block-row of matrix B there is only one nonzero block

$$B_{i,i+1} = E_{s \times s} + \tau_i f_z^\top (t_i, z_i, u_i, \xi), \quad i = 1, \ldots, N-1 \ .$$

Thus, the Euler discretization of this problem results in a very simple structure of the adjoint system matrix - its nonzero blocks are located just above the diagonal. The examination of expressions (39.16)-(39.21) leads to the following conclusions. After the $\Omega(u)$ gradient determination, to define an adjoint problems concerning calculation of $d^2\Omega/du^2$ it is necessary to define only the right hand sides of systems (39.19) and (39.20) determined by matrices $\tilde{F}_u^\top$, $(B \cdot p)_z^\top$, $(B \cdot p)_u^\top$ (denote them by Bu, BPz, and BPu, respectively). The matrix Bu has the same structure as B. The matrices BPz and BPu have block-diagonal structure with vanishing N-th block-row. For each l-th column of $d^2\Omega/du^2$ ($l = s \cdot (i_0 - 1) + j_0$), the Lagrange multipliers $\nu_1, \ldots, \nu_{i_0}$ are equal to zero. Since the matrix $d^2\Omega/du^2$ is symmetric, it is sufficient to determine its lower left triangle. Because of (39.21), to define the l-th column in this triangle it is enough to know only $\mu_{i_0+1}, \ldots, \mu_N$ amongst all $\mu_1, \ldots, \mu_N$.

Therefore, to define the l-th column of the triangle mentioned, it is enough to solve not the $2N \cdot s$ equations of systems (39.19) and (39.20), but only $2(N - i_0) \cdot s$ equations. In general, to define the whole of this triangle it is necessary to solve not the $2N \cdot s \cdot (N-1) \cdot m$ equations of adjoint systems (39.19) and (39.20), but only $N \cdot s \cdot (N-1) \cdot m$ equations, i.e. half as many. Hence one may say that on average, determining one column of $d^2\Omega/du^2$ requires the solution of one linear system similar to the adjoint system to be solved to calculate the gradient of $\Omega(u)$.

To take into account the constraints (39.23), we introduce additional functions (for example, a penalty function, the Lagrangian, the modified Lagrangian, and so on). According to (39.16)-(39.21), the formulas for determining all derivatives will be similar to the ones obtained if we replace W and its derivatives by these functions and their derivatives.

For some other finite-difference approximations of the optimal control problem, we obtain the following structures of the matrix B, which determine the corresponding adjoint systems.

Runge-Kutta order 4   Adams scheme order 4   Implicit scheme

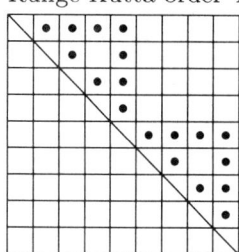
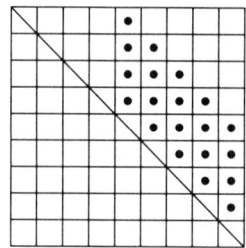
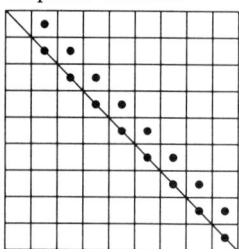

Nonzero blocks are marked with small circles, and the formulas for calculating them are simple. There are many identical blocks amongst them.

39.4   Numerical results

We applied our method to some problems of optimal control of a process governed by ordinary differential equations (39.22). The original problem was reduced by an Euler approximation to a nonlinear programming problem, which was solved by Newton's method. The gradient and the second order derivatives were computed by formulas (39.16)-(39.21). We considered the following test problems. Problem 1.a, Problem 1.b and Problem 1.c are variants of the problem

$$f^{\top}(z(t), u(t), \xi) = [z^1(t), (1 - z^0(t)^2)z^1(t) - z^0(t) + u^0(t),$$
$$z^0(t)^2 + z^1(t)^2 + u^0(t)^2],$$
$$T_0 \le t \le T_f, \qquad s = 3, \quad m = 1,$$
$$W(T_0, T_f, z(T_f), u(T_f), \xi) = z^2(T_f).$$

with the initial control $u(\cdot) \equiv 0$ and boundary conditions.

Problem 1.a. $z(T_0)^{\top} = [3, 0, 0]$, $T_0 = 0$, $T_f = 10$, $-10^{10} \le u^0(\cdot) \le 10^{10}$.

Problem 1.b. $z(T_0)^{\top} = [0, 1, 0]$, $T_0 = 0$, $T_f = 5$, $-10^{10} \le u^0(\cdot) \le 10^{10}$.

Problem 1.c. $z(T_0)^{\top} = [0, 1, 0]$, $T_0 = 0$, $T_f = 5$, $-0.8 \le u^0(\cdot) \le 0.8$.

Problem 2. Catalyst for two successive chemical reactions ($s = 2$ and $m = 1$).

$$f^{\top}(z(t), u(t), \xi) = [u^0(t)(10z^1(t) - z^0(t)), -u^0(t)(10z^1(t) - z^0(t))$$
$$- (1 - u^0(t))z^1(t)],$$
$$z(T_0)^{\top} = [1, 0], \qquad T_0 = 0, \quad T_f = 1,$$
$$W(T_0, T_f, z(T_f), u(T_f), \xi) = z^0(T_f) + z^1(T_f), \qquad 0 \le u^0(\cdot) \le 1,$$

and the initial control $u(\cdot) \equiv 0.5$.

Problem 3. Chemical reaction between gases ($s = 2$ and $m = 1$)

$$f^0(z(t), u(t), \xi) = -2c^1 z^0(t) u^0(t)/(2c^0 + z^1(t)),$$
$$f^1(zt), u(t), \xi) = -2f^0(z(t), u(t), \xi) - c^2 (u^0(t))^2 (2z^1(t)/(2c^0 + z^1(t)))^2,$$
$$c^0 = 1.475 \times 10^{-2}, \quad c^1 = 1.8725688803 \times 10^7, c^2 = 4.5304 \times 10^{-2},$$
$$z(T_0)^\top = [0.0105, 0.0085], \quad T_0 = 0, \quad T_f = 8,$$
$$W(T_0, T_f, z(T_f), u(T_f), \xi) = -100z^1(T_f), \quad 0 \le u^0(\cdot) \le 1,$$

and the initial control $u(\cdot) \equiv 0$.

All these problems are detailed in [231]. Calculations on Problems 1.a, 1.b, 1.c, and 2 showed that Newton's method finds solutions in 1/3 to 1/2 the number of iterations of the conjugate gradient method, but requires approximately the same computation time. Problem 3 is "stiff," and therefore was considered solely in following numerical experiment. We measured times required for computing the cost function, the gradient of the composite function $\Omega(u)$, and the Hessian $d^2\Omega/du^2$, denoted by t_f, t_{gr}, t_{hess}, respectively. The ratios t_{gr}/t_f and t_{hess}/t_f are almost independent of the initial conditions and control, rise insignificantly as the number of grid points (dimension of control) increases, but depend on the problem. On average, the ratio t_{gr}/t_f is 2.2 for Problem 3, 2.6 for Problem 2, and 2.75 for Problem 1. In all cases, it did not exceed 3. On average, the ratio t_{hess}/t_f is $0.52r$ for Problem 3 (r - dimension of control), $1.08r$ for Problem 2, and $1.97r$ for Problem 1. In all considered cases it did not exceed $2r$.

40

Application of Higher Order Derivatives to Parameterization

Jean-Daniel Beley, Stephane Garreau, Frederic Thevenon and Mohamed Masmoudi

ABSTRACT Research on automatic differentiation is mainly motivated by gradient computation and optimization. However, in the optimal design area, it is quite difficult to use optimization tools. Some constraints (e.g., aesthetics constraints, manufacturing constraints) are quite difficult to describe by mathematical expressions. In practice, the optimal design process is a dialog between the designer and the analysis software (structural analysis, electromagnetism, computational fluid dynamics, etc.). One analysis may take a while. Hence, parameterization tools such as design of experiments (D.O.E.) and neural networks are used. The aim of those tools is to build surrogate models. We present a parameterization method based on higher order derivatives computation obtained by automatic differentiation.

40.1 Introduction

The most popular parameterization methods are neural networks, design of experiments, and resolution of simplified models. We present a new design optimization approach based on automatic differentiation (AD). We first compute higher order derivatives of full models with respect to some parameters. Then we build a surrogate model using a Taylor expansion or a Padé approximation (rational expression).

Higher order derivatives have been used to solve stiff differential equations (see [240, 242]). We will give some industrial applications of this method showing that the higher order derivatives method is far more efficient than classical parameterization methods. One unexpected application of higher order derivatives is to catch singularities, poles, and the corresponding modes in acoustics, electromagnetism, and structural analysis.

40.2 Presentation of the Method

Let us denote by $P := \Pi_{i=1}^{n}[p_i^{min}, p_i^{max}]$ the parameters set and $p_0 \in P$ the reference parameter. An industrial CAE (Computer Aided Engineering) environment is based on several communicating codes. Figure 40.1 gives the general architecture of a CAE environment. Here are some constraints to be considered:

- The CAD, preprocessing, and postprocessing modules are commercial codes, whose source code is not generally available. There are still some in-house solvers, but the trend is to use commercial solvers.

- The solvers are huge Fortran programs.

The solvers are based on finite element methods, finite volume methods, finite differences methods, integral equations, etc. Behind these approximation methods are partial differential equations models. The solution of a partial differential equation usually depends analytically on the design parameters [262]. This analyticity is not transmitted systematically to the approximated problem, so some precautions are needed to preserve this analyticity. Our goal is to compute higher order derivatives of the solution $U(p)$ with respect to design parameters p. In the following subsections we will describe each component and the corresponding parameterization technique.

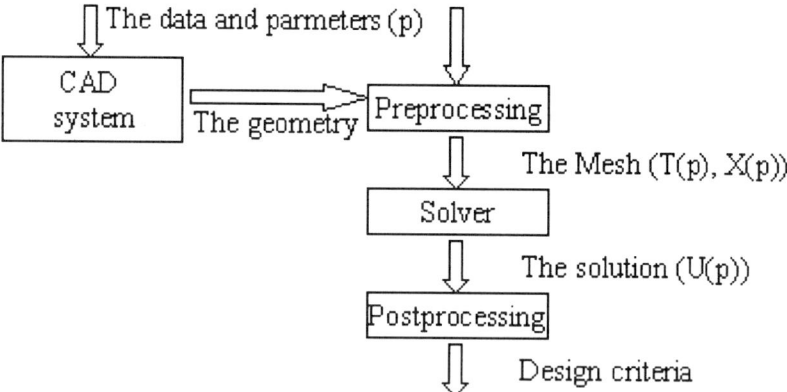

FIGURE 40.1. The CAE environment

40.2.1   The CAD system and the preprocessing step

The CAD system creates the geometry of the model. Modern CAD systems handle geometry parameterization. We will use this native parameterization to build mesh perturbations. The physical model is defined in the preprocessing environment (boundary conditions, material properties, etc.). The mesh (decomposition of the domain in small elements) is generated at this step. A mesh is defined by two tables. Typically for a tetrahedral element we have

- Nodes table $X = (x_i, y_i, z_i)_{i=1,...,N}$, where N is the number of nodes

- Elements table $T = (I_k^1, I_k^2, I_k^3, I_k^4)_{k=1,...,N_e}$, where N_e is the number of elements. The integers $1 \leq I_k^1, \ldots, I_k^4 \leq N$ are indices of the real table X.

For example $X(I_k^1)$ gives the coordinates of the first vertex of the kth tetrahedron. Two arbitrary closed sets of design parameters can produce completely different meshes with different N_e and N values. Thus the continuity of the map $p \longmapsto X(p)$ does not make sense.

Our idea is to build an analytical mesh. A reference mesh is created for p_0 using the preprocessing environment. The meshes for other values of $p \in P$ will be deduced as follows:

- $T(p) = T(p_0)$ for any $p \in P$, and

- The map $p \mapsto X(p)$ is analytic.

The software Adomesh of Cadoé creates these mesh perturbations that should fit with the geometry and satisfy some standard rules concerning the shape of the elements. The map $p \longmapsto X(p)$ is approximated by a multivariate polynomial by an interpolation technique. For each interpolation point, we use native geometry parameterization.

40.2.2   The solver

Most of the computing time is spent at this step. Our objective is to compute the parameterization of this step with a small additional cost.

The regularity of the parametric solution

For the sake of simplicity, we consider only the linear case:

$$A(p)U(p) = B(p),$$

where $A(p) = A(X(p))$ and $B(p) = B(X(p))$. Generally A and B are the sum of elementary matrices: $A = \sum_{i=1}^{Ne} A_i$ and $B = \sum_{i=1}^{Ne} B_i$.

Let us recall that f is meromorphic if and only if there exists two holomorphic functions h and g such that $f = h/g$. Now we can see that $p \longmapsto U(p)$ is meromorphic:

- by construction the map $p \longmapsto X(p)$ is analytic,

- the maps $X \longmapsto A_i(X)$ and $X \longmapsto B_i(X)$ are analytic (A_i and B_i are computed by a sequence of regular instructions independent of X),

- the maps $(A_i)_{i=1}^{Ne} \longmapsto A = \sum_{i=1}^{Ne} A_i$ and $(B_i)_{i=1}^{Ne} \longmapsto B = \sum_{i=1}^{Ne} B_i$ are analytic, and

- the map $(A, B) \longmapsto U = A^{-1}B$ is meromorphic if the matrix A is invertible.

The composition of those four maps is exactly the map $p \longmapsto U(p)$. Then it is natural to approximate this map by a Padé approximation: $U(p) \approx P/Q$, where P and Q are polynomials (Q has scalar coefficients).

The singularities of $A(p)$ can be defined by a couple (p_s, U_s) such that $A(p_s)U_s = 0$ and $U_s \neq 0$. The vector U_s is called the mode associated with p_s. If p_s is a zero of Q, then $U_s = P(p_s)$ is the corresponding mode. This remark gives an efficient way to catch singularities.

Higher order derivatives of the solution

Thanks to the analyticity we can compute higher order derivatives of $U(p)$. We first consider monovariate or directional derivatives: p is a scalar. The first variation U' of U with respect to p is the solution of the linear system $A(p_0)U' = B(p_0)' - A(p_0)'U$, and $U^{(d)}$ is the solution of the system

$$A(p_0)U^{(d)} = B^{(d)} - \sum_{i=1}^{d} \binom{d}{i} A^{(i)} U^{(d-i)}.$$

AD is used to compute higher order derivatives of A and B. We use our code Adogen to generate these derivatives. Adogen is designed to generate higher order derivatives of Fortran programs following the ideas of Griewank [233]. If N is the size of the problem, the space and time complexities behave like $O(N)$ for a given order.

Our method takes advantage of the matrix decomposition obtained in the first analysis. The computation of the Taylor expansion needs only to solve some already factorized systems. This method is particularly powerful for large scale problems: the rate

complexity of the factorization / complexity of the factorized system

increases with the dimension of the system. Even with iterative solvers, it is possible to save computing time using deflation methods.

40.2.3   The postprocessing step

Generally the most significant criteria are not differentiable. The maximum of the solution (e.g., maximum of elastic displacement, maximum of the velocity, maximum of the electrical field or current). Fortunately, these criteria can be evaluated easily and can be updated, in real time, for each $U(p)$.

40.2.4   The Adoc library

For a given parameter box P and a required accuracy, Adoc determines the derivative order associated with each variable $p_i, i = 1, \ldots, n$. Then cross derivatives of U are obtained from directional derivatives. The method proposed in [80] is adapted to our context: for each parameter p_i an appropriate derivative order is computed. We are studying the possibility of adapting an explicit method proposed in [238]. Then Adoc constructs a rational approximation [93] of the solution giving the best approximation of the solution.

40.3   Some Applications of the Method

40.3.1   Application to an electromagnetic problem

The tee filter presented in Figure 40.2 is a component of a spatial antenna feeding system. Thanks to the uniform thickness of the waveguide, this industrial example can be represented by a 2D finite element model. The model depends on the frequency $9.5 \leq f \leq 13.5 \; Ghz$ and on a geometry parameter $9 \leq p \leq 13 \; mm$ called D113 on Figures 40.2 and 40.4. The mesh perturbations presented in Figure 40.3 are obtained by Adomesh.

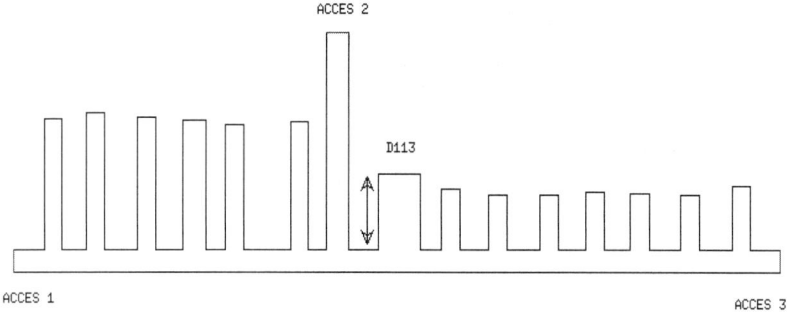

FIGURE 40.2. The shape of the waveguide

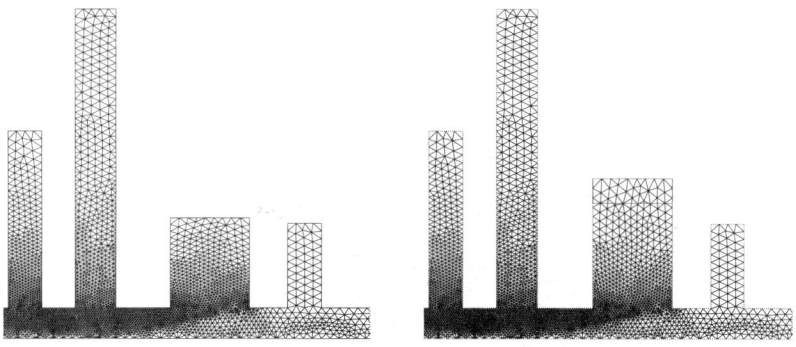

FIGURE 40.3. Mesh perturbations at p^{min} and p^{max}

For this frequency range and this quite complicated shape, the finite

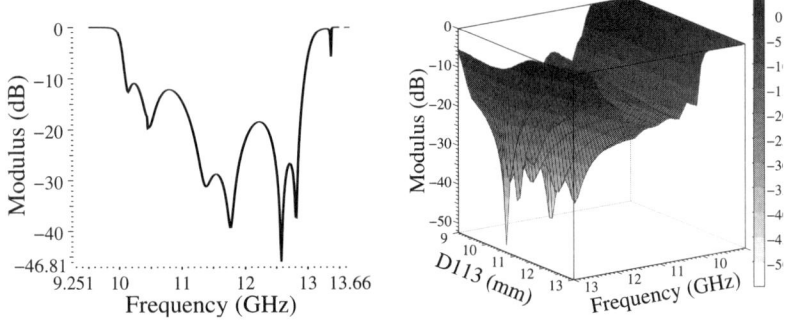

FIGURE 40.4. A multivariate parameterization for S-11

element model has 14821 elements and 21568 edges. A scalar complex unknown is associated with each edge.

The reflection coefficient $S_{11}(f, p)$ represents the part of loss energy (not
transmitted from access 1 to the other accesses). The software Adoc determined automatically the truncation of the Taylor expansion: at order 30
with respect to f and at order 5 with respect to p. The curve in Figure
40.4 is obtained on the reference geometry ($p_0 = 11.5$ mm). The singularities of the response $S_{11}(f, p_0)$ give interesting information on the physical
behaviour of the component. It is quite difficult to catch these singularity
by classical methods.

A standard run takes 10 sec. The parametric analysis takes 95 sec.

40.3.2 Application to a structural analysis problem

In this second example we compute the frequency response $U(f)$ of a car
body. The complex vector $U(f)$ is the solution to a parametric system
$A(f)U(f) = B(f)$. To take into account the damping effect, the map $f \longmapsto$
$A(f)$ is nonlinear.

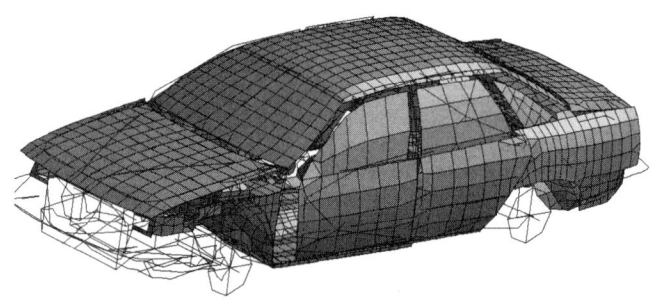

FIGURE 40.5. The car body

The mesh is relatively coarse, having 7132 nodes, 7304 elements, and 43075 unknowns. Two loads are applied vertically in phase on the two front wheels. There are two observation points at the center of the drive wheel (node 41080) and under the driver's seat (node 176700). The standard frequency response (one factorization and one resolution per point) requires 800 seconds for 160 points. The parametric response required 440 seconds for 80 derivatives. In Figure 40.6 we can see a good agreement between the two computations.

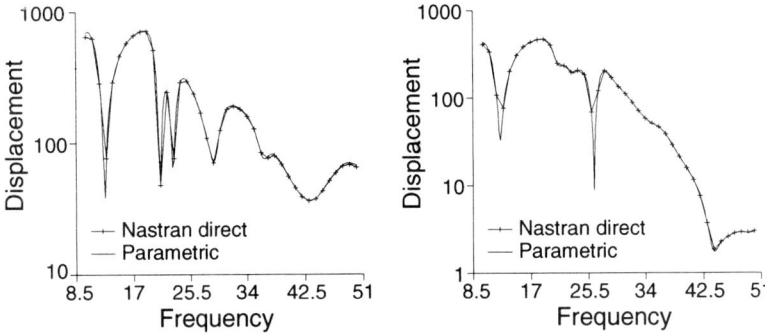

FIGURE 40.6. Frequency responses for nodes 41080 (left) and 176700 (right)

40.4   Conclusion

These two examples demonstrate the ability of our method to extend the parameterization over the singularities. It has many other advantages:

- it gives an efficient way to catch singularities,

- it saves human time since the designer creates only one finite element model (one mesh), and

- it compresses information; n derivatives of U give better parameterization than n computations of U.

41

Efficient High-Order Methods for ODEs and DAEs

Jens Hoefkens, Martin Berz and Kyoko Makino

ABSTRACT We present methods for the high-order differentiation through ordinary differential equations (ODEs), and more importantly, differential algebraic equations (DAEs). First, methods are developed that assert that the requested derivatives are really those of the solution of the ODE, and not those of the algorithm used to solve the ODE. Next, high-order solvers for DAEs are developed that in a fully automatic way turn an n-th order solution step of the DAEs into a corresponding step for an ODE initial value problem. In particular, this requires the automatic high-order solution of implicit relations, which is achieved using an iterative algorithm that converges to the exact result in at most $n + 1$ steps. We give examples of the performance of the method.

41.1 Introduction

Under certain conditions, the solutions of ordinary differential equations (ODEs) and differential algebraic equations (DAEs) can be expanded in Taylor series. In these cases, we can obtain good approximations of the solutions by computing the respective Taylor series [118, 119]. Here we show how high-order methods of AD can be used to obtain these expansions in an automated way and how these methods allow differentiation through the solutions. The method can be applied to explicit and implicit ODEs. Together with a structural analysis of Pryce [408, 425, 426], it can be extended to obtain Taylor series expansions of the solutions of DAEs.

To use methods of integrating explicit ODEs to solve DAEs, we perform a structural analysis of the DAE system to convert it into an equivalent set of implicit ODEs, which is converted into an explicit system by using differential algebraic methods for the high-order inversion of functional relations. For n-th order computations, these methods are guaranteed to converge to the exact result in at most $n + 1$ steps and return explicit ODEs that are order n equivalent to the original DAE problem. The resulting system of ODEs is then integrated to obtain the one-step solution of the DAE system. The method can be extended to multiple time steps by projecting the final coordinates of a particular time-step onto the constraint manifold and using the projections as consistent initial conditions for the next iteration.

This chapter is divided in three parts: differential algebraic AD tools used to solve explicit ODEs and convert implicit ODEs into explicit ones are presented in §41.2. §41.3 summarizes the structural analysis method suggested by Pryce that allows the automatic conversion of DAEs into implicit ODEs, and §41.4 presents an example that has been computed with the high-order code COSY Infinity [64]; it demonstrates how the combined method can successfully handle high-index DAE problems.

41.2   Efficient Differential Algebra Methods

For purposes of notation, we consider the set $\mathcal{C}^{n+1}(U, \mathbb{R}^w)$ (where $U \subset \mathbb{R}^v$ is an open set containing the origin) and equip it with the following relation: for $f, g \in \mathcal{C}^{n+1}(U, \mathbb{R}^w)$ we say $f =_n g$ (f equals g up to order n) if $f(0) = g(0)$, and all partial derivatives of orders up to n agree at the origin. This gives an equivalence relation on $\mathcal{C}^{n+1}(U, \mathbb{R}^w)$. The resulting equivalence classes are called DA vectors, and the class containing $f \in \mathcal{C}^{n+1}(U, \mathbb{R}^w)$ is denoted by $[f]_n$. For functions expressible via finitely many intrinsics, the class $[f]_n$ corresponds to the evaluation of f with any n-th order AD tool (see e.g. [433]). Our goal here is to determine $[f]_n$ when f is the solution of a DAE. The collection of these equivalence classes is called $_nD_v$ [47, 52]. When equipped with appropriate definitions of the elementary operations, the set $_nD_v$ becomes a differential algebra [52]. We have implemented efficient methods of using DA vectors in the arbitrary order code COSY Infinity [64].

Definition 1 *For $[f]_n \in {}_nD_v$, the depth $\lambda([f]_n)$ is defined to be the order of first non-vanishing derivative of f if $[f]_n \neq 0$, and $n+1$ otherwise.*

Definition 2 *Let $\mathcal{O}$ be an operator on $M \subset {}_nD_v$. $\mathcal{O}$ is contracting on M, if any $[f]_n, [g]_n \in M$ satisfy $\lambda(\mathcal{O}([f]_n) - \mathcal{O}([g]_n)) \geq \lambda([f]_n - [g]_n)$ with equality iff $[f]_n = [g]_n$.*

This definition has a striking similarity to the corresponding definitions on regular function spaces. A theorem that resembles the Banach Fixed Point Theorem can be established on $_nD_v$. However, unlike in the case of the Banach Fixed Point theorem, in $_nD_v$ the sequence of iterates is guaranteed to converge in at most $n+1$ steps.

Theorem 1 (DA Fixed Point Theorem) *Let $\mathcal{O}$ be a contracting operator and self-map on $M \subset {}_nD_v$. Then $\mathcal{O}$ has a unique fixed point $a \in M$. Moreover, for any $a_0 \in M$ the sequence $a_k = \mathcal{O}(a_{k-1})$ converges in at most $n+1$ steps to a.*

An extensive proof is given in [52]. In the following sections we demonstrate how this theorem can be used for high-order integration of ODEs and DAEs.

41.2.1   Integration of ODEs

Within $_nD_v$, high-order integration of ODEs can be accomplished by using the antiderivation operator and the DA Fixed Point Theorem.

Proposition 1 (Antiderivation is Contracting) *For $k \in \{1, \ldots, v\}$, the antiderivation $\partial_k^{-1} : {}_nD_v \to {}_nD_v$ is a contracting operator on $_nD_v$.*

The proof of this assertion is based on the fact that if $a, b \in {}_nD_v$ agree up to order l, the first non-vanishing derivative of $\partial_k^{-1}(a - b)$ is of order $l+1$. Using antiderivation, we rewrite the ODE initial value problem $\dot{x}(t) = f(x, t), x(t_0) = x_0$ in its fixed-point form

$$\mathcal{O}(x(t)) = x_0 + \int_{t_0}^{t} f(x, \tau) d\tau.$$

According to Proposition 1, this defines a contracting operator $\mathcal{O}$ on $_nD_{v+1}$. Thus, ODEs can be integrated very efficiently by iterating a relatively simple operator. The iteration is guaranteed to converge to the n-th order Taylor expansion of the solution in at most $n + 1$ steps.

Moreover, in the framework of DA, it is possible to replace the fixed initial value x_0 and additional parameters by additional DA variables. Thus, one not only obtains the solution of a particular initial value problem, but the method finds a Taylor expansion of the flow of the ODE as a function of the dependent variable t and all initial conditions and parameters.

41.2.2   Inversion of functional relations

In this section we show how the Implicit Function Theorem can be used in the DA framework to convert an implicit ODE to an explicit system. The following theorem shows that it is possible to compute a representative of the equivalence class $[f^{-1}]_n$ from a representative $\mathcal{M}$ of the class $[f]_n$.

Theorem 2 (Inversion by Iteration) *Assume that $f \in \mathcal{C}^{n+1}(U, \mathbb{R}^v)$ is an origin-preserving map, and let $\mathcal{M}$ be a representative of the equivalence class $[f]_n$. Write $\mathcal{M} = M + \mathcal{N}_\mathcal{M}$, where $\mathcal{N}_\mathcal{M}$ denotes the purely nonlinear part of $\mathcal{M}$. Assume furthermore that $M = \mathbf{D}f(0)$ is regular. Then*

$$\mathcal{O}(a) = M^{-1} \circ (\mathcal{I} - \mathcal{N}_\mathcal{M} \circ a)$$

is a well defined contracting operator on $_nD_v$. Moreover, for any $a_0 \in {}_nD_v$, the sequence of iterates converges in at most $n+1$ steps to a representative of the class $[f^{-1}]_n$.

The proof of this theorem uses the fact that if a and $b \in {}_nD_v$ agree up to order l, the compositions with $\mathcal{N}_\mathcal{M}$ agree to order $l + 1$, since $\lambda(\mathcal{N}_\mathcal{M}) > 1$. See [52] for a full proof of this theorem.

Together with the Implicit Function Theorem, the Inversion by Iteration Theorem allows the efficient computation of explicit expressions of an implicitly described function g. Given $f \in \mathcal{C}^{n+1}(U \times V, \mathbb{R}^w)$ $(U \subset \mathbb{R}^v$

and $V \subset \mathbb{R}^w$) and $x \in \mathbb{R}^v$ and $y \in \mathbb{R}^w$, we write $f(x, y)$. Suppose that $f(0, 0) = 0$ and that $f_y(0, 0)$ has rank w. Then, by the Implicit Function Theorem, there exists a unique $C^{n+1}(U, V)$ such that $g(0) = 0$, and in a neighborhood of $0 \in \mathbb{R}^v$, $f(x, g(x)) = 0$. Define $\Phi : \mathbb{R}^{v+w} \to \mathbb{R}^{v+w}$ by

$$\Phi \begin{pmatrix} x \\ y \end{pmatrix} = \begin{pmatrix} x \\ f(x, y) \end{pmatrix} = \begin{pmatrix} x \\ 0 \end{pmatrix}$$

and obtain an explicit expression for $g(x)$ from the last w components of $\Phi^{-1}(x, 0)$, because $(x, y) = (x, g(x)) = \Phi^{-1}(x, 0)$.

The technique of augmenting implicit systems has been used extensively to first order in bifurcation theory [339] and to high order for symplectic integration [49]. Other combinations of the Implicit Function Theorem and first order AD in the field of control theory are discussed by Evtushenko [175].

Finally, for a general function f with $f(x_0) = y_0$, one can extend the methods presented here using the fact that an origin-preserving function can be obtained via $\tilde{f}(x) = f(x + x_0) - y_0$.

41.3   Structural Analysis of DAEs

The structural analysis of DAEs is based on methods developed by Pryce [425, 426]. The signature method (Σ-method) is used to decide whether a given DAE possesses a unique solution and to transform it into an equivalent system of implicit ODEs.

We consider the DAE problem $f_1(\ldots) = \ldots = f_k(\ldots) = 0$ with the scalar independent variable t, the k dependent variables $x_j = x_j(t)$, and sufficiently smooth functions

$$f_i \left(x_1, \ldots, x_1^{(\xi_{i1})}, \ldots, x_k, \ldots, x_k^{(\xi_{ik})}, t \right) = 0.$$

For a given j, the i-th equation of this system does not necessarily depend on x_j, the derivatives $x_j^{(\eta)}$ for $\eta < \xi_{ij}$, or even any of its derivatives. However, if there is a dependence on at least one of the derivatives (including the 0-th derivative $x_j^{(0)} = x_j$), we denote the order of that highest derivative by ξ_{ij}. Using this notation, we define the $k \times k$ matrix $\Sigma = (\sigma_{ij})$ by

$$\sigma_{ij} = \begin{cases} -\infty & \text{if the } j\text{-th variable doesn't occur in } f_i \\ \xi_{ij} & \text{otherwise.} \end{cases}$$

Let P_k be the set of permutations of length k, and consider the assignment problem (AP) [46] of finding a maximal transversal $T \in P_k$ as defined by

$$\text{Maximize } \|T\| = \sum_{i=1}^{k} \sigma_{i,T(i)} \text{ with } \sigma_{i,T(i)} \geq 0.$$

The DAE is ill-posed if the AP is not regular. On the other hand, if such a maximal transversal exists, it is possible to compute the (uniquely determined [426]) smallest offsets $(c_i), (d_j) \in \mathbb{Z}^k$ that satisfy the requirements

$$\text{Minimize } \bar{z} = \sum d_j - \sum c_i \text{ with } d_j - c_i \geq \sigma_{ij} > -\infty \text{ and } c_i \geq 0 .$$

By differentiation it is possible to derive a k-dimensional system of ODEs from the original DAEs:

$$f_1^{(c_1)} = \ldots = f_k^{(c_k)} = 0.$$

Moreover, if an initial condition (x_0, t_0) satisfies the system of intermediate equations

$$f_1^{(0)} = \ldots = f_1^{(c_1 - 1)} = \ldots = f_k^{(0)} = \ldots = f_k^{(c_k - 1)} = 0,$$

and the system Jacobian

$$\mathbf{J} = J_{ij} = \partial f_i / \partial \left(x_j^{(d_j - c_i)} \right)$$

(with $J_{ij} = 0$ if the derivative is not present or if $d_j < c_i$) is non-singular at that point, the original DAE system has a unique solution in a neighborhood of t_0. The solution can be obtained as the unique solution of the derived ODE, and in a neighborhood of (x_0, t_0), the system has $\bar{z}$ degrees of freedom. Finally, the differentiation index ν_d of the system satisfies

$$\nu_d \leq \max c_i + \begin{cases} 0 & \text{if all } d_j > 0 \\ 1 & \text{otherwise.} \end{cases} \tag{41.1}$$

While it has been suggested that the index ν_d might equal the given expression, [436] shows that (41.1) is only an upper bound for it.

41.4   Example

Consider a pair of coupled pendulums in a frictionless environment and assume that the pendulums are massless and inextensible, with point masses on the ends (Figure 41.1). Denote the tensions in the strings by λ_1 and λ_2, respectively. Then the equations of motion expressed in Cartesian coordinates x_1, y_1, x_2, y_2 are

$$m_1 \ddot{x}_1 + \lambda_1 x_1 / l_1 - \lambda_2 (x_2 - x_1) / l_2 = 0$$
$$m_2 \ddot{y}_1 + \lambda_1 y_1 / l_1 - \lambda_2 (y_2 - y_1) / l_2 - m_1 g = 0$$
$$m_2 \ddot{x}_2 + \lambda_2 (x_2 - x_1) / l_2 = 0$$
$$m_2 \ddot{y}_2 + \lambda_2 (y_2 - y_1) / l_2 - m_2 g = 0$$
$$x_1^2 + y_1^2 - l_1^2 = 0$$
$$(x_2 - x_1)^2 + (y_2 - y_1)^2 - l_2^2 = 0.$$

The corresponding Σ matrix of this DAE is shown below, with the entries forming a maximal transversal in bold face.

$$\Sigma = \begin{pmatrix} 2 & -1 & 0 & -1 & \mathbf{0} & 0 \\ -1 & \mathbf{2} & -1 & 0 & 0 & 0 \\ 0 & -1 & 2 & -1 & -1 & \mathbf{0} \\ -1 & 0 & -1 & \mathbf{2} & -1 & 0 \\ \mathbf{0} & 0 & -1 & -1 & -1 & -1 \\ 0 & 0 & 0 & 0 & -1 & \mathbf{-1} \end{pmatrix}$$

Further analysis gives the offsets $c = (0,0,0,0,2,2)$ and $d = (2,2,2,2,0,0)$ and reveals that the system has four degrees of freedom and a differentiation index ν_d of at most three.

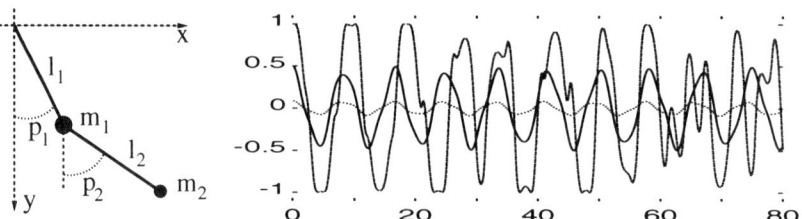

FIGURE 41.1. Coupled pendulum and x_1 as function of time.

The second plot in Figure 41.1 shows the time evolution of x_1 for the special case $g = 1, l_1 = l_2 = 1, m_1 = m_2 = 1$ and $p_1 = p_2 = 5°, 30°$ and $90°$ (Integrated to order 6, $0 \leq t \leq 80$, and $\Delta t = 0.1$), obtained with the DAE solver discussed above. The results were checked by studying energy conservation.

41.5   Conclusion

By combining existing AD methods, the Σ-method for the structural analysis of DAE systems and DA operations, we derived and demonstrated a high-order scheme for the efficient integration of high-index DAE problems. The method involves an automated structural analysis, the inversion of functional dependencies, and the high-order integration of an ODE system derived from the original DAE problem. Using techniques for the verified inversion of functional relations modeled by Taylor models [279, 58] and techniques for the verified integration of ODEs [364], the methods presented can readily be extended to verified high-order integration of DAEs.

The method presented of integrating DAEs can be simplified by realizing that in the framework of DA and Taylor models, the antiderivation is not fundamentally different from other elementary operations. Thus, the solution of ODEs and DAEs can be obtained by an inversion process. This could lead to a unified integration scheme for ODEs and DAEs that would also allow for high-order verification via Taylor models.

Part X

Error Estimates and Inclusions

42

From Rounding Error Estimation to Automatic Correction with Automatic Differentiation

Thierry Braconnier and Philippe Langlois

ABSTRACT Using automatic differentiation (AD) to estimate the propagation of rounding errors in numerical algorithms is classic. We propose a new application of AD to roundoff analysis providing an automatic correction of the first order effect of the elementary rounding errors. We present the main characteristics of this method and significant examples of its application to improve the accuracy of computed results and/or the stability of the algorithm.

42.1 Introduction

Rounding error analysis and automatic differentiation (AD) share at least 25 years' common history [35]. Iri's survey [300] emphasizes this topic. He describes rounding error estimates simply computable with reverse mode AD. These classical approaches yield either an *absolute bound* or a *probabilistic estimate* of the first-order error of the final calculated value with respect to the elementary rounding errors introduced by the computation of intermediate variables [300, 336].

We propose a new linear approach, the CENA method. The difference from previous methods is that elementary rounding errors are neither bounded nor modeled with random variables but computed. Thus we compute *a correcting term* to improve the accuracy of the finite precision result, together with *a bound on the residual error* of the corrected value.

In this chapter, we focus on absolute bound limitations, the main features of the CENA method, automatic bounding of rounding errors in the AD process (in the reverse mode), and significant applications of the correcting method. A complete presentation of the method is proposed in [341].

42.2 Linearization Methods and Bound Limitations

Let us consider $\widehat{x}_N = fl\left(f(X)\right)$ the floating point computation of $x_N = f(X)$, where f is a real function of the floating point vector $X = (\widehat{x}_1, \ldots, \widehat{x}_n)$. This computation introduces intermediate variables $\widehat{x}_k$ and associated elementary rounding errors δ_k that satisfy for $k = n + 1, \ldots, N$,

$$\widehat{x}_k = fl\left(\widehat{x}_i \circ \widehat{x}_j\right) = \widehat{x}_i \circ \widehat{x}_j + \delta_k = x_k + \delta_k, \qquad (42.1)$$

where $1 \leq i, j < k \leq N$, $\circ \in \{+, -, \times, /, \sqrt{\ }\}$, $\widehat{x}_j \neq 0$ when $\circ = /$, and $\widehat{x}_i = 0$, $\widehat{x}_j \geq 0$ when $\circ = \sqrt{\ }$. Hence, $\widehat{x}_N = \widehat{f}(X, \delta)$, i.e., a function of data X and elementary rounding errors $\delta = (\delta_{n+1}, \ldots, \delta_N)$. Let Δ_L denote the first order approximant with respect to δ of the global forward error to $\widehat{x}_N - x_N$, and E_L the linearization error associated to this approximant Δ_L. We have

$$\widehat{x}_N - x_N = \sum_{k=n+1}^{N} \frac{\partial \widehat{f}}{\partial \delta_k}(X, \delta) \cdot \delta_k - E_L = \Delta_L - E_L. \qquad (42.2)$$

Linearization methods use relation (42.2) to bound or estimate the global rounding error neglecting the linearization error E_L, e.g., [300, 356]. This linearization error describes the rounding error effects of order higher than one on the accuracy of the computed value $\widehat{x}$. Taking into account E_L may yield more accurate applications of relation (42.2). Alas, this linearization error E_L is generally unknown.

Classical deterministic application of relation (42.2) introduces another limitation when δ_k is bounded by (a function of) the floating point precision $\mathbf{u}$. Unsuitable bounds for $|\Delta_L|$ may be computed as the compensation of elementary rounding errors and exactly computed results are not taken into account. We respectively illustrate these two cases in IEEE-754 single precision computing in the natural evaluation order the expressions $E_1 = (2^{50} + 1) - 1$, and $E_2 = (2^{25} \times 2^{25}) - 2^{50}$. In both cases, the actual global error is equal to zero, but absolute bounding yields $|\Delta_L| \leq 2^{27}$. Here $|\delta_k| \leq |\widehat{x}_k|\,\mathbf{u}$, with $\mathbf{u} = 2^{-24}$, and absolute bounding satisfies $|\widehat{E}_i - E_i| = (2^{50} + 2^{50})\mathbf{u}$, for $i = 1, 2$. Automatic bounding may yield pessimistic results after very few floating point operations.

42.3   The CENA Method

42.3.1   Principles, hypothesis and applications

With the CENA method, we do not bound but *compute* Δ_L to yield a *corrected result* $\overline{x_N}$ more accurate than initial $\widehat{x}_N$. This corrected result suffers from both the truncation error E_L and rounding errors we denote by E_C. We *validate* the accuracy of $\overline{x_N}$ yielding a dynamic bound of this residual error for an *identified class of algorithms* that satisfies $E_L = 0$. In such conditions, we apply this automatic correction to *final* or *selected intermediate* computed results. This latter case is interesting for iterative algorithms (actual accuracy depends on the iterate) or when the algorithm stability depends on the accuracy of some intermediate variables.

42.3.2   Linear correction and linear algorithms

We *compute* the correcting factor $\widehat{\Delta}_L = fl(\Delta_L)$ and define the *corrected result* $\overline{x_N} = fl(\widehat{x}_N - \widehat{\Delta}_L)$. The correcting factor satisfies $\widehat{\Delta}_L = \Delta_L + E_C$, where E_C represents the rounding error introduced by the computation of the correction. The smaller the *residual error* $\overline{x_N} - x_N = -(E_L + E_C)$, the more efficient is the correcting method, *i.e.*, the more accurate is the corrected result $\overline{x_N}$.

We define a *linear algorithm* as one that only contains the operations $\{+, -, \times, /, \sqrt{\ }\}$ and such that

- every multiplication $fl(\widehat{x}_i \times \widehat{x}_j)$ satisfies $\widehat{x}_i = x_i$ or $\widehat{x}_j = x_j$,
- every division $fl(\widehat{x}_i / \widehat{x}_j)$ satisfies $\widehat{x}_j = x_j$, and
- every square root $fl(\sqrt{\widehat{x}_j})$ satisfies $\widehat{x}_j = x_j$.

Linear algorithms are interesting because $\widehat{x}_N$ computed with a linear algorithm satisfies $E_L = 0$ [343]. Examples of linear algorithms are classical summation algorithms, inner product computation, polynomial evaluation using Horner's method, and the solution of triangular linear systems.

42.3.3   Computing the correcting term $\widehat{\Delta}_L$

The correcting term $\widehat{\Delta}_L$ involves computing partial derivatives and elementary rounding errors defined by relation (42.2). Merging these two computations is the original feature of the CENA method. The computation of the partial derivatives $\partial \widehat{f}/\partial \delta_k$ is classic; the reverse mode of AD is applied as in [300] or [433], for example. Elementary rounding errors δ_k are well known in the computer arithmetic community. In the following relations, we note $\delta_k[\circ, i, j] = fl(x_k) - x_k$ with $x_k = x_i \circ x_j$, and right-side expressions are computed ones (we drop hats and fl-s).

The elementary rounding error associated with addition, subtraction and multiplication is a computable floating point number [155, 331]. We have
$$\delta_k[+, i, j] = (x_k - x_i) - x_j \text{ when } |x_i| \geq |x_j|, \text{ and}$$
$$\delta_k[\times, i, j] = x_k - (x_i^U \times x_j^U) - [(x_i^U \times x_j^L) + (x_i^L \times x_j^U)] - (x_i^L \times x_j^L).$$
In this last computation, each factor x is split into x^U and x^L, its upper and lower "half-length" parts that depend on floating point arithmetic characteristics. For division and square root, approximate elementary rounding error $\widehat{\delta}_k$ and corresponding truncation bounds for $|\widehat{\delta}_k - \delta_k|$ are also computable [342, 417]. We have
$$\widehat{\delta}_k[/, i, j] = \{(x_k \times x_j) - x_i - \delta_k[\times, k, j]\}/x_k, \text{ and}$$
$$\widehat{\delta}_k[\sqrt{\ }, i, j] = \{(x_k \times x_k) - x_i + \delta_k[\times, k, k]\}/(2 \times x_k).$$

42.3.4   Bounding the computing error E_C

The error E_C associated with the computation of $\widehat{\Delta}_L$ contains three sources of error. Truncation errors happen when $\widehat{\delta}_k$ is computed, *i.e.*, for division and square root. Rounding errors corrupt both the final inner product in relation (42.2) and the AD process for the $\partial \widehat{f}/\partial \delta$ themselves. We dynamically bound each of these errors using Wilkinson's running error analysis [502].

We consider here the computation of the bound ε_k of the rounding errors introduced in the AD process, i.e., $|fl(D_k) - D_k|$, where $D_k = \partial \widehat{x}_N/\partial \delta_k$ and $fl(D_k)$ is the corresponding value computed with AD in the reverse mode. The following algorithm describes the simultaneous computation of partial derivatives $fl(D_k)$ (lines 3 and 4) and the value $\varepsilon_k = \mathbf{u}e_k$ that bounds $|fl(D_k) - D_k|$ when the precision of computation is $\mathbf{u}$ (line 5 and output).

Input: $\partial x_k/\partial x_i$ $(k = n+1, N), Dx_N = 1,$ *others* $= 0$
  1: **for** $k = N$ **to** $n+1$ **do**
  2:     **for** $i \in$ Vertices_To(x_k) **do**
  3:         $C_{ki} = \partial x_k/\partial x_i$
  4:         $Dx_i = Dx_i + Dx_k \times C_{ki}$
  5:         $e_i = e_i + |Dx_k \times C_{ki}| + |Dx_i| + e_k \times |C_{ki}| + f_{ik} \times |Dx_k|$
  6:     **end for**
  7: **end for**
Output: $\partial \widehat{x}_N/\partial \delta_k = Dx_k,\ \varepsilon_k = \mathbf{u} \times e_k$ $(k = n+1, N)$

For intermediate and output variables x_k such that $x_k = fl(x_i \circ x_j)$, Vertices_To(x_k) returns intermediate or input variables x_i and x_j; f_{ik} bounds the elementary rounding error in the corresponding computation of $\partial x_k/\partial x_i$ that may be different from zero when $\circ = /$ or $\sqrt{\ }$. These values are naturally computed in the construction stage of the reverse mode.

42.4   Accuracy Improvement and Algorithm Stabilization

In this final section, we propose two significant applications of the CENA method. The *final* correction and accuracy improvement are illustrated with an inner product computation. We also stabilize Newton's iteration used to compute a polynomial multiple root correcting sensitive *intermediate* variables. Computed and corrected values are evaluated using IEEE-754 single precision [299]. In both cases, we compare the efficiency of the correcting method to the classical use of higher precision (that corresponds here to IEEE-754 double precision).

42.4.1   Correcting the final result

We compute the inner product $X^T Y$ with

$$X = [1, 2, \cdots, 2^n, -1, -2, \cdots, -(2^n)]^T \text{ and } Y = [1, 1, \cdots, 1]^T.$$

The chosen bound n guarantees no overflow during this computation. The exact result satisfies $X^T Y = 0.0$, but IEEE-754 single and double precision

TABLE 42.1. Inner product $X^T Y$ in IEEE-754 single precision arithmetic

$X_k Y_k$	Computed $\sum_k X_k Y_k$	Error	δ_k	Correction
1	$1.000\ldots00\ 2^0$	0	0	0
2	$1.100\ldots00\ 2^1$	0	0	0
⋮	⋮	⋮	⋮	⋮
2^i	$1.111\ldots00\ 2^i$	0	0	0
⋮	⋮	⋮	⋮	⋮
2^{23}	$1.111\ldots11\ 2^{23}$	0	0	0
2^{24}	$1.000\ldots00\ 2^{25}$	-1	-1	-1
2^{25}	$1.000\ldots00\ 2^{26}$	-1	0	-1
2^{26}	$1.000\ldots00\ 2^{27}$	-1	0	-1
⋮	⋮	⋮	⋮	⋮
2^i	$1.000\ldots00\ 2^{i+1}$	-1	0	-1
⋮	⋮	⋮	⋮	⋮
2^n	$1.000\ldots00\ 2^{n+1}$	-1	0	-1
-1	$1.000\ldots00\ 2^{n+1}$	-2	-1	-2
-2	$1.000\ldots00\ 2^{n+1}$	-4	-2	-4
⋮	⋮	⋮	⋮	⋮
-2^i	$1.000\ldots00\ 2^{n+1}$	-2^{i+1}	-2^i	-2^{i+1}
⋮	⋮	⋮	⋮	⋮
-2^{n-24}	$1.000\ldots00\ 2^{n+1}$	-2^{n-23}	-2^{n-24}	-2^{n-23}
-2^{n-23}	$1.111\ldots11\ 2^n$	-2^{n-23}	0	-2^{n-23}
-2^{n-22}	$1.111\ldots01\ 2^n$	-2^{n-23}	0	-2^{n-23}
⋮	⋮	⋮	⋮	⋮
-2^{n-1}	$1.000\ldots01\ 2^n$	-2^{n-23}	0	-2^{n-23}
-2^n	$1.000\ldots00\ 2^{n-23}$	-2^{n-23}	0	-2^{n-23}

computations suffer from a forward error that increases as 2^{n-n_u} when $n > n_u$, a threshold that depends on the precision (Figure 42.1).

Table 42.1 shows the partial sum and forward error, the elementary error δ_k during the accumulations of the inner product and the correcting factor in IEEE-754 single precision arithmetic (24 bits of mantissa). The correcting factor in relation (42.2) here verifies $\widehat{\Delta}_L = \sum_k \delta_k$. For the first 24 terms, the partial sums are accumulated without error, so $\delta_k = 0$. Applying the "round to the nearest even" rounding rule of IEEE arithmetic, the next addition yields $\delta_{25} = -1$, the first contribution to the correcting factor. Afterwards the accumulation of positive $X_k Y_k$ is exact (provided no overflow occurs) and the correcting factor does not change. As the first subtractions do not modify the partial sum, the resulting errors are captured in

δ_k and accumulated in the correcting factor. This inaccurate computation continues until we get a large enough $X_k Y_k$, that is the last 24 terms of the computation. Then the subtraction is again exact and the correcting factor describes the exact forward error in the computed inner product since this latter is a power of the arithmetic base in this case.

Applying the correcting method to the single precision computation, we have $\overline{X^T Y} = 0.0$ (that does not appear on the log-scale Figure 42.1). Therefore the corrected inner product is exact here for every considered n. It is worth noting that the computed B_{E_C} that bounds the computing error E_C proves that single precision results are *false* without knowing the exact result since $|\overline{X^T Y} - X^T Y| = |X^T Y| \leq B_{E_C} \ll |fl(X^T Y)|$.

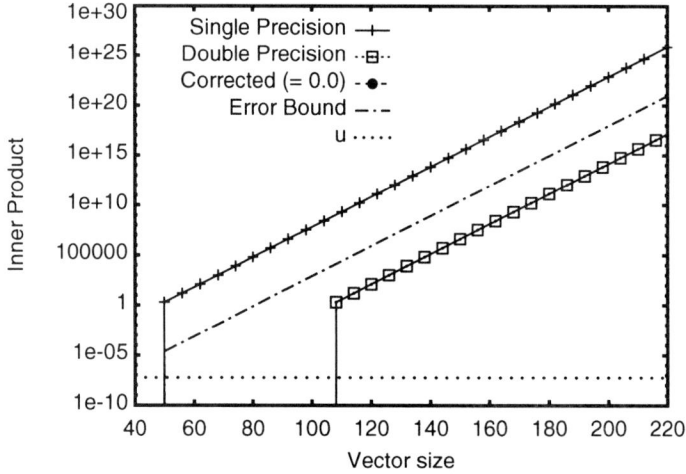

FIGURE 42.1. Computing $X^T Y = 0.0$.

42.4.2   Correcting sensitive intermediate variables

Rounding errors destabilize Newton's iteration when it is applied to compute a polynomial root of multiplicity $m > 1$ [430]. Correcting the final result of an iterative process is inefficient, but correcting well-chosen intermediate variables may stabilize the iteration.

We consider the polynomial $P(x) = (x - 1)^6$. Starting with $x_0 = 2.0$, we compute Newton's iterate x_k until the absolute residual $|P(x_k)| < \sigma'$. Here, we *a priori* choose $\sigma' = O(\sigma) = 1.0 \times 10^{-37}$, where $\sigma \approx 1.2 \times 10^{-38}$ is the smallest non-zero positive number in IEEE-754 single precision.

After a reasonable beginning of convergence, both single and double precision iterations become unstable until stopping fortunately with a zero absolute residual (breakdown is observed for other choices of initial x_0).

When polynomials P and P' are evaluated using Horner's scheme, the

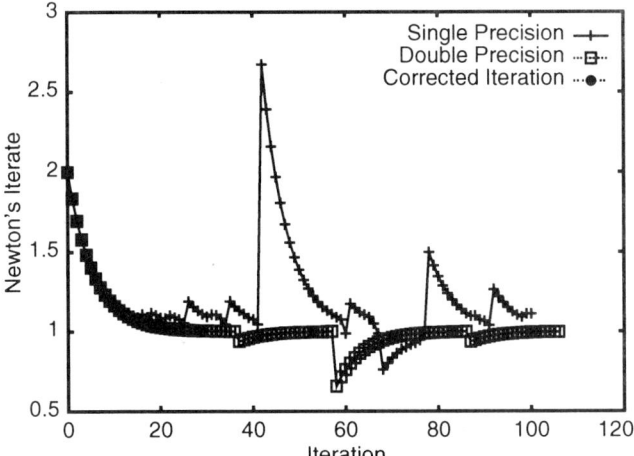

FIGURE 42.2. Newton's iterates until termination.

corrected Newton's iteration $x_{k+1} = x_k - \overline{P(x_k)}/\overline{P'(x_k)}, (k = 0, 1, \dots)$ is stable and converges with the mathematically expected geometrical order $(m-1)/m$. The computed approximate x_k is more accurate than the classical estimation in $O(\mathbf{u}^m)$ where $\mathbf{u}$ is the precision of the computation. Figures 42.2 and 42.3 illustrate these properties.

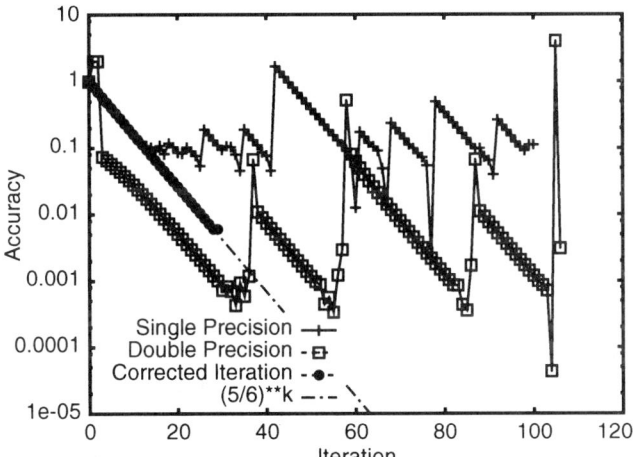

FIGURE 42.3. Accuracy of Newton's iterates until termination.

43

New Applications of Taylor Model Methods

Kyoko Makino and Martin Berz

ABSTRACT Taylor model methods unify many concepts of high-order computational differentiation with verification approaches covering the Taylor remainder term. Not only do they provide local multivariate derivatives, they also allow for highly efficient and sharp verification. We present several recent results obtained with Taylor model methods, including verified optimization, verified quadrature and verified propagation of extended domains of initial conditions through ODEs, approaches towards verified solution of DAEs and PDEs. In all cases, the methods allow the development of new numeric-analytic tools that efficiently capitalize on the availability of derivatives and sharp inclusions over extended ranges. Applications of the methods are given, including global optimization, very high-dimensional numeric quadrature, particle accelerators, and dynamics of near-earth asteroids.

43.1 Taylor Model Arithmetic

The remainder of any $(n + 1)$ times continuously partially differentiable function f approximated by the nth order Taylor polynomial $P_{n,f}$ at the expansion point $\vec{x}_0$ can be bounded by an interval $I_{n,f}$ satisfying

$$\forall \vec{x} \in [\vec{a}, \vec{b}], \quad f(\vec{x}) \in P_{n,f}(\vec{x} - \vec{x}_0) + I_{n,f} \tag{43.1}$$

that scales with $|\vec{x} - \vec{x}_0|^{n+1}$. In practice, over reasonable box sizes, the remainder bound interval $I_{n,f}$ can be made very small. A pair $(P_{n,f}, I_{n,f})$ satisfying (43.1) is called a Taylor model of f and denoted by

$$T_{n,f} = (P_{n,f}, I_{n,f}).$$

Any computer representable function $f(\vec{x})$ can be modeled by Taylor models if the function f satisfies the above mentioned mathematical conditions in $[\vec{a}, \vec{b}]$. The expansion point $\vec{x}_0$ and the order n specify the Taylor polynomial part $P_{n,f}(\vec{x} - \vec{x}_0)$ uniquely with coefficients described by floating point numbers on a computer. The remainder interval part $I_{n,f}$ further depends on the domain $[\vec{a}, \vec{b}]$ and the details of the algorithm to compute it.

The dependency problem in interval arithmetic [8, 338, 384, 385, 386] is typically caused by cancellation effects. For example, if $I = [a, b]$, then

$I - I$, if not recognized to represent the same number, is computed as $[a, b] - [a, b] = [a, b] + [-b, -a] = [a - b, b - a]$, resulting in a width that is not zero, but twice as large as before. In Taylor models, the bulk of the functional dependency is kept in the Taylor polynomial part, and the cancellation of the dependency happens there, thus the dependency problem in interval computations is suppressed except for the small remainder bound interval part. Thus, not only do Taylor models provide local multivariate derivatives, they also allow for highly efficient and sharp verification. The benefit of the sharpness becomes dramatic for multi-dimensional problems, which otherwise require an unrealistically large number of subdivision of the interested multi-dimensional domain box.

The tools to calculate Taylor models for standard computer representable functions have been developed and implemented in the code COSY Infinity, starting from sums and products and covering intrinsic functions [51, 364, 365]. The arithmetic starts from preparing the variables of the function, $\vec{x}$, represented by Taylor models; the polynomial part is $\vec{x}_0 + (\vec{x} - \vec{x}_0)$, and there is no error involved. Then, Taylor model arithmetic is carried through binary operations and intrinsic functions which compose the function f sequentially. Because the resulting objects represent functions, it is very advantageous to use the antiderivation operation

$$\partial_i^{-1}(P_n, I_n) = \left(\int P_{n-1} dx_i \,, \ (B(P_n - P_{n-1}) + I_n) \cdot |b_i - a_i| \right), \quad (43.2)$$

where B denotes the bounds of the argument, as a new intrinsic function in the spirit of the differential algebraic approach [364].

43.2   Sharp Enclosure of Multivariate Functions and Global Optimization

The straightforward Taylor model computation of a function $f(\vec{x})$ starting from the identity functions $i_i(\vec{x}) = x_i$ gives a resulting Taylor model of the function f, which carries the information on the derivatives as well as the sharp verification of the range enclosure of the function. This can be used directly for the exclusion procedure of domain decomposition approaches in verified global optimization methods.

To increase the accuracy of range enclosures, in general, the very first step should be a subdivision of the domain of interest. In naive interval arithmetic, the accuracy of the range enclosures increases linearly with the width of the argument; while in Taylor model arithmetic, the accuracy of the remainder intervals increases with $(n + 1)$st power, so the increase of the order of computation also could be the very first step.

Technically there are three questions for practical computation. First, the method requires an efficient mechanism to compute multivariate Taylor polynomials with floating point coefficients. The tools supplied in the

TABLE 43.1. Widths of the local bounds of Gritton's function around $x_0 = 1.5$ by non-verified rastering, the naive interval method, and the 10th order Taylor model method as well as the widths of the remainder intervals.

Subdomain Width	Widths of Local Bounds			TM Remainder
	Rastering	Interval	TM 10th	TM 10th
0.4000	0.2323	144507.	0.7775	2.998×10^{-6}
0.1000	1.854×10^{-2}	24555.	3.349×10^{-2}	6.360×10^{-13}
0.0250	5.478×10^{-3}	5788.	6.000×10^{-3}	1.472×10^{-19}

code COSY Infinity for more than ten years, have been used for various practical problems mostly in the field of beam physics [48, 51, 64]. The implementation of the Taylor model computation uses the existing multivariate Taylor polynomial computation tools [364]. The second problem is the rigorous estimation of the cumulative computational errors in the floating point arithmetic of the Taylor coefficients, which are all lumped into the remainder bound.

Finally, if the purpose of computation is to bound the range of a function, it is necessary to bound sufficiently sharply the range of the Taylor polynomial part. A variety of verified methods exist, and several methods are implemented in COSY Infinity, beginning with centered Horner methods. Because of the inherent dominance of lower order terms in the Taylor polynomial, it is possible to develop special-purpose tools that are sufficiently accurate while still being very fast. The key idea is to combine exact range solvers for the lower order parts of the polynomials with interval-based tools to enclose the higher order terms [364]. In most cases, however, already the majority of the benefit of the Taylor model approach can be achieved by very simple bounding techniques based on mere interval evaluation of the Taylor polynomial via Horner's scheme. In order not to distract the flow of the argument by discussing bounding techniques, we use this approach throughout the remainder of the chapter.

We show some challenging example problems for verified global optimization. The first example is Gritton's second problem from chemical engineering, addressed by Kearfott [319], which is known to suffer from a severe dependency problem. The function is an 18th order polynomial in one dimension, having 18 roots in the range $[-12, 8]$. The function varies roughly from -4×10^{13} to 6.03×10^{14}, and all the local maxima and minima have different magnitudes. As an illustration of its complicated structure, we note that there are four local extrema in the range $[1.4, 1.9]$ with function values varying only between around -0.1 and 0.1. Mere interval computation shows a severe blow up due to cancellation as summarized in Table 43.1. For intervals, to achieve the comparable result to the Taylor models in the 0.1 width subdomain, the domain would have to be cut to a practically hard to achieve size of 10^{-7}.

The pictures in Figure (1) show the absolute value of the function in

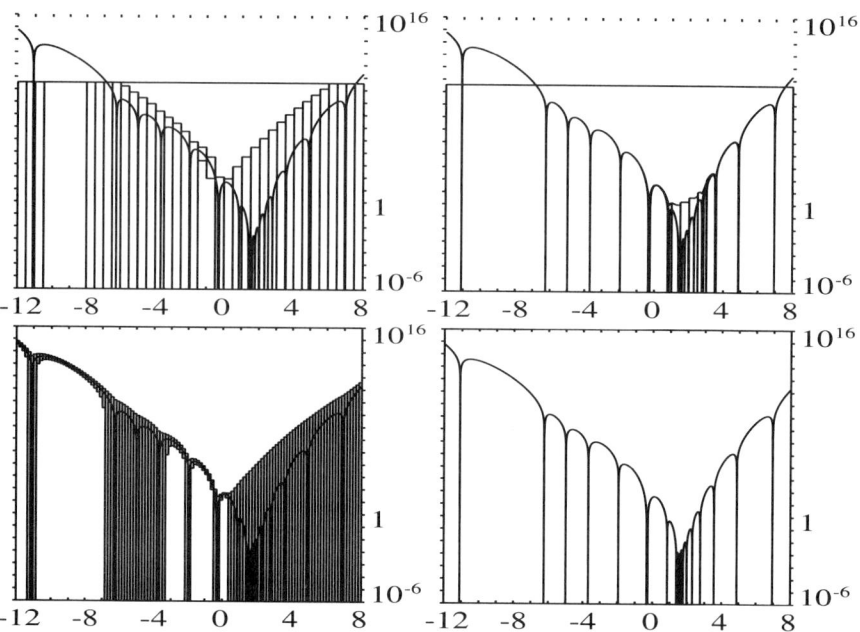

FIGURE 43.1. Gritton's function in $[-12, 8]$ evaluated by 40 (left top) and 120 (left bottom) subdivided intervals, and by 40 subdivided Taylor models with fourth order (right top) and eighth order (right bottom). The pictures show the absolute value of the function in logarithmic scale.

logarithmic scale. Any of the eighteen visible dips to the lower picture frame corresponds to a zero of the function. The left two pictures show enclosures by subdivided intervals; when the local bound interval contains zero, the lower end of the interval range box reaches the lower frame. In the pictures, the advantage of smaller subdivisions is hardly visible for $x > 0$. The right two pictures are obtained with 40 subdivided Taylor models. At order four (right top), there still remains a visible band width of the Taylor models in the range $0.5 \leq x \leq 3$, but at order eight (right bottom), the verified enclosure of the original function reaches printer resolution.

Another challenging example is the pseudo-Lyapunov function of weakly nonlinear dynamical systems [52]. The function is a six dimensional polynomial up to roughly 200th order which involves a large number of local minima and maxima, but the function value itself is almost zero. Hence, there is a large amount of cancellation, and the problem is a substantial challenge for interval methods. The study showed the Taylor model in sixth order already gives fairly tight enclosure of the function comparable to the rastering result in the domain box with the width 0.02 in each dimension. To achieve the similar sharpness in naive interval methods requires 10^{24} subdivided domains [364, 366].

TABLE 43.2. Bounds estimates of an eight dimensional integral, where the analytical answer is ≈ 99.4896438

Division	Step Rule	Trapezoidal	10th Taylor Models
1^8	`[8E-15,164.73]`	80.25	`[99.105950,99.910689]`
2^8	`[53.06,137.03]`	94.61	`[99.489371,99.489920]`
4^8	`[77.25,119.51]`	98.28	`[99.489643,99.489644]`

Points	Monte-Carlo Method
1	72.406666
100	95.614737
10000	99.865452
1000000	99.503242

43.3 Multidimensional Verified Quadratures

Using the handily available antiderivation operator (43.2), quadratures are computed with sharp verification straightforwardly, yet lead to a powerful method especially for the multi-dimensional case [62], where otherwise the Monte Carlo method often represents the only other viable approach.

Based on the following double definite integral

$$\int_0^{\frac{\pi}{2}} \int_0^{\frac{\pi}{2}} \frac{\sin y \sqrt{1 - k^2 \sin^2 x \sin^2 y}}{1 - k^2 \sin^2 y} \, dx \, dy = \frac{\pi}{2\sqrt{1 - k^2}},$$

we construct an eight dimensional integral, having as integrand the summation of four terms [62], which is to have the value of $\pi^7/(32\sqrt{1 - k^2})$, or approximately 99.4896438 for $k^2 = 0.1$. Even the evaluation with the simple trapezoidal rule without verification is quite expensive. The results are shown in Table 43.2 by the step rule with verification, the trapezoidal rule without verification, the tenth order Taylor model, and the non-verified Monte Carlo method with some numbers of subdivisions and sampling points. The tenth order Taylor model computation without any domain division already gives a remarkably good bound estimate.

43.4 Verified Integration of ODEs

ODE solvers in the Taylor model methods start from the integral form of the ODE to use the antiderivation operator (43.2), then bring it to a fixed point problem, where the nth order polynomial part P_n can be found in at most $(n + 1)$ steps. The rest involves tasks to check the inclusion of intervals, which are trivially done [61]. Our Taylor model ODE solver carries the functional dependency of the solutions to the initial conditions in the frame work of Taylor models. Thus it can optimally eliminate the wrapping effect, which has been the most challenging issue in verified ODE solvers [55]. Here again, in the Taylor model approach, the fact that the bulk of the functional dependency is kept in the polynomial part is key. The suppression of the wrapping effect allows the method to deal with larger domains of initial conditions. When combined with methods for verified solutions of constraint conditions over extended domains using Taylor models, the ODE solver forms a natural basis of a verified DAE solver [280].

An important application of the method is the dynamics of near-earth asteroids, addressed by Moore [387] and described by the six dimensional ODEs

$$\ddot{\mathbf{r}} = G \sum_i \frac{m_i (\mathbf{r}_i - \mathbf{r})}{r_i^3} \left\{ 1 - \frac{2(\beta + \gamma)}{c^2} G \sum_j \frac{m_j}{r_j} - \frac{2\beta - 1}{c^2} G \sum_{j \neq i} \frac{m_j}{r_{ij}} \right.$$

$$+ \frac{\gamma |\dot{\mathbf{r}}|^2}{c^2} + \frac{(1 + \gamma)|\dot{\mathbf{r}}_i|^2}{c^2} - \frac{2(1 + \gamma)}{c^2} \dot{\mathbf{r}} \cdot \dot{\mathbf{r}}_i - \frac{3}{2c^2} \left[\frac{(\mathbf{r} - \mathbf{r}_i) \cdot \dot{\mathbf{r}}_i}{r_i} \right]^2$$

$$\left. + \frac{1}{2c^2} (\mathbf{r}_i - \mathbf{r}) \cdot \ddot{\mathbf{r}}_i \right\} + \frac{3 + 4\gamma}{2c^2} G \sum_i \frac{m_i \ddot{\mathbf{r}}_i}{r_i}$$

$$+ \frac{1}{c^2} G \sum_i \frac{m_i}{r_i^3} \left\{ [\mathbf{r} - \mathbf{r}_i] \cdot [(2 + 2\gamma)\dot{\mathbf{r}} - (1 + 2\gamma)\dot{\mathbf{r}}_i] \right\} (\dot{\mathbf{r}} - \dot{\mathbf{r}}_i),$$

where $\mathbf{r}_i$ is the solar-system barycentric position of body i, including the sun, the planets, the moon and the five major asteroids; $r_i = |\mathbf{r}_i - \mathbf{r}|$; β and γ are the parametrized post-Newtonian parameters [309, 449]. The problem is challenging since initial conditions for asteroids are usually not very well known. To perform verified integrations it is thus necessary to transport a large box over an extended period of time. Hence the system is very susceptible to wrapping effect problems, but poses no difficulty for the Taylor model based integrator. Refer to, for example, [63], which discusses the resulting Taylor models for the position of the asteroid 1997XF11 obtained via verified integration over a period of about 3.47 years, with relative overestimation of the size of the resulting domain of less than a magnitude of 10^{-5}, showing the far-reaching avoidance of the wrapping effect.

Acknowledgments: We thank many colleagues for various stimulating discussions. The work was supported in part by the US Department of Energy and an Alfred P. Sloan Fellowship.

Taylor Models in Deterministic Global Optimization

R. Baker Kearfott and Alvard Arazyan

ABSTRACT Deterministic global optimization requires a global search with rejection of subregions. To reject a subregion, bounds on the range of the constraints and objective function can be used. Although often effective, simple interval arithmetic sometimes gives impractically large bounds on the ranges. However, Taylor models as developed by Berz et al. may be effective in this context. Efficient incorporation of such models in a general global optimization package is a significant project. Here, we use the system COSY Infinity by Berz et al. to study the bounds on the range of various order Taylor models for certain difficult test problems we have previously encountered. Based on that, we conclude that Taylor models may be useful for some, but not all, problems in verified global optimization. Forthcoming improvements in the COSY Infinity interface will help us reach stronger conclusions.

44.1 Deterministic Global Optimization

Deterministic global optimization involves exhaustive search over the domain. The domain is subdivided ("branching"), and those subdomains that cannot possibly contain global minimizers are rejected. For example, if the problem is the unconstrained problem

$$\text{Enclose the minimizers of} \quad \phi(x) \atop \text{subject to} \quad x \in \boldsymbol{x}, \tag{44.1}$$

then evaluating ϕ at a particular point x gives an upper bound for the global minimum of ϕ over the region $\boldsymbol{x}$. Some method is then used to bound the range of ϕ over subregions $\tilde{\boldsymbol{x}} \subset \boldsymbol{x}$. If the lower bound $\underline{\phi}$, so obtained, for ϕ over $\tilde{\boldsymbol{x}}$ has $\underline{\phi} > \phi(x)$, then $\tilde{\boldsymbol{x}}$ may be rejected as not containing any global optima; see Figure 44.1 for the situation in one dimension.

A related problem is that of finding all roots within a given region,

$$\text{Enclose all } x \text{ with} \quad f(x) = 0 \atop \text{subject to} \quad x \in \boldsymbol{x}. \tag{44.2}$$

In equation (44.2), bounds on f over a subregion $\tilde{\boldsymbol{x}} \subset \boldsymbol{x}$ are obtained; denote the interval vector representing such bounds by $\boldsymbol{f}(\boldsymbol{x})$. If $0 \notin \boldsymbol{f}(\boldsymbol{x})$, that is, unless the lower bound for each component of f is less than zero and the upper bound is greater than zero, then there cannot be a solution of $f(x) = 0$ in $\tilde{x}$, and $\tilde{x}$ can be rejected.

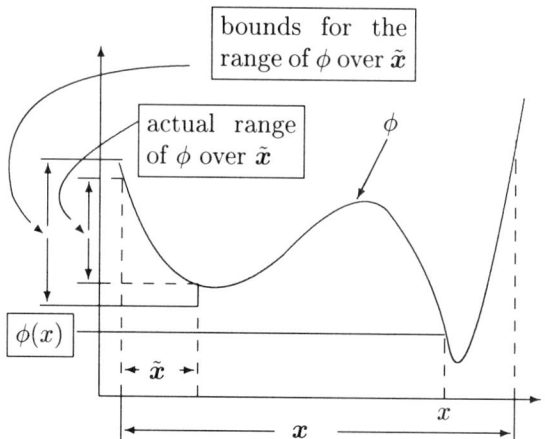

FIGURE 44.1. Rejecting $\tilde{x}$ because of a high objective value

A simple interval evaluation $\phi(\tilde{x})$ (or $f(\tilde{x})$) of ϕ (or of f) over $\tilde{x}$ is some-times a practical way of obtaining the lower bound. (See [268, 316, 402], or a number of other introductory expositions.) However, there are some func-tions for which interval evaluation gives an extreme overestimation, and other techniques are necessary. One such function arises from Gritton's second problem, a chemical engineering model that J. D. Seader previously pointed out to us.

Example 1 *(Gritton's second problem) The eighteen real solutions of* $f(x) = 0$ *in* $x = [-12, 8]$ *are sought, where* f *is defined by*

$$
\begin{aligned}
f(x) = & - 371.93625x^{18} - 791.2465656x^{17} + 4044.944143x^{16} \\
& + 978.1375167x^{15} - 16547.8928x^{14} + 22140.72827x^{13} \\
& - 9326.549359x^{12} - 3518.536872x^{11} + 4782.532296x^{10} \\
& - 1281.47944x^9 - 283.4435875x^8 + 202.6270915x^7 \qquad (44.3) \\
& - 16.17913459x^6 - 8.88303902x^5 + 1.575580173x^4 \\
& + 0.1245990848x^3 - 0.03589148622x^2 \\
& - 0.0001951095576x + 0.0002274682229.
\end{aligned}
$$

This example can be treated by careful domain subdivision and use of point evaluations, as explained in [317]. However, sharper bounds on the range would provide power that would make the code much simpler. Taylor models have shown promise for this.

44.2   Taylor Models and Global Optimization

An interval Taylor model in COSY Infinity [54, 64] for $\phi : \mathbb{R}^n \to \mathbb{R}$ is of the form

$$\phi(x) \in P_d(x - x_0) + \boldsymbol{I}_d, \tag{44.4}$$

where $P_d(x)$ is a degree-d polynomial in the n variables $x \in \mathbb{R}^n$, x_0 is a base point (often the midpoint of the interval vector $\boldsymbol{x}$), and $\boldsymbol{I}_d$ is an interval that encompasses the truncation error over the interval vector $\boldsymbol{x}$ and possible roundoff errors in computing the coefficients of P_d. Early work in interval computations did not indicate that Taylor models were promising. In particular, if one merely evaluated P_m with interval arithmetic over a box (i.e. over an interval vector) $\boldsymbol{x}$, then the difference between the width of $P_m(\boldsymbol{x}) + \boldsymbol{I}_d$ and the width of the actual range of ϕ over $\boldsymbol{x}$ decreases no faster than the square of the widths of $\boldsymbol{x}$, a rate that can already be achieved with $m = 2$. A higher convergence order can be achieved if the range of P_m can be estimated accurately, but computing such an estimation is NP-complete in the length of the expression defining P_m; see [334, Ch. 3 and Ch. 4].

However, Berz et al. have found Taylor models to be highly effective at computing low-overestimation enclosures of the range of functions [366]. Berz' group has applied such models successfully to the analysis of stability of particle beams in accelerators [59], and has advocated its use for global optimization in general [366]. This, among other applications, is discussed in [367]. In an informal communication, Berz and Makino illustrated that the overestimation in Gritton's problem can be reduced by many orders of magnitude simply by approximating the degree-18 function with its degree-5 Taylor polynomial.

In summary, Taylor models do not work in principle, but are effective in practice. The effectiveness can be viewed as a type of symbolic preconditioning of the algebraic expression: If the domain widths are not excessive, then interval dependencies are reduced when the original expression is replaced by a Taylor model. This indicates that additional careful study of Taylor models in general deterministic global optimization algorithms is warranted.

44.3   Scope and Purpose of This Preliminary Study

During the past few years, with support from a SunSoft Cooperative Research and Development contract, we have gained experience with various practical problems within the GlobSol [148, 316] interval global optimization package. Although successful with some problems, GlobSol, and another package Numerica [484] based on similar principles, cannot solve certain problems without subdividing the region $\boldsymbol{x}$ into an impractically large number of subregions. To solve such problems via verified global optimization, we need to identify the cause or causes of this algorithmic failure. These causes may include the following (among possibly other reasons).

(a) The stopping criteria for the subdivision process are inappropriate;

(b) the way that the boxes are subdivided (such as the method of selecting the coordinate to bisect in a bisection process) is inappropriate;

(c) the bounds on the ranges have an excessive amount of overestimation;

(d) there is something inherent in the mathematics of the equations, such as coupling between the components, that causes problems.

We have considered stopping criteria (item (a) above) in [318], and feel we understand the mechanisms in most cases. We [316, §4.3.2] and others (e.g. [45, 151, 435]) have studied the criteria for subdividing the box (item (b) above). We have also determined that such subdivision criteria are best if consistent with the stopping criteria; such integrated subdivision and stopping criteria have already been implemented in GlobSol.

Certain inherent conditions, such as manifolds of solutions, can result in an impractically large number of subregions. In these cases, problem reformulation is probably necessary for higher dimensions, although use of more powerful equipment (for example, a sufficient number of parallel or distributed processors) may be appropriate if the number of variables is not too large. These cases may be hard to distinguish from overestimation on the bounds (item (c)), and probably need to be studied individually.

Overestimation on bounds (item (c)) potentially can be handled using various tools within the algorithm itself. As mentioned in §44.2, §44.4, and in [366, 367], Taylor models can sometimes reduce overestimation considerably. However, the computational differentiation techniques in Taylor models, such as described in [48, 60], must be done efficiently to be practical in general global optimization algorithms. This differentiation process involves indexing schemes to reference the (generally) sparsely occurring non-zero terms from among the $(d + \nu)!/(d!\nu!)$ possible terms of a polynomial of degree d in ν variables [48]. This and other implementation details make quality implementation of Taylor model computations a major task.

Berz et al. have a good Taylor model implementation in COSY Infinity [54, 64]. Objective functions can be coded in the special COSY language, translated and interpreted within the package. In the experiments reported here, due to technical and licensing limitations, the objective had to be coded the COSY language, and called in a stand-alone mode. However, Jens Hoefkens has recently developed a Fortran 90 module for access to the COSY Infinity package. Future experiments will be easier and more comprehensive with this module.

A thorough test of Taylor arithmetic for general global optimization will need to integrate COSY Infinity computations within the global optimization algorithm, since the effect of range bound overestimation is different at different points (say, near a solution and far away from one), and since it is probably often advantageous to use local Taylor models specific to smaller subregions; this will be done with Hoefkens' module. However, because of the aforementioned limitations, we have proceeded in this chapter as follows:

1. We have provided a simple translator that translates GlobSol's "code list" to the COSY-INFINITY language.

2. We have identified two interesting problems we have tried to solve with GlobSol.

3. We have computed Taylor model ranges for several expansion points and interval vector widths.

The goal of this study is to evaluate the potential usefulness of Taylor models in verified global optimization. In particular, we wish to know what order and degree are necessary. Can the same benefits be gotten by just implementing lower-order, or is there a benefit of full generality? Also, how much of an impact is there on particular problems? Orders of magnitude difference in widths of range bounds for larger boxes would be useful, but small differences (perhaps less than a factor of 2) would be unimpressive, since Taylor arithmetic is more expensive than ordinary interval arithmetic.

44.4   Our Results

Our investigations to date have been with two examples.

44.4.1   Gritton's second problem

We initially tried Gritton's Example (1). One troublesome point is $\check{x} \approx 1.381$ with $f(\check{x}) = 0$. The range bounds over subintervals near this point should not contain zero, for such subintervals to be efficiently rejected. We tried base point $x_0 = 1.36$, and focused on the interval $\tilde{x} = [1.35, 1.37]$, experimenting also with different widths. The result is Table 44.4.1. "Rastering" is a heuristic method COSY Infinity uses to obtain inner bounds on the range: In these tables, the functions are evaluated at the end points of the component intervals and at three random values in the interior of each component interval; the minimum and maximum value so-obtained give a heuristic estimate for the range.

For Example (1), the Taylor model is definitely helpful. In particular, a straightforward interval calculation over $\tilde{x} = [1.35, 1.37]$ gives $f(x) \in [-1381, 1384]$, whereas the Taylor model of degree 5 gives $f(x) \in [0.009023, 0.0431]$, close enough to the actual range of $[0.0111, 0.0431]$ to determine that there is no zero of f in $\tilde{x}$. This contrasts sharply with the simple interval value, which is roughly 100,000 times too wide to be of use. In

TABLE 44.1. Widths of enclosing intervals for Example 1

Width	2	0.2	0.02
Degree 1	$3.46 \times 10^{+6}$	$1.76 \times 10^{+3}$	$1.27 \times 10^{+1}$
Degree 2	$1.08 \times 10^{+6}$	$7.03 \times 10^{+1}$	9.25×10^{-2}
Degree 3	$3.63 \times 10^{+5}$	$4.03 \times 10^{+0}$	4.29×10^{-2}
Degree 4	$1.16 \times 10^{+5}$	9.52×10^{-1}	4.27×10^{-2}
Degree 5	$4.68 \times 10^{+4}$	8.16×10^{-1}	4.27×10^{-2}
Degree 10	$3.20 \times 10^{+3}$	8.09×10^{-1}	4.27×10^{-2}
Simple interval	$4.10 \times 10^{+6}$	$3.22 \times 10^{+5}$	$2.65 \times 10^{+2}$
Rastering	$2.64 \times 10^{+3}$	5.62×10^{-1}	4.03×10^{-2}

fact, the interval evaluation at $[1.3599995, 1.3600005]$, an interval of length 10^{-6}, contains the interval $[-.0436, .0939]$, whereas the interval evaluation at $[1.3599999, 1.3600001]$, an interval of length 2×10^{-7}, is contained in the interval $[.01137, .03882]$. Thus, intervals on the order of 2×10^{-7} are needed to reject portions of the region as far out as 10^{-2} from the root, whereas an interval of length 2×10^{-2} (or perhaps larger) can be rejected with a degree-5 Taylor model.

44.4.2   A six-dimensional quartic

Neither GlobSol nor Numerica could solve the following six-dimensional polynomial system, whose components are of degree 4.

Example 2 *Find a_1, a_2, a_3, x_1, x_2, and x_3 such that $c_i = 0$, $i = 1, \ldots, 6$, where*

$$c_1 = 0.08413r + 0.2163q_1 + 0.0792q_2 - 0.1372q_3,$$
$$c_2 = -0.3266r - 0.57q_1 - 0.0792q_2 + 0.4907q_3$$
$$c_3 = 0.2704r + 0.3536\left(a_1(x_1 - x_3) + a_2(x_1^2 - x_3^2) + a_3(x_1^3 - x_3^3) \right.$$
$$\left. + x_1^4 - x_3^4\right)$$
$$c_4 = 0.02383p_1 - 0.01592r - 0.08295q_1 - 0.05158q_2 + 0.0314q_3$$
$$c_5 = -0.04768p_2 - 0.06774r - 0.1509q_1 + 0.1509q_3$$
$$c_6 = 0.02383p_3 - 0.1191r - 0.0314q_1 + 0.05158q_2 + 0.08295q_3, \quad where$$
$$r = a_1 + a_2 + a_3 + 1, \quad and$$
$$p_i = a_1 + 2a_2x_i + 3a_3x_i^2 + 4x_i^3, \quad i = 1, 2, 3,$$
$$q_i = a_1x_i + a_2x_i^2 + a_3x_i^3 + x_i^4, \quad i = 1, 2, 3.$$

In Example 2, we took center point $(a, x) = (1, 1, 1, 1, 1, 1)$, and took intervals of widths 1, 0.1, and 0.01 (equal widths in each direction) about this center point. The results appear in Table 44.4.2. In Table 44.4.2, we only

TABLE 44.2. Widths of enclosing intervals for c_1 for Example 2

Domain Width	1	0.1	0.01
Degree 1	9.6	0.54	0.0509
Degree 2	8.52	0.53	0.0508
Degree 3	8.52	0.53	0.0508
Degree 4	8.49	0.53	0.0508
Simple interval	6.85	0.59	0.0588
Rastering	6.24	0.51	0.0505

display the results for c_1; the widths of the range bounds for the other five components behave similarly. From Table 44.4.2, it is clear for this problem that

- The simple interval computations do not have excessive overestimation, at least for this problem.

- There is probably not an advantage to using the Taylor model representations to compute interval range bounds on this particular problem, since they apparently do not lead to narrower-width enclosures.

- The Taylor model representations show order-1 convergence as the interval widths are decreased.

All in all, there is evidence that, in Example 2, the difficulties of GlobSol are probably not due to overestimation in function range bounds. Nonetheless application of Taylor models reveals behavior of Taylor models not apparent in Gritton's problem (Example 1) or in Makino's example [366].

Despite the lack of advantage of interval computations in range estimation for the components in Example 2, Taylor models may still be useful for such problems. In particular, coupling between the equations (i.e. c_j and c_k both depend strongly on x_i for the same index i) plays a role in the difficulty. The equations are uncoupled with preconditioning. Hansen and others have pointed out that, for larger independent variable widths, the preconditioning is much more effective if it is done *symbolically*, before an interval evaluation is attempted. To symbolically precondition a system, the component functions need to be represented in terms of a basis and coefficients. The multivariate Taylor form provides such a basis.

Some additional remarks

The version of COSY Infinity currently available [54] evaluates the polynomial part $P_d(x - x_0)$ by substituting the interval vector $\tilde{x}$ into the expression and performing simple interval arithmetic. Since that version of COSY Infinity does not support the power function for all data types, x^2 is evaluated as $x \times x$, so that e.g., $[-1, 1]^2$ evaluates to $[-1, 1]$ rather than $[0, 1]$. We have observed slightly sharper (but non-rigorous) Taylor bounds from computations with Mathematica compared with COSY Infinity; we attribute the difference to COSY's present treatment of the power function. Nonetheless, as evidenced in Makino's example (ibid.) and both examples presented here, in at least some cases, the Taylor model approach is either powerful without sharp bounding of the polynomial part or else sharply bounding the polynomial part would not help much. The COSY-INFINITY development group is testing better bounding procedures, including sharp treatment of even powers, use of linear trends, etc.

44.5   Some Conclusions

It is difficult to draw definitive conclusions from our still limited experience. However, it is apparent that Taylor models are sometimes helpful and sometimes not helpful in verified global optimization. Given a hard problem, a user of verified global optimization software should probably first determine whether or not the intractability is due to overestimation in range bounds or due to some other reason. Heuristic tools for this purpose include stand-alone Taylor model evaluators (such as COSY Infinity) and rastering schemes.

Although costing a significant implementation effort, Taylor model capability would be useful if bundled with verified global optimization software. However, it should either be a well-documented user-controlled option or automatically chosen with a good heuristic, since there is evidence that it sometimes will provide great benefit and sometimes could make the algorithm significantly slower. Taylor models also may be helpful in putting systems of equations into a form in which we can symbolically precondition.

Actual implementation of Taylor models in a global optimization scheme may be somewhat different from the current COSY Infinity system, since models for first and second-order derivatives, in addition to the function itself, would be useful.

Acknowledgments: We wish to thank Martin Berz, who organized and supported a mini-conference at the National Superconducting Cyclotron Center for us to discuss and become familiar with COSY Infinity and its potential uses. We also wish to thank Kyoko Makino, who has been very helpful during our initial experiments with COSY Infinity.

45

Towards a Universal Data Type for Scientific Computing

Martin Berz

ABSTRACT Modern scientific computing uses an abundance of data types. Besides floating point numbers, we routinely use intervals, univariate Taylor series, Taylor series with interval coefficients, and more recently multivariate Taylor series. Newer are Taylor models, which allow verified calculations like intervals, but largely avoid many of their limitations, including the cancellation effect, dimensionality curse, and low-order scaling of resulting width to domain width. Another more recent structure is the Levi-Civita numbers, which allow viewing many aspects of scientific computation as an application of arithmetic and analysis with infinitely small numbers, and which are useful for a variety of purposes including the assessment of differentiability at branch points. We propose new methods based on partially ordered Levi-Civita algebras that allow for a unification of all these various approaches into one single data type.

45.1 Introduction

In this chapter we attempt to combine various data types used in scientific computation into a single one. Our primary interest lies in the ability to compute derivatives of as wide a class of functions representable on a computer as possible; in addition, we strive to make our arguments rigorous by employing interval techniques. To this end, we provide a generalization of the Levi-Civita numbers that on the one hand covers the multidimensional case, and on the other hand employs interval techniques to keep all arguments rigorous. The interval technique in particular will also allow answering a variety of questions concerned with proper exception handling merely within the one-dimensional case.

The Levi-Civita field $\mathcal{R}$ can be viewed as a set of functions from the rational numbers into the real numbers that are zero except on a set of rational numbers forming at most a strictly monotonically diverging sequence, as described in [50]. The structure forms an ordered field that is algebraically closed and admits infinitely small and large numbers [53]. In [50] and [454, 455] it is shown that for functions that can be represented as a computational graph consisting of elementary operations and the common intrinsics, derivatives can be obtained up to infinitely small error by evaluation of the difference quotient $(f(x + \Delta x) - f(x))/\Delta x$ for any infinitesimally small Δx; similar formulae hold for higher derivatives. Thus, the method retroactively justifies the thinking of Newton and Leibnitz, and allows treatment of infinitely small numbers even on the computer. Additionally, the algebraic closure implies that derivatives can also be calculated and proven to exist in many cases of branch points, and even under the presence of non-differentiable pieces as long as the overall graph represents a differentiable function [454].

The method of [50] is limited to functions of one variable. This allowed the treatment of partial derivatives by repeated evaluation, but no assessment of differentiability over entire multidimensional neighborhoods. Below the structures are extended to the canonical treatment of v independent differentials. One obtains functions from Q^v into R such that the sequence formed by adding the v rational numbers $q_1 + ... + q_n$ at nonzero entries forms a strictly diverging sequence. Most favorable properties are retained, in particular the structure still admits roots for most numbers, and similar to above it is now possible to obtain partial derivatives of any order as the respective difference quotients with respect to different infinitesimals. Again the methods also hold under branching and in the presence of certain non-differentiable pieces.

In a practical implementation requiring mathematical rigor, it is also necessary to account for possible floating point errors. This can be achieved by embedding the structures into functions from Q^v into the set of floating point intervals. Using this method, the presence/absence of critical points can be detected in a rigorous way if appropriate.

In practice, first and higher order partial derivatives are often used in the framework of sensitivity analysis. We note that it is possible to develop a further extension of the previous numbers that even support rigorous bounds of the errors made in a sensitivity representation over a pre-specified domain. This approach provides a generalization of the Taylor Model approach [365, 367] that now also rigorously accounts for non-differentiability. For reasons of space, we have to refer to a forthcoming paper for details.

Altogether, the newly proposed data type unites differentiation data types for any order and number of variables, interval methods, the sharper approach of Taylor models, and has the ability to assert differentiability even if not all pieces are differentiable.

45.2   The Levi-Civita Numbers $\mathcal{R}_v$

We begin our discussion with a group of definitions.

Definition 1 *(Diagonal Finiteness) We say a subset M of the rational v-tuples Q^v is diagonally finite if for any m, there are only finitely many elements $(q_1, ..., q_v) \in M$ that satisfy $q_1 + ... + q_v \leq m$.*

Diagonal finiteness is a generalization of the left-finiteness introduced for the Levi-Civita numbers [50]. In a similar way, we now introduce a set of numbers as follows:

Definition 2 (*Extended Levi-Civita Numbers*) *We define the extended Levi-Civita numbers $\mathcal{R}_v$ to be the set of functions x from Q^v into the real numbers R such that the points $\mathbf{q} = (q_1, ..., q_v)$ in the support of x, i.e. those points that satisfy $x[\mathbf{q}] \neq 0$, are diagonally finite.*

For two extended Levi-Civita Numbers x and y in $\mathcal{R}_v$, we define $(x + y)$ and $(x \cdot y)$ via

$$(x + y)[\mathbf{q}] = x[\mathbf{q}] + y[\mathbf{q}] \tag{45.1}$$

$$(x \cdot y)[\mathbf{q}] = \sum_{\mathbf{q=r+s}} x[\mathbf{r}] \cdot y[\mathbf{s}] \tag{45.2}$$

where for a given $\mathbf{q}$, the sum is carried over only those $\mathbf{r}$ and $\mathbf{s}$ that satisfy $\mathbf{q} = \mathbf{r} + \mathbf{s}$ and $x[\mathbf{r}] \neq 0$, $y[\mathbf{s}] \neq 0$, of which there are only finitely many.

Definition 3 *We define the interval Levi-Civita Numbers $\mathcal{R}_v^I$ to be the set of functions from Q^v into intervals with diagonally finite support. The arithmetic for the intervals (45.1) and (45.2) follows standard practice ([8, 338, 384, 385, 386]). The arithmetic for the $\mathbf{q}$ is performed as exact rational arithmetic with check for overflow beyond a maximum q_o and underflow beyond a minimum q_u.*

For a generalization of the Taylor model approach [364, 365, 367], one would map into a product space (I_1, I_2) of two intervals, the first of which agrees with the previous one, and the second of which would represent a bound of the remainder error for truncation at the respective order.

We note that implementation of the Levi-Civita numbers $\mathcal{R}_v$ is a straightforward generalization of high-order multivariable Taylor methods; to any given Levi-Civita number, one decides a depth l to which support points are kept. For the finitely many support points below l, the common denominator is determined, and the numerators are manipulated just as in high-order multivariate automatic differentiation [48].

Definition 4 (*Degree, Finiteness, Core and Pyramidality*) *To a number nonzero $x \in \mathcal{R}_v$, we define the degree*

$$\lambda(x) = \min\{q_1 + ... + q_v \mid x[q_1, ..., q_v] \neq 0\}, \tag{45.3}$$

i.e. the minimum of the support of x. We also define $\lambda(0) = +\infty$. If $\lambda(x) = 0$, > 0, or < 0, we say x is finite, infinitely small, or infinitely large, respectively. Any point $(q_1^c, ..., q_v^c)$ that satisfies $x[q_1^c, ..., q_v^c] \neq 0$ and $q_1 + ... + q_v = \lambda(x)$ is called a core of x. We say x is pyramidal if it has only one core.

The most important case, representing the case of conventional automatic differentiation, corresponds to finite numbers with support points $(q_1, ..., q_v)$ that satisfy $q_i \geq 0$. These numbers are pyramidal with core $(0, ..., 0)$.

Definition 5 *(Positive Numbers, Ordering)* *Let $x \in R_v$ be pyramidal, and let $\mathbf{q}_c$ be the core of x. We say x is positive if x if $x[\mathbf{q}_c] > 0$, and negative if $x[\mathbf{q}_c] < 0$. We say $x > y$ if $(x - y)$ is pyramidal and positive.*

Apparently the ordering is only partial, as some numbers are neither positive nor negative. Altogether, we arrive at the following situation.

Theorem 7 *(Structure of R_v)* *The Levi-Civita Numbers R_v form an extension of the real numbers R by virtue of the embedding*

$$r \in R \to x \in R_v \text{ with } x[0] = r \text{ and } x[\mathbf{q}] = 0 \text{ for } \mathbf{q} \neq 0 \qquad (45.4)$$

which preserves arithmetic and order of R.

They form a partially ordered real algebra, and the order is compatible with the algebraic operations. The order is non-Archimedean; for example, the elements d_v given by

$$d_v[e_v] = 1, \text{ and } 0 \text{ else} \qquad (45.5)$$

where e_v is the v-th Cartesian basis vector, are infinitely small, and their multiplicative inverses are infinitely large.

Pyramidal elements admit inverses and odd roots, and positive pyramidal elements admit even roots.

The proofs of these assertions follow rather parallel to the corresponding ones in the case of R presented in [50]. Specifically, let $(q_1^c, ..., q_v^c)$ be the (unique) core of the number x. We then write

$$x = x[q_1^c, ..., q_v^c] \cdot (1 + \tilde{x})$$

and observe that by the definition of pyramidality, $\lambda(\tilde{x}) > 0$. This allows for the treatment of $(1 + \tilde{x})$ through a (converging) power series for the root, in a similar way as in [50].

45.3 Intrinsic Functions for Finite and Infinite Arguments

In [50, 53, 455], the common intrinsic functions were extended to R based on a theory of power series on R and their power series representation. The concept of weak convergence on which the theory of power series on R is based can be extended directly to R_v in the same way as before based on the family of seminorms

$$||x||_r = \max_{q_1 + ... + q_v \leq r} (|x[q_1, ..., q_v]|).$$

In this way, one obtains that the conventional power series all converge in the same way as in $\mathcal{R}$, and thus all conventional intrinsic functions are available in the same fashion.

However, the canonical treatment of intrinsics via power series only applies to arguments within the finite radius of convergence, and thus for example does not cover infinitely large numbers, limiting their use for cases where also in all intermediate computations, the arguments stay finite, and precludes the study of cases like $f(x) = x^2 \cdot \log(x)$ $(x \neq 0)$, $f(0) = 0$ at the origin. Within the framework of $\mathcal{R}$ and $\mathcal{R}_v$, it is however also desirable to assign values to intrinsics for non-finite numbers.

We now extend the common intrinsic functions to $\mathcal{R}_v$ beyond the domains in which they can be represented by power series by representing their asymptotic behavior in terms of intervals, and as a consequence we will be able to study cases like $x^2 \log(x)$.

Definition 6 (Intrinsic Functions)

$$\sin(x) = \begin{cases} \sum_{i=0}^{\infty} (-1)^{i+1} \frac{x^{2i+1}}{(2i+1)!} & \lambda(x) \geq 0 \\ [-1,1] & x \text{ infinitely large} \end{cases}$$

$$\exp(x) = \begin{cases} \sum_{i=0}^{\infty} \frac{x^i}{i!} & \lambda(x) \geq 0 \\ [0,\infty] \cdot d_1^{q_o} \cdot \ldots \cdot d_v^{q_o} & x \text{ positive and infinitely large} \\ [0,\infty] \cdot d_1^{q_u} \cdot \ldots \cdot d_v^{q_u} & x \text{ negative and infinitely large} \end{cases}$$

$$\log(x) = \begin{cases} \log(\mathrm{Re}(x)) + \sum_{i=0}^{\infty} \frac{(x - \mathrm{Re}(x))^i}{i \cdot \mathrm{Re}(x)^i} & x \text{ positive and } \lambda(x) = 0 \\ [-\infty, 0] & x \text{ positive and infinitely small} \\ [0, \infty] & x \text{ positive and infinitely large} \end{cases}$$

where q_o and q_u are the largest and smallest representable rational numbers introduced above, and $\mathrm{Re}(x)$ denotes the real part $x[0, ..., 0]$ introduced above.

Apparently these definitions have a slight similarity with assigning things such as "NAN" etc. for floating point exceptions, where here in addition it is possible to distinguish different speed of divergence by assigning "different kinds of infinity." The sine stays bounded asymptotically, the exponential diverges faster than any rational power, and the logarithm, while diverging, does so slower than any rational power. With these and similar definitions for the other intrinsics not explicitly listed, it is now possible to treat many different kinds of branch points and exceptions, as exemplified below.

We have similar theorems as in $\mathcal{R}$:

Theorem 8 (Continuity and Differentiability)
    The function f is infinitely often partially differentiable at $\mathbf{x}$ in R^v, and is totally differentiable if it can be evaluated for some true interval inclusion that has $\mathbf{x}$ in its interior without tripping a code branch during its evaluation. In this case, all partial derivatives of f can be evaluated up to

infinitely small error by the corresponding divided difference formulae with infinitely small differences.

The function f is continuous in the direction of unit vector $\mathbf{e}_v$ if it can be evaluated for an interval that may have $\mathbf{x}$ on its boundary, as well as $\mathbf{x} \pm d_v$, and if the evaluation $f(\mathbf{x})$ and $f(\mathbf{x} \pm d_v)$ agree up to infinitely small error, even if $\mathbf{x}$ and $\mathbf{x} \pm d_v$ lie on different code branches.

The function f is partially differentiable in direction $\mathbf{e}_v$ if it can be evaluated for an interval that may have $\mathbf{x}$ on its boundary as well as $\mathbf{x} \pm d_v$ and if the difference quotients $(f(x + d_v) - f(x))/d_v$ and $(f(x) - f(x - d_v))/d_v$ agree up to infinitely small error, even if $\mathbf{x}$ and $\mathbf{x} \pm d_v$ lie on different code branches.

The function f is higher-order partially differentiable in direction $\mathbf{e}_v$ if it can be evaluated for an interval that may have $\mathbf{x}$ on its boundary as well as $\mathbf{x} + \sum c_j d_j$ for suitable coefficients c_j and if the corresponding higher-order difference quotients agree up to infinitely small error, even if $\mathbf{x}$ and the $\mathbf{x} + \sum c_j d_j$ lie on different code branches.

In this view, the treatment of branch conditions in IF structures etc. through intervals I are to be viewed in a set theoretical sense with a partial order $I_1 < I_2$ if $x_1 < x_2$ for all $x_1 \in I_1$ and $x_2 \in I_2$. This is similar to other interval-based treatment of branches [36, 315]. In the first case, if there is a true inclusion that is "safe" for evaluation, this already implies that all the arguments in the divided difference schemes can be evaluated. The other cases parallel the situation in $\mathcal{R}$ [50], except that now also partial derivatives are possible, and there is no reliance on the accuracy of floating point arithmetic.

45.4   Critical Points of Computer Functions and Their Distribution

It has been widely recognized [36, 237, 315] that the detection of critical points during code execution is a difficult task without relying on verified techniques such as interval methods or the interval-Levi Civita numbers introduced above. However, it is frequently at least assumed that any critical points that can occur in computer code to be processed by automatic differentiation tools through branching or inherent singularities of intrinsic functions are isolated and hence easily distinguishable. However, this intuitive notion is indeed rather misleading, as we shall illustrate with a few examples. In particular, it is possible to construct computer functions with critical points at nearly every computer number in a certain range. Consider for example the function f and its critical points

$$f(x) = 1/\sin(\exp(1/x)) \tag{45.6}$$
$$x_n^{crit} = 1/\ln(n\pi) \tag{45.7}$$

Apparently the critical points of f are spread all over the interval $[0, 1]$, and they get closer together as we approach 0. Table 45.1 shows the distribution of the critical points of f in $(0, 1)$. Furthermore, any point in the interval

TABLE 45.1. Distribution of the critical points of f in $(0, 1)$

Subinterval	Number of critical points of f
$\left(\frac{1}{16}, \frac{1}{8}\right]$	$2, 827, 587$
$\left(\frac{1}{32}, \frac{1}{16}\right]$	$25, 134, 688, 039, 947$
$\left(\frac{1}{64}, \frac{1}{32}\right]$	more than 1.9×10^{27}

$(0, 0.03]$ is less than 10^{-16} away from a critical point of f. In passing we note that the actual size of the interval is rather immaterial since through a simple appropriate re-scaling we can apparently obtain the same result in $(0, 1]$ rather than in $(0, 0.03]$. Between any two computer numbers in $(0, 0.03]$, there lies a critical point of f, and thus critical points cannot computationally be distinguished from non-critical points.

To illustrate this point, we investigate what happens if we evaluate f at the critical point $x = 1/\ln\left(10^{15}\pi\right) \in (0, 0.03]$. We first evaluate x numerically with 14, 16, 18, 20 digits precision and then we evaluate f with the same precision, using Maple. The results are given in Table 45.2 The results

TABLE 45.2. Critical points cannot computationally be distinguished from non-critical points

Digits	$x = 1/\ln\left(10^{15}\pi\right)$	$f(x)$
14	$2.\,802\,415\,189\,056\,6 \times 10^{-2}$	$-1.\,410\,081\,582\,760\,7$
16	$2.\,802\,415\,189\,056\,638 \times 10^{-2}$	$1.\,000\,812\,471\,855\,461$
18	$2.\,802\,415\,189\,056\,637\,88 \times 10^{-2}$	$-9.\,237\,866\,180\,822\,379\,43$
20	$2.\,802\,415\,189\,056\,637\,879\,1 \times 10^{-2}$	$-791.\,989\,471\,416\,785\,333\,8$

apparently do not reflect the fact that x is a critical point of f, but each individual evaluation looks rather inconspicuous. The situation is rather similar for other choices of critical points. With only minor additional effort, the situation can become even much more complicated; for example, consider the function $g(x)$ with critical points $x^c_{m,n}$ given by

$$g(x) = 1/\sin(f(x)) = 1/\sin\left(1/\sin(\exp(1/x))\right);$$

$$x^c_{m,n} = \frac{1}{\ln\left(n\pi + \arcsin\left(\frac{1}{m\pi}\right)\right)} \quad \text{where } m, n \text{ are nonzero integers.}$$

Thus, to every critical point of f in $(0, 1)$ there corresponds a whole family of infinitely many critical points of g. All these difficulties that are not transparent in a floating point environment are overcome more or less directly when working in any kind of rigorous interval environment. This is also the case for the interval Levi-Civita numbers R_v^I.

45.5   Examples

To conclude, we illustrate the behavior of the method for various pathological cases, some of which have been considered previously [36, 236, 237, 273, 315, 353].

Problem 1 $f_1(x) = 1$ *for* $x^2 > 2$, *and* 0 *otherwise, at* $x = \sqrt{2}$.

Depending on whether the floating point result of the calculation of $\sqrt{2}$ is above or below the true value of $\sqrt{2}$, using floating point methods, the value of f_1 would be chosen as 1 or 0. The derivative would be returned as 0. Evaluation with interval arithmetic would recognize the branch and conclude that it cannot decide differentiability. Depending on implementation of the IF structure and the possibility of launching branch threads, an implementation may even return the result $[0, 1]$. Evaluation in R_v^1 would recognize the branch and conclude that it cannot decide differentiability.

Problem 2 $f_2(x) = x \cdot \exp(1/x)$ *for* $x \neq 0$ *and* $f_2(0) = 0$, *at* $x = 0$

Evaluation R_v^1 for $f(0)$ and $f(0+d)$ yields $[0, 0] \cdot d^0$ and $d \cdot [0, \infty] \cdot d_1^{q_o} \cdot \ldots \cdot d_v^{q_o}$. Since $f(0)$ and $f(0+d)$ differ by more than an infinitely small amount, we conclude that f_2 is discontinuous at the origin.

Problem 3 $f_3(x) = x \cdot \sin(1/\sqrt{|x|})$ *for* $x \neq 0$ *and* $f_3(0) = 0$, *at* $x = 0$

Since $f_3(0) = 0$, $f_3(-d) = -d \cdot [-1, 1]$, $f_3(d) = d \cdot [-1, 1]$ all differ only by infinitely small amounts, we conclude f_3 is continuous at 0. Since $(f_3(d) - f_3(0))/d = [-1, 1]$ is finite, we know f is not differentiable at 0.

Problem 4 $f_4(x) = |x|^{3/2} \cdot \log(|x|)$ *for* $x \neq 0$ *and* $f_4(0) = 0$, *at* $x = 0$

Since $f_4(0) = 0$, $f_4(-d) = d^{3/2} \cdot [-\infty, 0]$, $f_4(d) = d^{3/2} \cdot [-\infty, 0]$ all differ only by infinitely small amounts, we conclude f_4 is continuous at 0. Since $(f_4(d) - f_4(0))/d = (f_4(0) - f_4(-d))/d = d^{1/2} \cdot [-\infty, 0]$ are equal and thus differ by not more than infinitely small amounts, f_4 is differentiable at 0.

Problem 5 $f_5(x, y) = 0$ *if* $x^2 = y$ *and* $(x, y) \neq (0, 0)$, $f_5(0, 0) = 2$, *and* $f_5(x, y) = (2x + 1)(|y| + 2)$ *otherwise, at* $x = y = 0$.

Evaluating the function at any small interval containing the origin triggers a code branch, avoiding the affirmative answer that the function is totally differentiable at the origin. Moreover, if within the implementation of code transformation, all branches can be followed, it is concluded that the resulting interval has width of at least $[0, 2]$, and hence the function is concluded not to be totally differentiable. Testing partial differentiability in the x direction, we have $f_5(0,0) = 2$, $f_5(d_x,0) = 2 + 4d_x$, and $f_5(-d_x,0) = 2 - 4d_x$. Hence $(f_5(d_x,0) - f_5(0,0))/d_x = (f_5(-d_x,0) - f_5(0,0))/(-d_x) = 4$, and we conclude that f is partially differentiable in the x direction with derivative 4. On the other hand, evaluating $f_5(0,0) = 2$, $f_5(0,d_y) = 2 + d_y$, and $f_5(0,-d_y) = 2 + d_y$, we have $(f_5(0,d_y) - f_5(0,0))/d_y = 1$, while $(f_5(0,-d_y) - f_5(0,0))/(-d_y) = -1$, and so we conclude f_5 is not partially differentiable in the y direction.

Acknowledgments: For performing the calculations on the distribution of critical points of the example functions I would like to thank Khodr Shamseddine. For various fruitful conversations about critical points, I would like to thank George Corliss and Baker Kearfott.

Bibliography of Automatic Differentiation

George F. Corliss

ABSTRACT   This is a bibliography of work related to automatic differentiation. It represents the compilation of all works cited by all chapters in this volume, *Automatic Differentiation: From Simulation to Optimization*, George Corliss, Christèle Faure, Andreas Griewank, Laurent Hascoët, and Uwe Naumann (eds.), Springer, New York, 2001 [144]. Authors compiled bibliographies for their own chapters. The separate bibliographies were merged into a single BibTex database. Because it includes all of the works cited by any chapter in this book, it includes most widely cited AD work and many citations that are not directly related to AD. For example, it includes basic references in optimization, symbolic algebra systems, and several applications areas.

Comments

This bibliography builds on collected bibliographies:

- [145] Corliss, Automatic Differentiation Bibliography, in *Automatic Differentiation of Algorithms: Theory, Implementation, and Application* [239] (Breckenridge, 1991).

- [505] Yang and Corliss, Bibliography of Computational Differentiation, in *Computational Differentiation: Techniques, Applications, and Tools* [57] (Sante Fe, 1996).

This collection includes all chapters from those two volumes, but not all the entries from those two bibliographic collections. For getting started in automatic differentiation, my favorite surveys include:

- [431] Rall, *Automatic Differentiation: Techniques and Applications*, Springer, 1981.

- [234] Griewank, The Chain Rule Revisited in Scientific Computing, *SIAM News*, 1991.

- [433] Rall and Corliss, An Introduction to Automatic Differentiation, in *Computational Differentiation: Techniques, Applications, and Tools*, [57], SIAM, 1996.

- [238] Griewank, *Evaluating Derivatives: Principles and Techniques of Algorithmic Differentiation*, SIAM, 2000.

- See www.mcs.anl.gov/autodiff.

Web URL's given here may change. These are the best known at the time of publication.

The editors hope that this collection will help new (and veteran) researchers find their way through the extensive literature of automatic differentiation. This bibliography is available from NETLIB (netlib.bell-labs.com).

[1] Jason Abate, Steve Benson, Lisa Grignon, Paul D. Hovland, Lois C. McInnes, and Boyana Norris. Integrating AD with object-oriented toolkits for high-performance scientific computing. In Corliss et al. [144], chapter 20, pages 173–178.

[2] Jason Abate, Christian Bischof, Alan Carle, and Lucas Roh. Algorithms and design for a second-order automatic differentiation module. In *International Symposium on Symbolic and Algebraic Computing (ISSAC)*, pages 149–155, Philadelphia, Penn., 1997. SIAM.

[3] John Abbott and André Galligo. Reversing a finite sequence. Preprint, 1991.

[4] Anatole Abragam and Berbis Bleaney. *Résonance paramagnétique électronique des ions de transition*, chapter 3.2. L'effet de l'anisotropie sur le facteur g, formule (3,10). Presses Universitaires de France, Paris, 1972.

[5] Milton Abramowitz and Irene A. Stegun, editors. *Handbook of Mathematical Functions With Formulas, Graphs, and Mathematical Tables*. U.S. Government Printing Office, Washington, D.C., 20402, 1964.

[6] Alfred V. Aho, Ravi Sethi, and Jeffrey D. Ullman. *Compilers: Principles, Techniques, and Tools*. Addison-Wesley, Reading, Mass., 1995.

[7] A. F. Albou, V. I. Gorbunov, and V. I. Zubov. Optimal control of the process of melting. *Computational Mathematics and Mathematical Physics*, 40(4):491–504, 2000.

[8] Götz Alefeld and Jürgen Herzberger. *Introduction to Interval Computations*. Academic Press, New York, 1983.

[9] J. Allen. Computer optimisation of cornering line. Master's thesis, School of Mechanical Engineering, Cranfield University, 1997. See www.cranfield.ac.uk/cils/library.

[10] R. Altpeter and Wolfram Klein. Numerical solutions in stationary thermodynamic calculations of steam generators. In *Proceedings of GAMM 2000*, 2000.

[11] AMPL. See `www.ampl.com`.

[12] E. Anderson, Z. Bai, Christian Bischof, John Demmel, Jack Dongarra, J. Du Croz, A. Greenbaum, Sven Hammarling, A. McKenney, S. Ostrouchov, and Danny Sorensen. *LAPACK User's Guide, 2nd edn.* SIAM, Philadelphia, Penn., 1995.

[13] W. Kyle Anderson and Daryl L. Bonhaus. Airfoil design on unstructured grids for turbulent flows. *AIAA J.*, 37(2):185–191, 1999.

[14] W. Kyle Anderson, William D. Gropp, Dinesh K. Kaushik, David E. Keyes, and Barry F. Smith. Achieving high sustained performance in an unstructured mesh CFD application. In *Proceedings of SC'99*. IEEE Computer Society, 1999. Gordon Bell Prize Award Paper in the "Special" Category.

[15] T. E. Arbetter, J. A. Curry, and J. A. Maslank. Effects of rheology and ice thickness distribution in a dynamic-thermodynamic sea-ice model. *J. Phys. Oceanog.*, 29:2656–2670, 1999.

[16] Uri M. Ascher, Robert M. Mattheij, and Robert D. Russell. *Numerical Solution of Boundary Value Problems.* Prentice Hall, Englewood Cliffs, N.J., 1988.

[17] Pierre Aubert. meta::Expr⟨T_emplate⟩ a library for building efficient numerical analysis classes in C++. *Journal of Computational and Applied Mathematics*, 2000. Submitted. See `pierre.aubert. free.fr/research/meta.ps`.

[18] Pierre Aubert and Nicolas Di Césaré. Expression templates and forward mode automatic differentiation. In Corliss et al. [144], chapter 37, pages 311–315.

[19] Pierre Aubert, Nicolas Di Césaré, and Olivier Pironneau. Automatic differentiation in C++ using expression templates and application to a flow control problem. *Computing and Visualisation in Sciences*, 2000. Accepted. See `pierre.aubert.free.fr/research/fad.ps`.

[20] Pierre Aubert and Bernard Rousselet. Sensitivity computations and shape optimization for a nonlinear arch model with simple instabilities. *IJNME*, 42:15–48, 1998.

[21] Brett M. Averick, Richard G. Carter, Jorge J. Moré, and Guo-Liang Xue. The MINPACK-2 test problem collection. Preprint MCS–P153–0692, ANL/MCS–TM–150, Rev. 1, Mathematics and

Computer Science Division, Argonne National Laboratory, Argonne, Ill., 1992. See `ftp://info.mcs.anl.gov/pub/MINPACK-2/tprobs/P153.ps.Z`.

[22] Brett M. Averick and Jorge J. Moré. User guide for the MINPACK-2 test problem collection. Technical Memorandum ANL/MCS-TM-157, Argonne National Laboratory, Argonne, Ill., 1991. Also issued as Preprint 91-101 of the Army High Performance Computing Research Center at the University of Minnesota.

[23] Brett M. Averick, Jorge J. Moré, Christian H. Bischof, Alan Carle, and Andreas Griewank. Computing large sparse Jacobian matrices using automatic differentiation. *SIAM J. Sci. Comput.*, 15(2):285–294, 1994.

[24] David F. Bacon, Susan L. Graham, and Oliver J. Sharp. Compiler transformations for high-performance computing. *ACM Computing Surveys*, 26(4):345–420, 1994.

[25] Satish Balay, William D. Gropp, Lois C. McInnes, and Barry F. Smith. Efficient management of parallelism in object oriented numerical software libraries. In E. Arge, A. M. Bruaset, and Hans Petter Langtangen, editors, *Modern Software Tools in Scientific Computing*, pages 163–202. Birkhauser Press, 1997.

[26] Satish Balay, William D. Gropp, Lois C. McInnes, and Barry F. Smith. The portable extensible toolkit for scientific computing PETSc 2.0 users manual. Technical Report ANL-95/11 - Revision 2.0.28, Argonne National Laboratory, Mar. 2000. See `www.mcs.anl.gov/petsc`.

[27] William E. Ball. *Material and Energy Balance Computations*, pages 560–566. John Wiley, 1969.

[28] L. Baratchart, José Grimm, J. Leblond, M. Olivi, F. Seyfert, and F. Wielonsky. Identification d'un filtre hyperfréquences par approximation dans le domaine complexe. Technical Report RT-0219, INRIA, 1998.

[29] Michael C. Bartholomew-Biggs. Using forward accumulation for automatic differentiation of implicitly-defined functions. *Computational Optimization and Applications*, 9:65–84, 1998.

[30] Michael C. Bartholomew-Biggs. A two-dimensional search used with a non-linear least-squares solver. *Journal of Optimization Theory and Applications*, 104(1):215–234, 2000.

[31] Michael C. Bartholomew-Biggs, Steve Brown, Bruce Christianson, and Laurence C. W. Dixon. Automatic differentiation of algorithms. *Journal of Computational and Applied Mathematics*, 124:171–190, 2000.

[32] John J. Barton and Lee R. Nackman. *Scientific and Engineering C++*. Addison-Wesley, Reading, Mass., 1994.

[33] Douglas M. Bates and Donald G. Watts. Relative curvature measures of nonlinearity. *Journal of the Royal Statistical Society B*, 42(1):1–25, 1980.

[34] A. Battermann and M. Heinkenschloss. Preconditioners for Karush-Kuhn-Tucker matrices arising in optimal control of distributed systems. In W. Desch, F. Kappel, and K. Kunisch, editors, *Optimal Control of Partial Differential Equations (Vorau 1996)*, pages 15–32. Birkheuser Verlag, Basel, 1998.

[35] Friedrich L. Bauer. Computational graphs and rounding errors. *SIAM J. Numer. Anal.*, 11(1):87–96, 1974.

[36] T. Beck and Herbert Fischer. The IF-problem in automatic differentiation. *Journal of Comput. Appl. Math.*, 50:119–131, 1995.

[37] Jean-Daniel Beley, Stephane Garreau, Frederic Thevenon, and Mohamed Masmoudi. Application of higher order derivatives to parameterization. In Corliss et al. [144], chapter 40, pages 335–341.

[38] Adel Ben-Haj-Yedder, Eric Cances, and Claude Le Bris. Optimal laser control of chemical reactions using AD. In Corliss et al. [144], chapter 24, pages 205–211.

[39] Jochen Benary. Parallelism in the reverse mode. In Berz et al. [57], pages 137–147.

[40] Martin P. Bendsøe. *Optimization of Structural Topology, Shape and Material*. Springer, Berlin, 1995.

[41] Claus Bendtsen and Ole Stauning. FADBAD, a flexible C++ package for automatic differentiation. Technical Report IMM-REP-1996-17, Technical University of Denmark, IMM, Departement of Mathematical Modeling, Lyngby, 1996.

[42] Steve Benson, Lois C. McInnes, and Jorge J. Moré. GPCG: A case study in the performance and scalability of optimization algorithms. Technical Report ANL/MCS-P768-0799, Mathematics and Computer Science Division, Argonne National Laboratory, 1999.

[43] Steve Benson, Lois C. McInnes, and Jorge J. Moré. TAO users manual. Technical Report ANL/MCS-TM-242, Mathematics and Computer Science Division, Argonne National Laboratory, 2000. See www.mcs.anl.gov/tao.

[44] Claude Berge. *Graphs and Hypergraphs*. North Holland, Amsterdam, 1976.

[45] Sonja Berner. A parallel method for verified global optimization. In Götz Alefeld, Andreas Frommer, and Bruno Lang, editors, *Scientific Computing and Validated Numerics*, Mathematical Research, pages 200–206, Berlin, 1996. Akademie Verlag.

[46] Dimitri P. Bertsekas. *Linear Network Optimization*. MIT Press, Massachusetts and London, 1991.

[47] Martin Berz. Differential algebraic description of beam dynamics to very high orders. *Particle Accelerators*, 24:109, 1989.

[48] Martin Berz. Forward algorithms for high orders and many variables with application to beam physics. In Griewank and Corliss [239], pages 147–156.

[49] Martin Berz. Symplectic tracking in circular accelerators with high order maps. In *Nonlinear Problems in Future Particle Accelerators*, page 288. World Scientific, 1991.

[50] Martin Berz. Calculus and numerics on Levi-Civita fields. In Berz et al. [57], pages 19–35.

[51] Martin Berz. COSY INFINITY Version 8 reference manual. Technical Report MSUCL–1088, National Superconducting Cyclotron Laboratory, Michigan State University, East Lansing, MI 48824, 1997. See cosy.nscl.msu.edu.

[52] Martin Berz. *Modern Map Methods in Particle Beam Physics*. Academic Press, San Diego, 1999.

[53] Martin Berz. Analytical and computational methods for the Levi-Civita fields. In *Proc. Sixth International Conference on Nonarchimedean Analysis*, pages 21–34. Marcel Dekker, New York, 2000.

[54] Martin Berz. COSY INFINITY web page, 2000. See cosy.nscl.msu.edu.

[55] Martin Berz. Higher order verified methods and applications. In *Procedings of SCAN 2000*. Kluwer Academic Publishers, Dordrecht, Netherlands, 2001.

[56] Martin Berz. Towards a universal data type for scientific computing. In Corliss et al. [144], chapter 45, pages 373–381.

[57] Martin Berz, Christian Bischof, George Corliss, and Andreas Griewank, editors. *Computational Differentiation: Techniques, Applications, and Tools*. SIAM, Philadelphia, Penn., 1996.

[58] Martin Berz and Jens Hoefkens. Verified inversion of functional dependencies and superconvergent interval Newton methods. *Reliable Computing*, 7(5), 2001.

[59] Martin Berz and Georg H. Hoffstätter. Exact estimates of the long term stability of weakly nonlinear systems applied to the design of large storage rings. *Interval Computations*, 2:68–89, 1994.

[60] Martin Berz and Georg H. Hoffstätter. Computation and application of Taylor polynomials with interval remainder bounds. *Reliable Computing*, 4(1):83–97, 1998.

[61] Martin Berz and Kyoko Makino. Verified integration of ODEs and flows with differential algebraic methods on Taylor models. *Reliable Computing*, 4:361–369, 1998.

[62] Martin Berz and Kyoko Makino. New methods for high-dimensional verified quadrature. *Reliable Computing*, 5:13–22, 1999.

[63] Martin Berz, Kyoko Makino, and Jens Hoefkens. Verified integration of dynamics in the solar system. *Nonlinear Analysis*, 2001, in print.

[64] Martin Berz, Kyoko Makino, Khodr Shamseddine, Georg H. Hoffstätter, and Weishi Wan. COSY INFINITY and its applications to nonlinear dynamics. In Berz et al. [57], pages 363–365.

[65] Dieter Bestle and Michael Zeitz. Canonical form observer design for non-linear time-variable systems. *Int. J. Control*, 38(2):419–431, 1983.

[66] K. Beulker. Berechnung der Kegelradzahngeometrie. FVA-Heft 548, Forschungsvereinigung Antriebstechnik, 1998.

[67] George Biros and Omar Ghattas. Parallel Newton-Krylov methods for PDE-constrained optimization. In *Proceedings of SC99*. IEEE Computer Society, 1999.

[68] George Biros and Omar Ghattas. A Lagrange-Newton-Krylov-Schur method for PDE-constrained optimization. *SIAG/OPT Views-and-News*, 11(2):1–6, 2000.

[69] Christian Bischof, Peyvand Khademi, Ali Bouaricha, and Alan Carle. Efficient computation of gradients and Jacobians by dynamic exploitation of sparsity in automatic differentiation. *Optim. Methods Software*, 7:1–39, 1996.

[70] Christian H. Bischof. Issues in parallel automatic differentiation. In Griewank and Corliss [239], pages 100–113.

[71] Christian H. Bischof. Availability of ADIFOR software, 1999. See www.mcs.anl.gov/autodiff/ADIFOR.

[72] Christian H. Bischof. A collection of automatic differentiation tools, Dec. 1999. See www.mcs.anl.gov/autodiff/AD_Tools.

[73] Christian H. Bischof, Ali Bouaricha, Peyvand Khademi, and Jorge J. Moré. Computing gradients in large-scale optimization using automatic differentiation. *INFORMS J. Computing*, 9:185–194, 1997.

[74] Christian H. Bischof, H. Martin Bücker, and Dieter an Mey. A case study of computational differentiation applied to neutron scattering. In Corliss et al. [144], chapter 6, pages 69–74.

[75] Christian H. Bischof, Martin Bücker, and Paul D. Hovland. On combining computational differentiation and toolkits for parallel scientific computing. Technical Report ANL/MCS-P797-0200, Mathematics and Computer Science Division, Argonne National Laboratory, 2000. In Proceedings of EuroPar 2000, Springer LNCS 1900, p. 86–94.

[76] Christian H. Bischof and Alan Carle. Users' experience with ADIFOR 2.0. In Berz et al. [57], pages 385–392.

[77] Christian H. Bischof, Alan Carle, George F. Corliss, Andreas Griewank, and Paul D. Hovland. ADIFOR: Generating derivative codes from Fortran programs. *Scientific Programming*, 1:11–29, 1992.

[78] Christian H. Bischof, Alan Carle, Paul D. Hovland, Peyvand Khademi, and Andrew Mauer. ADIFOR 2.0 user's guide (Revision D). Technical report, Mathematics and Computer Science Division Technical Memorandum no. 192 and Center for Research on Parallel Computation Technical Report CRPC-95516-S, 1998. See www.mcs.anl.gov/adifor.

[79] Christian H. Bischof, Alan Carle, Peyvand Khademi, and Andrew Mauer. ADIFOR 2.0: Automatic differentiation of Fortran 77 programs. *IEEE Computational Science & Engineering*, 3(3):18–32, 1996.

[80] Christian H. Bischof, George F. Corliss, and Andreas Griewank. Structured second- and higher-order derivatives through univariate Taylor series. *Optimization Methods and Software*, 2:211–232, 1993.

[81] Christian H. Bischof, Larry Green, Ken Haigler, and T. Knauff. Calculation of sensitivity derivatives for aircraft design using automatic differentiation. In *Proceedings of the 5th AIAA/NASA/USAF/ISSMO Symposium on Multidisciplinary Analysis and Optimization, AIAA 94-4261*, pages 73–84. American Institute of Aeronautics and Astronautics, 1994. Also appeared as Argonne National Laboratory, Mathematics and Computer Science Division, Preprint MCS-P419-0294.

[82] Christian H. Bischof and Mohammad R. Haghighat. Hierarchical approaches to automatic differentiation. In Berz et al. [57], pages 83–94.

[83] Christian H. Bischof, Peyvand M. Khademi, A. Bouaricha, and Alan Carle. Efficient computation of gradients and Jacobians by dynamic exploitation of sparsity in automatic differentiation. *Optimization Methods and Software*, 7:1–39, 1997.

[84] Christian H. Bischof, Lucas Roh, and Andrew Mauer. ADIC — An extensible automatic differentiation tool for ANSI-C. *Software– Practice and Experience*, 27(12):1427–1456, 1997. See `www-fp.mcs. anl.gov/division/software`.

[85] Christian H. Bischof and P. Wu. Time-parallel computation of pseudo-adjoints for a leapfrog scheme. Preprint MCS-P639-0197, Argonne National Laboratory Mathematics and Computer Science Division, Argonne, Ill., 1997.

[86] Joakim O. Blanch, William W. Symes, and Roelof Versteeg. A numerical study of linear viscoacoustic inversion. In Robert G. Keys and Douglas J. Foster, editors, *Comparison of Seismic Inversion Methods on a Single Real Data Set*, pages 13–44. Society of Exploration Geophysicists, Tulsa, Oklahoma, USA, 1998.

[87] H. G. Bock. Randwertprobleme zur Parameteridentifizierung in Systemen nichtlinearer Differentialgleichungen. Bonner Mathematische Schriften 183, Universität Bonn, 1987.

[88] François Bodin, Y. M/'evel, and R. Quiniou. A user level program transformation tool. In *Proceedings of the International Conference on Supercomputing (Melbourne, Australia)*, 1998.

[89] François Bodin and Antoine Monsifrot. Performance issues in automatic differentiation on superscalar processors. In Corliss et al. [144], chapter 4, pages 51–57.

[90] Jeff Borggaard and Arun Verma. An AD technique for computing approximations to the continuous sensitivity equation. Presented at the Third International Conference on Automatic Differentiation (Nice), 2000.

[91] Ali Bouaricha and Jorge J. Moré. Impact of partial separability on large-scale optimization. *Comp. Optim. Appl.*, 7(1):27–40, 1997.

[92] Thierry Braconnier and Philippe Langlois. From rounding error estimation to automatic correction with AD. In Corliss et al. [144], chapter 42, pages 351–357.

[93] Claude Brezinski. *Padé Type Approximation and General Orthogonal Polynomials*. Birkhäuser Verlag, 1980.

[94] Peter N. Brown and Youcef Saad. Hybrid Krylov methods for nonlinear systems of equations. *SIAM J. Sci. Stat. Comput.*, 11(3):450–481, 1990.

[95] Peter N. Brown and C. S. Woodward. Preconditioning strategies for fully implicit radiation diffusion with material-energy transfer. Technical Report UCRL–JC–139087, Lawrence Livermore National Laboratory, 2000.

[96] Richard H. Byrd, Jorge Nocedal, and Robert B. Schnabel. Representations of quasi-Newton matrices and their use in limited-memory methods. *Math. Prog., Ser. A*, 63:129–156, 1994.

[97] George D. Byrne and Alan C. Hindmarsh. PVODE, an ODE solver for parallel computers. *Int. J. High Perf. Comput. Appl.*, 13:354–365, 1999.

[98] Xiao-Chuan Cai. Some domain decomposition algorithms for non-selfadjoint elliptic and parabolic partial differential equations. Technical Report 461, Courant Institute, New York, 1989.

[99] Xiao-Chuan Cai and David E. Keyes. Nonlinearly preconditioned inexact Newton algorithms. *SIAM J. Sci. Comput.*, 2000, submitted. See www.icase.edu/~keyes/papers/aspin.pdf.

[100] Zhiqiang Cai, Raytcho D. Lazarov, Thomas A. Manteuffel, and Stephen F. McCormick. First-order least squares for partial differential equations: Part I. *SIAM J. Numer. Anal.*, 31:1785–1799, 1994.

[101] Zhiqiang Cai, Thomas A. Manteuffel, and Stephen F. McCormick. First-order least squares for partial differential equations: Part II. *SIAM J. Numer. Anal.*, 34:425–454, 1997.

[102] Jean-Baptiste Caillau. Continuation technique for a weakly controlled satellite. In *Nonlinear Analysis 2000*, New-York, 2000. Invited poster session.

[103] Jean-Baptiste Caillau, Joseph Gergaud, and Joseph Noailles. Trajectoires optimales à poussée continue. Contract Report R & T A3006, CNES (French Space Agency) / ENSEEIHT–IRIT, July 1998. (in French).

[104] Jean-Baptiste Caillau and Joseph Noailles. Geometric study of time optimal orbit transfer. Technical Report RT/APO/00/1, ENSEEIHT-IRIT, March 2000.

[105] Jean-Baptiste Caillau and Joseph Noailles. Wavelets for adaptive solution of boundary value problems. In *Proceedings of the 16th IMACS World Congress*, Lausanne, Switzerland, 2000.

[106] Jean-Baptiste Caillau and Joseph Noailles. Optimal control sensitivity analysis with AD. In Corliss et al. [144], chapter 11, pages 109–115.

[107] Jean-Baptiste Caillau and Joseph Noailles. Sensitivity analysis for time optimal orbit transfer. *Optimization*, 49(4):327–350, 2001.

[108] Stephen L. Campbell and Richard Hollenbeck. Automatic differentiation and implicit differential equations. In Berz et al. [57], pages 215–227.

[109] Bernard Cappelaere, David Elizondo, and Christèle Faure. Odyssée versus hand differentiation of a terrain modelling application. In Corliss et al. [144], chapter 7, pages 75–82.

[110] Alan Carle. Personal communication, 1997.

[111] Alan Carle and Mike Fagan. Improving derivative performance for CFD by using simplified recurrences. In Berz et al. [57], pages 343–351.

[112] Alan Carle and Mike Fagan. ADIFOR 3.0 overview. Technical Report CAAM-TR-00-02, Rice University, 2000.

[113] Alan Carle and Mike Fagan. Automatically differentiating MPI-1 datatypes: The complete story. In Corliss et al. [144], chapter 25, pages 215–222.

[114] Alan Carle, Mike Fagan, and Larry L. Green. Preliminary results from the application of automated code generation to CFL3D. Conference paper Paper 98-4807, AIAA, 1998.

[115] David L. Carroll. FORTRAN genetic algorithm driver, March 1999. See www.staff.uiuc.edu/~carroll/ga.html.

[116] Daniele Casanova. Performance optimisation of formula one cars. Talk presented at Technical University Dresden, December 2000.

[117] Daniele Casanova, Robin S. Sharp, Mark Final, Bruce Christianson, and Pat Symonds. Application of automatic differentiation to race car performance optimisation. In Corliss et al. [144], chapter 12, pages 117–124.

[118] Y. F. Chang and George F. Corliss. Solving ordinary differential equations using Taylor series. *ACM Trans. Math. Software*, 8:114–144, 1982.

[119] Y. F. Chang and George F. Corliss. ATOMFT: Solving ODEs and DAEs using Taylor series. *Computers and Mathematics with Applications*, 28:209–233, 1994.

[120] W. L. Chapman, W. J. Welch, K. P. Bowman, J. Sacks, and J. E. Walsh. Arctic sea-ice variability: Model sensitivities and a multi-decadal simulation. *J. Geophys. Res.*, 99:919–935, 1994.

[121] Bruce W. Char. Computer algebra as a toolbox for program generation and manipulation. In Griewank and Corliss [239], pages 53–60.

[122] Isabelle Charpentier and Noël Jakse. Exact derivatives of the pair-correlation function of simple liquids using the tangent linear method. *J. Chem. Phys*, 114:2284–2292, 2001.

[123] Isabelle Charpentier, Noël Jakse, and Fabrice Veersé. Second order exact derivatives to perform optimization on self-consistent integral equations problems. In Corliss et al. [144], chapter 22, pages 189–195.

[124] Guy Chavent, Jérôme Jaffré, Sophie Jégou, and Jun Liu. A symbolic code generator for parameter estimation. In Berz et al. [57], pages 129–136.

[125] Bruce Christianson. Reverse accumulation and accurate rounding error estimates for Taylor series coefficients. *Optimization Methods and Software*, 1(1):81–94, 1991. Also appeared as Tech. Report No. NOC TR239, The Numerical Optimisation Centre, University of Hertfordshire, U.K., July 1991.

[126] Bruce Christianson. Reverse accumulation and attractive fixed points. *Optimization Methods and Software*, 3:311–326, 1994.

[127] Bruce Christianson. Reverse accumulation and implicit functions. *Optimization Methods and Software*, 9(4):307–322, 1998.

[128] Bruce Christianson. Automatic differentiation and Pantoja's algorithm. *Optimization Methods and Software*, 10(1):729–743, 1999.

[129] Bruce Christianson and Michael C. Bartholomew-Biggs. Globalization of Pantoja's optimal control algorithm. In Corliss et al. [144], chapter 13, pages 125–130.

[130] Bruce Christianson, Laurence C. W. Dixon, and Steven Brown. Sharing storage using dirty vectors. In Berz et al. [57], pages 107–115.

[131] Bruce Christianson, Andreas Griewank, and Uwe Naumann. Automated analysis and enhancement of applications code. Pre-proposal submitted to the European Science Foundation, 1999.

[132] Thomas F. Coleman and Jin-Yi Cai. The cyclic coloring problem and estimation of sparse Hessian matrices. *SIAM J. Alg. Disc. Meth.*, 7(2):221–235, 1986.

[133] Thomas F. Coleman, Burton S. Garbow, and Jorge J. Moré. Software for estimating sparse Jacobian matrices. *ACM Trans. Math. Software*, 10(3):329–345, 1984.

[134] Thomas F. Coleman, Burton S. Garbow, and Jorge J. Moré. Software for estimating sparse Hessian matrices. *ACM Trans. Math. Software*, 11(4):363–377, 1985.

[135] Thomas F. Coleman and Aiping Liao. An efficient trust region method for unconstrained discrete-time optimal control problems. *Computational Optimization and Applications*, 4:47–66, 1995.

[136] Thomas F. Coleman and Jorge J. Moré. Estimation of sparse Jacobian matrices and graph coloring problems. *SIAM J. Numer. Anal.*, 20(1):187–209, 1983.

[137] Thomas F. Coleman and Jorge J. Moré. Estimation of sparse Hessian matrices and graph coloring problems. *Math. Programming*, 28:243–270, 1984.

[138] Thomas F. Coleman and Arun Verma. Structure and efficient Jacobian calculation. In Berz et al. [57], pages 149–159.

[139] Thomas F. Coleman and Arun Verma. ADMAT: An automatic differentiation toolbox for MATLAB. Technical report, Computer Science Department, Cornell University, 1998.

[140] Thomas F. Coleman and Arun Verma. The efficient computation of sparse Jacobian matrices using automatic differentiation. *SIAM J. Sci. Comput.*, 19(4):1210–1233, 1998.

[141] Thomas F. Coleman and Arun Verma. ADMIT-1: Automatic differentiation and MATLAB interface toolbox. *ACM Trans. Math. Softw.*, 26(1):150–175, 2000.

[142] R. L. Colony and I. Rigor. International Arctic buoy program data report for 1 January 1992 31 December 1992. Tech. Memo. APL-UW TM29-93, Appl. Phys. Lab., Univ. of Washington, Seattle, 1993.

[143] D. Conforti and Marco Mancini. A curvilinear search algorithm for unconstrained optimization by automatic differentiation. *Optimization Methods and Software*, 15r(3-4):283–297, 2001.

[144] George Corliss, Christèle Faure, Andreas Griewank, Laurent Hascoët, and Uwe Naumann, editors. *Automatic Differentiation: From Simulation to Optimization*. Computer and Information Science. Springer, New York, 2001.

[145] George F. Corliss. Automatic differentiation bibliography. In Griewank and Corliss [239], pages 331–353.

[146] George F. Corliss. Overloading point and interval Taylor operators. In Griewank and Corliss [239], pages 139–146.

[147] George F. Corliss. ADIFOR case study : VODE + ADIFOR. Technical Memorandum ANL/MCS–TM–168, Mathematics and Computer Science Division, Argonne National Laboratory, Argonne, Ill., 1992.

[148] George F. Corliss. Globsol entry page, 1998. See `www.mscs.mu.edu/~globsol`.

[149] George F. Corliss. Automatic differentiation bibliography. In Corliss et al. [144], chapter 46, pages 383–425.

[150] Beatrice Creusillet and F. Irigoin. Interprocedural array region analysis. Rapport cri, A-282, Ecole des Mines de Paris, FRANCE, January 1996.

[151] Tibor Csendes and Dietmar Ratz. Subdivision direction selection in interval methods for global optimization. *SIAM J. Numer. Anal.*, 34(3):922–938, 1997.

[152] A. R. Curtis, Michael J. D. Powell, and John K. Reid. On the estimation of sparse Jacobian matrices. *J. Inst. Math. Appl.*, 13:117–119, 1974.

[153] Michel J. Dayde and Iain S. Duff. The RISC BLAS: A blocked implementation of level 3 BLAS for RISC processors. *ACM Trans. Math. Software*, 25(3):316–340, 2000.

[154] Bram de Jager. The use of symbolic computation in nonlinear control: Is it viable? *IEEE Trans. on Automatic Control*, AC-40(1):84–89, 1995.

[155] Theodorus J. Dekker. A floating-point technique for extending the available precision. *Numer. Math.*, 18:224–242, 1971.

[156] Denk, U. Feldmann, and C. Hammer et al. Erweiterung eines Standard-Schaltungs Simulators in Richtung VDHL-AMS. In *Proceedings of ITG Workshop Analog '99, München*, 1999.

[157] John E. Dennis and Robert B. Schnabel. *Numerical Methods for Unconstrained Optimization and Non-Linear Equations*. Number 16 in Classics in Applied Mathematics. SIAM, Philadelphia, Penn., 1996.

[158] Florian Dignath, Peter Eberhard, and Axel Fritz. Analytical aspects and practical pitfalls in technical applications of AD. In Corliss et al. [144], chapter 14, pages 131–136.

[159] Claude Dion. *Dynamique de L'alignement et de L'orientation Moleculair'e Induite Par Laser. Simulations Numériques sur HCN En Champ Infrarouge*. PhD thesis, Université de Sherbrooke et Université de Paris-Sud, 1999.

[160] Laurence C. W. Dixon. On the impact of automatic differentiation on the relative performance of parallel truncated Newton and variable metric algorithms. *SIAM J. Optim.*, 1:475–486, 1991.

[161] Laurence C. W. Dixon. Use of automatic differentiation for calculating Hessians and Newton steps. In Griewank and Corliss [239], pages 114–125.

[162] John J. Doherty. Strategy for design optimisation. Customer Report DERA/MSS5/CR980389/1.0, Defence Evaluation & Research Agency, September 1998.

[163] Elizabeth D. Dolan and Jorge J. Moré. Benchmarking optimization software with COPS. Technical Memorandum ANL/MCS-TM-246, Argonne National Laboratory, Argonne, Ill., 2000.

[164] Kevin Dowd and Charles R. Severence. *High Performance Computing*. O'Reilly, 1998.

[165] Norman R. Draper and Harry Smith. *Applied Regression Analysis*. Wiley Series in Probability and Mathematical Statistics. John Wiley and Sons, Inc., New York, second edition, 1981.

[166] M. Dryja and O. B. Widlund. An additive variant of the Schwarz alternating method for the case of many subregions. Technical Report 339, Department of Computer Science, Courant Institute, New York, 1987.

[167] Iain S. Duff, Roger G. Grimes, and John G. Lewis. Users' guide for the Harwell-Boeing sparse matrix collection (Release I). Technical Report tr/pa/92/86, CERFACS, 1992.

[168] Der-Ming Duh, Anthony Douglas, and John Haymet. Integral equation theory for uncharged liquids: The Lennard-Jones fluid and the bridge function. *J. Chem. Phys*, 103:2625–2633, 1995.

[169] J. C. Dunn and Dimitri P. Bertsekas. Efficient dynamic programming implementations of Newton's method for unconstrained optimal control problems. *Journal of Optimization Theory and Applications*, 63(1):23–38, 1989.

[170] Roland Dutschk. *Geometrische Probleme bei Herstellung und Eingriff bogenverzahnter Kegelräder*. PhD thesis, Technical University of Dresden, 1994.

[171] Peter Eberhard. Adjoint variable method for the sensitivity analysis of multibody systems interpreted as continuous, hybrid form of automatic differentiation. In Berz et al. [57], pages 319–328.

[172] Peter Eberhard. *Kontaktuntersuchungen an hybriden Mehrkörpersystemen, Habilitation*. University of Stuttgart, 2000.

[173] Peter Eberhard and Christian H. Bischof. Automatic differentiation of numerical integration algorithms. *Mathematics of Computation*, 68:717–731, 1999.

[174] Jonathon Elliott and Jaime Peraire. Practical 3D aerodynamic design and optimization using unstructured meshes. *AIAA J.*, 35(9):1479–1485, 1997.

[175] Yuri G. Evtushenko. Automatic differentiation viewed from optimal control. In Griewank and Corliss [239], pages 25–30.

[176] Yuri G. Evtushenko. Computation of exact gradients in distributed dynamic systems. *Optimization Methods and Software*, 9:45–75, 1998.

[177] Yuri G. Evtushenko, E. S. Zasuhina, and V. I. Zubov. Numerical optimization of solutions to Burgers' problems by means of boundary conditions. *Computational Mathematics and Mathematical Physics*, 37(12):1406–1414, 1997.

[178] Yuri G. Evtushenko, E. S. Zasuhina, and V. I. Zubov. FAD method to compute second order derivatives. In Corliss et al. [144], chapter 39, pages 327–333.

[179] Christèle Faure. Splitting of algebraic expressions for automatic differentiation. In Berz et al. [57], pages 117–127.

[180] Christèle Faure. Adjoining strategies for multi-layered programs. *Optimisation Methods and Software*, 2001. To appear. Also appeared as INRIA Rapport de recherche no. 3781, BP 105-78153 Le Chesnay Cedex, FRANCE, 1999.

[181] Christèle Faure. Memory limited adjoint code generation. *Flow, Turbulence and Combustion*, 2001, in press.

[182] Christèle Faure and Isabelle Charpentier. Comparing automatically generated and hand coded adjoints. Rapport de recherche 3811, INRIA, 1999.

[183] Christèle Faure and Patrick Dutto. Extension of Odyssée to the MPI library - Direct mode. Technical Report 3715, INRIA, June 1999.

[184] Christèle Faure and Patrick Dutto. Extension of Odyssée to the MPI library - Reverse mode. Technical Report 3774, INRIA, October 1999.

[185] Christèle Faure and Uwe Naumann. Minimizing the tape size. In Corliss et al. [144], chapter 34, pages 293–298.

[186] Christèle Faure and Yves Papegay. Odyssée User's Guide. Version 1.7. Rapport technique RT–0224, INRIA, Sophia-Antipolis, France, September 1998. See www.inria.fr/RRRT/RT-0224.html, and www.inria.fr/safir/SAM/Odyssee/odyssee.html.

[187] Christèle Faure, Edgar J. Soulié, and Théo Berclaz. Résonance paramagnétique électronique, optimisation et différentiation automatique. Rapport de recherche 3907, INRIA, 2000.

[188] William F. Feehery and Paul I. Barton. A differentiation-based approach to dynamic simulation and optimization with high-index differential-algebraic equations. In Berz et al. [57], pages 239–252.

[189] William F. Feehery, John E. Tolsma, and Paul I. Barton. Efficient sensitivity analysis of large-scale differential-algebraic equations. *Appl. Numer. Math.*, 25:41–54, 1997.

[190] Michael C. Ferris, Mike P. Mesnier, and Jorge J. Moré. NEOS and Condor: Solving optimization problems over the Internet. Preprint ANL/MCS-P708-0398, Mathematics and Computer Science Division, Argonne National Laboratory, Argonne, 1998.

[191] Mark Final. Automatic differentiation and non-linear optimisation. Master's thesis, University of Hertfordshire, College Lane, Hatfield, Herts AL10 9AB, UK, 2000. To appear.

[192] Herbert Fischer. Special problems in automatic differentiation. In Griewank and Corliss [239], pages 43–50.

[193] Harley Flanders. Automatic differentiation of composite functions. In Griewank and Corliss [239], pages 95–99.

[194] Harley Flanders. Application of ad to a family of periodic functions. In Corliss et al. [144], chapter 38, pages 319–326.

[195] Michel Fliess, Jean Lévine, Philippe Martin, and Pierre Rouchon. Flatness and defect of non-linear systems: Introductory theory and examples. *Int. J. Control*, 61:1327–1361, 1995.

[196] Otto Föllinger. Entwurf zeitvarianter Systeme durch Polvorgabe. *Regelungstechnik*, 26(6):189–196, 1978.

[197] Brian Ford and Françoise Chatelin, editors. *Problem Solving Environments for Scientific Computing*. North Holland, 1985.

[198] Shaun A. Forth. Automatic differentiation for flux linearisation. AMOR Report 98/1, Cranfield University (RMCS Shrivenham), Swindon SN6 8LA, England, 1998. Poster Presentation at *16th International Conference on Numerical Methods in Fluid Dynamics*, July 6-10, 1998, Arcachon, France.

[199] Shaun A. Forth and Trevor P. Evans. Aerofoil optimisation via ad of a multigrid cell-vertex Euler flow solver. In Corliss et al. [144], chapter 17, pages 153–160.

[200] D. A. S. Fraser. Regression analysis, nonlinear or nonnormal: Simple and accurate p values from likelihood analysis. *Journal of the American Statistical Association*, 94(448):1286–1295, 1999.

[201] Axel Fritz. Optimal control of a vehicle convoy. In J. Ambrosio and W. Schiehlen, editors, *Proceedings EUROMECH 404, Lissabon, Advances in Computational Multibody Dynamics*, pages 695–712, 1999.

[202] Pascale Fulcheri and Martine Olivi. Matrix rational H_2 approximation: A gradient algorithm based on Schur analysis. *SIAM J. Cont. Optim.*, 36(6):2103–2127, 1998.

[203] GAMS. See www.gams.com.

[204] Hin-Hark Gan and Byung Chan Eu. Ornstein-Zernike derivative relations and thermodynamic functions. *J. Chem. Phys*, 96:558–564, 1992.

[205] Oscar García. A system for the differentiation of Fortran code and an application to parameter estimation in forest growth models. In Griewank and Corliss [239], pages 273–286.

[206] David M. Gay. Automatic differentiation of nonlinear AMPL models. In Griewank and Corliss [239], pages 61–73.

[207] David M. Gay. More AD of nonlinear AMPL models: Computing Hessian information and exploiting partial seperability. In Berz et al. [57], pages 173–184.

[208] Sophie Geffroy, Richard Epenoy, and Joseph Noailles. Averaging techniques in optimal control for orbital low-thrust transfers and rendez-vous computation. In *Proceedings of the 11th International Astrodynamics Symposium*, pages 166–171, Gifu, Japan, 1996.

[209] Uwe Geitner, Jean Utke, and Andreas Griewank. Automatic computation of sparse Jacobians by applying the method of Newsam and Ramsdell. In Berz et al. [57], pages 161–172.

[210] E. Michael Gertz, Philip E. Gill, and Julia Muetherig. User's guide for SnadiOpt: A package adding automatic differentiation to Snopt. Report NA 01-1, Department of Mathematics, University of California, San Diego, 2000.

[211] Ralf Giering. *Tangent Linear and Adjoint Model Compiler, Users Manual*. Center for Global Change Sciences, Department of Earth, Atmospheric, and Planetary Science, MIT, Cambridge, MA, December 1997. Unpublished. Available at puddle.mit.edu/~ralf/tamc.

[212] Ralf Giering and Thomas Kaminski. Recipies for adjoint code construction. *ACM Trans. Math. Software*, 24(4):437–474, 1998. Also appeared as Max-Planck Institut für Meteorologie Hamburg, Technical Report No. 212, 1996.

[213] Ralf Giering and Thomas Kaminski. On the performance of derivative code generated by TAMC. Manuscript, FastOpt, Hamburg, Germany, 2000. Submitted to Optimization Methods and Software. See www.FastOpt.de/papers/perftamc.ps.gz.

[214] Ralf Giering and Thomas Kaminski. Generating recomputations in reverse mode AD. In Corliss et al. [144], chapter 33, pages 283–291.

[215] Jean-Charles Gilbert, Georges Le Vey, and John Masse. La différentiation automatique de fonctions représentées par des programmes. Rapport de Recherche 1557, INRIA, Le Chesnay, France, 1991.

[216] Jean-Charles Gilbert and Claude Lemaréchal. The modules M1QN3 and N1QN3. Version 2.0c. Technical report, INRIA, 1995.

[217] Michael B. Giles. On the use of Runge-Kutta time-marching and multigrid for the solution of steady adjoint equations. Technical Report NA00/10, Oxford University Computing Laboratory, 2000.

[218] Michael B. Giles. On the iterative solution of adjoint equations. In Corliss et al. [144], chapter 16, pages 145–151.

[219] Michael B. Giles and Niles A. Pierce. An introduction to the adjoint approach to design. *Flow, Turbulence and Combustion*, to appear.

[220] Philip E. Gill and Walter Murray. Algorithms for the solution of the nonlinear least-squares problem. *SIAM J. Numer. Anal.*, 15(5):977–992, 1978.

[221] Philip E. Gill, Walter Murray, and Michael A. Saunders. SNOPT: An SQP algorithm for large-scale constrained optimisation. Numerical Analysis Report 97-2, Department of Mathematics, University of California, San Diego, La Jolla, CA, 1997.

[222] Philip E. Gill, Walter Murray, and Margret H. Wright. *Practical Optimization*. Academic Press, New York, 1981.

[223] Mark S. Gockenbach. Understanding code generated by TAMC. Technical Report 00-30, Department of Computational and Applied Mathematics, Rice University, Houston, Texas, U.S.A, 2000.

[224] Mark S. Gockenbach, M. J. Petro, and William W. Symes. C++ classes for linking optimization with complex simulations. *ACM Trans. Math. Software*, 25(2):191–212, 1999.

[225] Mark S. Gockenbach, Daniel R. Reynolds, and William W. Symes. Automatic differentiation and the adjoint state method. In Corliss et al. [144], chapter 18, pages 161–166.

[226] Mark S. Gockenbach, Daniel R. Reynolds, and William W. Symes. Efficient and automatic implementation of the adjoint state method. *ACM Trans. Math. Software*, submitted.

[227] Donald Goldfarb and Philippe L. Toint. Optimal estimation of Jacobian and Hessian matrices that arise in finite difference calculations. *Mathematics of Computation*, 43(167):69–88, 1984.

[228] Victor V. Goldman and Gerard Cats. Automatic adjoint modelling within a program generation framework: A case study for a weather forecasting grid-point model. In Berz et al. [57], pages 185–194.

[229] Victor V. Goldman, J. Molenkamp, and J. A. van Hulzen. Efficient numerical program generation and computer algebra environments. In Griewank and Corliss [239], pages 74–83.

[230] Édouard Goursat and Earle R. Hedrick. *A Course in Mathematical Analysis I*. Ginn, 1904.

[231] N. I. Grachev and A. N. Filkov. Solution of optimal control problems in system DISO. Technical report, Computing Center of Russian Academy of Sciencesy, Moscow, 1986. (In Russian).

[232] Andreas Griewank. On automatic differentiation. In Masao Iri and Kunio Tanabe, editors, *Mathematical Programming: Recent Developments and Applications*, pages 83–108. Kluwer Academic Publishers, Dordrecht, 1989.

[233] Andreas Griewank. Automatic evaluation of first- and higher-derivative vectors. In R. Seydel, F. W. Schneider, T. Küpper, and H. Troger, editors, *Proceedings of the Conference at Würzburg, Aug. 1990, Bifurcation and Chaos: Analysis, Algorithms, Applications*, volume 97, pages 135–148. Birkhäuser Verlag, Basel, Switzerland, 1991.

[234] Andreas Griewank. The chain rule revisited in scientific computing. *SIAM News*, 24, 1991. no. 3, p. 20 & no. 4, p. 8.

[235] Andreas Griewank. Achieving logarithmic growth of temporal and spatial complexity in reverse automatic differentiation. *Optimization Methods and Software*, 1:35–54, 1992.

[236] Andreas Griewank. Automatic directional differentiation of nonsmooth composite functions. In *Recent Developments in Optimization, French-German Conference on Optimization*, Dijon, 1994.

[237] Andreas Griewank. Private communication, 2000.

[238] Andreas Griewank. *Evaluating Derivatives: Principles and Techniques of Algorithmic Differentiation*. Number 19 in Frontiers in Appl. Math. SIAM, Philadelphia, Penn., 2000.

[239] Andreas Griewank and George F. Corliss, editors. *Automatic Differentiation of Algorithms: Theory, Implementation, and Application*. SIAM, Philadelphia, Penn., 1991.

[240] Andreas Griewank, George F. Corliss, Petra Henneberger, Gabriella Kirlinger, Florian A. Potra, and Hans J. Stetter. High-order stiff ODE solvers via automatic differentiation and rational prediction. In *Numerical Analysis and Its Applications*, number 1196 in Lecture Notes in Computer Science, pages 114–125. Springer, Berlin, 1997. Also appeared as Technical Report IOKOMO 02 96, Technical University of Dresden.

[241] Andreas Griewank, David Juedes, H. Mitev, Jean Utke, Olaf Vogel, and Andrea Walther. ADOL-C: A package for the automatic differentiation of algorithms written in C/C++. Technical report, Technical University of Dresden, Institute of Scientific Computing and Institute of Geometry, 1999. Updated version of the paper published in *ACM Trans. Math. Software* 22, 1996, 131–167.

[242] Andreas Griewank, David Juedes, and Jean Utke. ADOL–C, a package for the automatic differentiation of algorithms written in C/C++. *ACM Trans. Math. Software*, 22(2):131–167, 1996. See ftp://info.mcs.anl.gov/pub/tech_reports/reports/TM162.ps.

[243] Andreas Griewank, David Juedes, and Jean Utke. A package for the automatic differentiation of algorithms written in C/C++. User manual. Technical report, Institute of Scientific Computing, Technical University of Dresden, Dresden, Germany, 1996. See www.math.tu-dresden.de/wir/project/wwwadolc/index.html.

[244] Andreas Griewank and Christo Mitev. Detecting Jacobian sparsity patterns by Bayesian probing. Preprint IOKOMO-04-2000, Technical University of Dresden, 2000. Submitted to Math. Progr.

[245] Andreas Griewank and Christo Mitev. Verifying jacobian sparsity. In Corliss et al. [144], chapter 32, pages 271–279.

[246] Andreas Griewank and George W. Reddien. Computation of cusp singularities for operator equations and their discretizations. *Journal of Computational and Applied Mathematics*, pages 133–153, 1989.

[247] Andreas Griewank and Shawn Reese. On the calculation of Jacobian matrices by the Markowitz rule. In Griewank and Corliss [239], pages 126–135.

[248] Andreas Griewank and Philippe L. Toint. On the unconstrained optimization of partially separable functions. In Michael J. D. Powell, editor, *Nonlinear Optimization 1981*, pages 301–312. Academic Press, New York, 1982.

[249] Andreas Griewank, Jean Utke, and Andrea Walther. Evaluating higher derivative tensors by forward propagation of univariate Taylor series. *Mathematics of Computation*, 69:1117–1130, 2000.

[250] Andreas Griewank and Andrea Walther. Algorithm 799: Revolve: An implementation of checkpoint for the reverse or adjoint mode of computational differentiation. *ACM Trans. Math. Software*, 26(1):19, 1999. Also appeared as Technical University of Dresden, Technical Report IOKOMO-04-1997.

[251] Anne K. Griffith and Nancy K. Nichols. Accounting for model error in data assimilation using adjoint models. In Berz et al. [57], pages 195–204.

[252] R. Griffiths. Minimum lap time simulation of a racing car. Master's thesis, School of Mechanical Engineering, Cranfield University, 1992.

[253] José Grimm. Rational approximation of transfer functions in the hyperion software. Rapport de recherche 4002, INRIA, September 2000.

[254] José Grimm. Complexity analysis of automatic differentiation in the hyperion software. In Corliss et al. [144], chapter 36, pages 305–310.

[255] José Grimm, Loïc Pottier, and Nicole Rostaing-Schmidt. Optimal time and minimum space-time product for reversing a certain class of programs. In Berz et al. [57], pages 95–106.

[256] William D. Gropp, Steven Huss-Lederman, Andrew Lumsdaine, Ewing Lusk, Bill Nitzberg, William Saphir, and Marc Snir. *MPI: The Complete Reference, Volume 2, The MPI-2 Extensions*. MIT Press, 1998.

[257] William D. Gropp, Dinesh K. Kaushik, David E. Keyes, and Barry F. Smith. Performance modeling and tuning of an unstructured mesh CFD application. In *Proceedings of SC2000*. IEEE Computer Society, 2000.

[258] William D. Gropp, David E. Keyes, Lois C. McInnes, and M. D. Tidriri. Globalized Newton-Krylov-Schwarz algorithms and software for parallel implicit CFD. *Int. J. High Performance Computing Applications*, 14:102–136, 2000.

[259] William D. Gropp, Ewing Lusk, Nathan Doss, and Anthony Skjellum. A high-performance portable implementation of the MPI message passing interface standard. *Parallel Computing*, 22:789–828, 1996.

[260] William D. Gropp, Ewing Lusk, and Anthony Skjellum. *Using MPI – Portable Parallel Programming with the Message Passing Interface*. MIT Press, Cambridge, 1994.

[261] John Guckenheimer. Bifurcations, automatic differentiation and computer generated proofs. In Berz et al. [57], pages 229–237.

[262] Philippe Guillaume and Mohamed Masmoudi. Computation of higher order derivatives in optimal shape design. *Numerische Mathematik*, 67:231–250, 1994.

[263] Gundolf Haase, Ulrich Langer, Ewald Lindner, and Wolfram Mühlhuber. Optimal sizing of industrial structural mechanics problems using ad. In Corliss et al. [144], chapter 21, pages 181–188.

[264] Raphael T. Haftka and Zafer Gürdal. *Elements of Structural Optimization.* Kluwer Academic Publishers, Dordrecht, 1992.

[265] Steve Hague and Uwe Naumann. Present and future scientific computation environments. In Corliss et al. [144], chapter 5, pages 59–66.

[266] Michael G. Hall. Cell-vertex multigrid solution of the Euler equations for transonic flow past aerofoils. Technical Report 84116, Royal Aircraft Establishment, December 1984.

[267] D. C. Hamilton. Confidence regions for parameter subsets in nonlinear regression. *Biometrika*, 73:57–64, 1986.

[268] Elden R. Hansen. *Global Optimization Using Interval Analysis.* Marcel Dekker, New York, 1992.

[269] Jean-Pierre Hansen. *Theory of Simple Liquids.* Academic Press, 1986.

[270] Laurent Hascoët. The data-dependence graph of adjoint programs. Research Report 4167, INRIA, Sophia-Antipolis, France, 2001. See www.inria.fr/rrrt/rr-4167.html.

[271] Laurent Hascoët, Stefka Fidanova, and Christophe Held. Adjoining independent computations. In Corliss et al. [144], chapter 35, pages 299–304.

[272] Jaroslav Haslinger and Pekka Neittaanmäki. *Finite Element Approximation for Optimal Shape Design: Theory and Applications.* John Wiley & Sons Ltd., Chichester, 1988.

[273] Eric Hassold and André Galligo. Automatic differentiation applied to nonsmooth convex optimization. In Berz et al. [57], pages 287–297.

[274] J. P. M. Hendrikx, T. J. J. Meijlink, and R. F. C. Kriens. Application of optimal control theory to inverse simulation of car handling. *Vehicle System Dynamics*, 26(6):449–462, 1996.

[275] W. D. Hibler, III. Modeling a variable thickness sea-ice cover. *Mon. Wea. Rev.*, 108:1943–1973, 1980.

[276] Kenneth E. Hillstrom. Users guide for JAKEF. Technical Memorandum ANL/MCS–TM–16, Mathematics and Computer Science Division, Argonne National Laboratory, Argonne, Ill., 1985.

[277] Alan C. Hindmarsh and Allan G. Taylor. User documentation for IDA, a differential-algebraic equation solver for sequential and parallel computers. Technical Report UCRL-MA-136910, Lawrence Livermore National Laboratory, 1999.

[278] Dorit S. Hochbaum, editor. *Approximation Algorithms for NP-Hard Problems.* PWS Publishing Company, 20 Park Plaza, Boston, 1997.

[279] Jens Hoefkens and Martin Berz. Verification of invertibility and charting of constraint manifolds in DAEs. *Reliable Computing*, to appear, 2001. Special issue: Proceedings of the Sixth International Conference on Applications of Computer Algebra.

[280] Jens Hoefkens, Martin Berz, and Kyoko Makino. Efficient high-order methods for ODEs and DAEs. In Corliss et al. [144], chapter 41, pages 343–348.

[281] D. M. Holland, L. A. Mysak, and D. K. Manak. Sensitivity study of a dynamic-thermodynamic sea-ice model. *J. Geophys. Res.*, 98:2561–2586, 1993.

[282] E. Horowitz and P. Sahni. *Fundamentals of Computer Algorithms.* Computer Science Press, 1978.

[283] Jim E. Horwedel. GRESS: A preprocessor for sensitivity studies on Fortran programs. In Griewank and Corliss [239], pages 243–250.

[284] Jim E. Horwedel, Brian A. Worley, E. M. Oblow, and F. G. Pin. GRESS version 1.0 users manual. Technical Memorandum ORNL/TM 10835, Martin Marietta Energy Systems, Inc., Oak Ridge National Laboratory, Oak Ridge, Tenn., 1988.

[285] A.K.M. Shahadat Hossain. *On the Computation of Sparse Jacobian Matrices and Newton Steps.* PhD thesis, Department of Informatics, University of Bergen, 1998. Technical Report 146.

[286] A.K.M. Shahadat Hossain and Trond Steihaug. Computing a sparse Jacobian matrix by rows and columns. *Optimization Methods and Software*, 10:33–48, 1998.

[287] Shahadat Hossain and Trond Steihaug. Reducing the number of AD passes for computing a sparse Jacobian matrix. In Corliss et al. [144], chapter 31, pages 263–270.

[288] Paul D. Hovland. *Automatic Differentiation of Parallel Programs.* PhD thesis, University of Illinois at Urbana-Champaign, Urbana, Ill., May 1997.

[289] Paul D. Hovland and Christian H. Bischof. Automatic differentiation of message-passing parallel programs. In *Proceedings of the First Merged International Parallel Processing Symposium and Symposium on Parallel and Distributed Processing*, pages 98–104, Los Alamitos, CA, 1998. IEEE Computer Society Press.

[290] Paul D. Hovland and Michael T. Heath. Adaptive SOR: A case study in automatic differentiation of algorithm parameters. Technical Report ANL/MCS-P673-0797, Mathematics and Computer Science Division, Argonne National Laboratory, 1997.

[291] Paul D. Hovland and Lois C. McInnes. Parallel simulation of compressible flow using automatic differentiation and PETSc. Technical Report ANL/MCS-P796-0200, Mathematics and Computer Science Division, Argonne National Laboratory, 2000. To appear in a special issue of *Parallel Computing* on "Parallel Computing in Aerospace".

[292] Paul D. Hovland, Bijan Mohammadi, and Christian H. Bischof. Automatic differentiation of Navier-Stokes computations. Technical Report MCS-P687-0997, Argonne National Laboratory, 1997.

[293] Paul D. Hovland, Boyana Norris, Lucas Roh, and Barry F. Smith. Developing a derivative-enhanced object-oriented toolkit for scientific computations. In *Proceedings of the SIAM Workshop on Object Oriented Methods for Inter-operable Scientific and Engineering Computing*, pages 129–137, Philadelphia, Penn., 1999. SIAM.

[294] Mark J. Huiskes. Virtual population analysis with the adjoint method. In F. Funk et al., editor, *Fishery Stock Assessment Models*, pages 639–658. Alaska Sea Grant College Program Report No. AK-SG-98-01, University of Alaska, Fairbanks, 1998.

[295] Mark J. Huiskes. Automatic differentiation for modern nonlinear regression. In Corliss et al. [144], chapter 8, pages 83–90.

[296] Mark J. Huiskes. Evaluation of parametric and structural uncertainty in stock assessment models with an application to North Sea herring. *Canadian Journal of Fisheries and Aquatic Sciences*, 2001. Submitted.

[297] E. C. Hunke and J. K. Dukowicz. An elastic-viscous-plastic model for sea-ice dynamics. *J. Phys. Oceanog.*, 27:1849–1867, 1997.

[298] Ulf Hutschenreiter. A new method for bevel gear tooth flank computation. In Berz et al. [57], pages 329–341.

[299] IEEE. *IEEE Standard for Binary Floating-Point Arithmetic, ANSI/IEEE Standard 754-1985*. Institute of Electrical and Electronics Engineers, New York, 1985. Reprinted in SIGPLAN Notices, 22(2):9–25, 1987.

[300] Masao Iri. History of automatic differentiation and rounding estimation. In Griewank and Corliss [239], pages 1–16.

[301] Masao Iri and Koichi Kubota. Methods of fast automatic differentiation and applications. Research Memorandum RMI 87 – 02, Department of Mathematical Engineering and Information Physics, Faculty of Engineering, University of Tokyo, 1987.

[302] Alberto Isidori. *Nonlinear Control Systems: An Introduction.* Springer-Verlag, Berlin, third edition, 1995.

[303] Machio Iwasaki. Second-order perturbation treatment of the general spin Hamiltonian in an arbitrary coordinate system. *Journal of Magnetic Resonance*, 16:417–423, 1974.

[304] R. Jacob. *Low Frequency Variability of the Atmosphere-Ocean System.* PhD thesis, Univ. of Wisconsin, Madison, 1997.

[305] D. H. Jacobson and D. Q. Mayne. *Differential Dynamic Programming.* American Elsevier, New York, 1970.

[306] Antony Jameson and John Vassberg. Studies of alternate numerical optimization methods applied to the brachistochrone problem. In *OptiCON '99 Conference*, 1999.

[307] Max E. Jerrell. Automatic differentiation and interval arithmetic for estimation of econometric functions. In Berz et al. [57], pages 265–272.

[308] R. Jessani and M. Putrino. Comparison of single- and dual-pass multiply-add fused floating-point units. *IEEE Trans. on Computers*, 47(9), 1998.

[309] Jet Propulsion Laboratory. JPL Solar System Dynamics. See `ssd.jpl.nasa.gov`.

[310] Bo-Nan Jiang. *The Least-Squares Finite Element Method, Theory and Applications in Computational Fluid Dynamics and Electromagnetics.* Scientific Computation. Springer, 1998.

[311] T. Jouhanique and Paul Rascle. A fifth equation to model the relative velocity in the 3–D thermal-hydraulic code THYC. In R. C. Block and F. Feiner, editors, *Proceedings of the Sixth Nuclear Reactor Thermal-Hydraulic (NURETH) Congress*. American Nuclear Society, Nuclear Regulatory Commission Publication., 1995.

[312] David W. Juedes. A taxonomy of automatic differentiation tools. In Griewank and Corliss [239], pages 315–329.

[313] David W. Juedes and Karthik Balakrishnan. Generalized neural networks, computational differentiation, and evolution. In Berz et al. [57], pages 273–285.

[314] Dan Kalman and Robert Lindell. Automatic differentiation in astrodynamical modeling. In Griewank and Corliss [239], pages 228–243.

[315] R. Baker Kearfott. Automatic differentiation of conditional branches in an operator overloading context. In Berz et al. [57], pages 75–81.

[316] R. Baker Kearfott. *Rigorous Global Search: Continuous Problems.* Kluwer Academic Publishers, Dordrecht, Netherlands, 1996.

[317] R. Baker Kearfott. Empirical evaluation of innovations in interval branch and bound algorithms for nonlinear algebraic systems. *SIAM J. Sci. Comput.*, 18(2):574–594, 1997.

[318] R. Baker Kearfott. On stopping criteria in verified nonlinear systems or optimization algorithms. *ACM Trans. Math. Software*, 26(3):373–389, 2000.

[319] R. Baker Kearfott and Alvard Arazyan. Taylor series models in deterministic global optimization. In Corliss et al. [144], chapter 44, pages 365–372.

[320] C. T. Kelley and David E. Keyes. Convergence analysis of pseudotransient continuation. *SIAM J. Numer. Anal.*, 35:508–523, 1998.

[321] David E. Keyes and William D. Gropp. A comparison of domain decomposition techniques for elliptic partial differential equations and their parallel implementation. *SIAM J. Sci. Stat. Comput.*, 8(2):s166–s202, 1987.

[322] David E. Keyes, Paul D. Hovland, Lois C. McInnes, and Widodo Samyono. Using automatic differentiation for second-order matrix-free methods in PDE-constrained optimization. In Corliss et al. [144], chapter 3, pages 35–50.

[323] Jong G. Kim and Paul D. Hovland. Sensitivity analysis and parameter tuning of a sea-ice model. In Corliss et al. [144], chapter 9, pages 91–98.

[324] D. E. Kirk. *Optimal Control Theory: An Introduction.* Prentice Hall, Englewood Cliffs, New Jersey, 1970.

[325] Wolfram Klein. Report on symbolic manipulators for automatic code generation. EG project Jessi AC 12: Analog expert design system. *Milestone Report*, 1995.

[326] Wolfram Klein. Comparisons of automatic differentiation tools in circuit simulation. In Berz et al. [57], pages 297–307.

[327] Wolfram Klein. Symbolic modeling in circuit simulation. In *Proceedings of IX. ECMI Conference, Kopenhagen (1996)*, 1996.

[328] Wolfram Klein, Andreas Griewank, and Andrea Walther. Differentiation methods for industrial strength problems. In Corliss et al. [144], chapter 1, pages 3–23.

[329] Wolfram Klein and Andrea Walther. Application of techniques of computational differentiation to a cooling system. *Optimization Methods and Software*, 13:65–78, 2000. Also appeared as Technical University of Dresden, preprint IOKOMO-05-1998.

[330] Dana A. Knoll and W. J. Rider. A multigrid preconditioned Newton-Krylov method. *SIAM J. Sci. Stat. Comput.*, 21:691–710, 2000.

[331] Donald E. Knuth. *The Art of Computer Programming, Vol. 2. Seminumerical Algorithms.* Addison-Wesley, Reading, Mass., third edition, 1981.

[332] Vamshi Mohan Koriva, Arthur C. Taylor III, Perry A. Newman, Gene W. Hou, and Henry E. Jones. An approximately factored incremental strategy for calculating consistent discrete aerodynamic sensitivity derivatives. *Journal of Computational Physics*, 113:336–346, 1994.

[333] D. Kraft. Algorithm 733: TOMP—Fortran modules for optimal control calculations. *ACM Trans. Math. Software*, 20:262–281, 1994.

[334] Vladik Kreinovich, Anatoly Lakeyev, Jiri Rohn, and Patrick Kahl. *Computational Complexity and Feasibility of Data Processing and Interval Computations.* Number 10 in Applied Optimization. Kluwer, Dordrecht, Netherlands, 1998.

[335] Koichi Kubota. PADRE2, a Fortran precompiler yielding error estimates and second derivatives. In Griewank and Corliss [239], pages 251–262.

[336] Koichi Kubota. PADRE2 - Fortran precompiler for automatic differentiation and estimates of rounding error. In Berz et al. [57], pages 367–374.

[337] James L. Kuester and Joe H. Mize. *Optimization Techniques with Fortran*, pages 240–250. McGraw-Hill, New-York, 1973.

[338] Ulrich W. Kulisch and Willard L. Miranker. *Computer Arithmetic in Theory and Practice.* Academic Press, New York, 1981.

[339] Peter Kunkel. Augmented systems for generalized turning points. In R. Seydel, F. W. Schneider, T. Küpper, and H. Troger, editors, *Proceedings of the Conference at Würzburg, Aug. 1990, Bifurcation and Chaos: Analysis, Algorithms, Applications*, pages 231–236, Basel, 1991. Birkhäuser.

[340] Stanislas Labik, Anatol Malijevski, and Peter Vonka. A rapidly convergent method of solving the OZ equation. *Mol. Phys.*, 56:709–715, 1985.

[341] Philippe Langlois. Automatic linear correction of rounding errors. *BIT*, 41(3):515–539, 2001.

[342] Philippe Langlois and Fabrice Nativel. Reduction and bounding of the rounding error in floating point arithmetic. *C.R. Acad. Sci. Paris, Série 1*, 327:781–786, 1998.

[343] Philippe Langlois and Fabrice Nativel. When automatic linear correction of rounding errors is exact. *C.R. Acad. Sci. Paris, Série 1*, 328:543–548, 1999. Erratum in 328:829, 1999.

[344] Hans Petter Langtangen. *Computational Partial Differential Equations: Numerical Methods and Diffpack Programming*. Number 2 in Lecture Notes in Computational Science and Engineering. Springer, Berlin, 1999.

[345] Charles L. Lawson. Automatic differentiation of inverse functions. In Griewank and Corliss [239], pages 87–94.

[346] J. Daniel Layne. Applying automatic differentiation and self-validating numerical methods in satellite simulations. In Griewank and Corliss [239], pages 211–217.

[347] Cong Tanh Le. *Contrôle optimal et transfert orbital en temps minimal*. PhD thesis, ENSEEIHT, Institut National Polytechnique de Toulouse, France, 1999.

[348] Claude Le Bris. Control theory applied to quantum chemistry: Some tracks. In *International Conference on Control of Systems Governed by PDEs (Nancy, March 1999)*, volume 8, pages 77–94. ESAIM PROC, 2000. See `cermics.enpc.fr/reports/CERMICS-99-174.ps.gz`.

[349] François-Xavier Le Dimet and Olivier Talagrand. Variational algorithms for analysis and assimilation of meteorological observations: Theoretical aspects. *Tellus*, 38A:97–110, 1986.

[350] Lloyd L. Lee. An accurate integral equation theory for hard spheres: Role of the zero-separation theorems in the closure relation. *J. Chem. Phys*, 103:9388–9396, 1995.

[351] Steven L. Lee, Alan C. Hindmarsh, and Peter N. Brown. User documentation for SensPVODE, a variant of PVODE for sensitivity analysis. Technical Report UCRL-MA-140211, Lawrence Livermore National Laboratory, 2000.

[352] Steven L. Lee and Paul D. Hovland. Sensitivity analysis using parallel ODE solvers and automatic differentiation in C: SensPVODE and ADIC. In Corliss et al. [144], chapter 26, pages 223–229.

[353] Claude Lemaréchal and R. Mifflin. Nonsmooth optimization. In *Proceedings of a IIASA Workshop*, Oxford, 1977. Pergamon Press.

[354] Shengtai Li and Linda Petzold. Design of new DASPK for sensitivity analysis. Technical Report TRCS99–28, University of California at Santa Barbara, 1999.

[355] H. Linhart and W. Zucchini. *Model Selection*. Wiley Series in Probability and Mathematical Statistics. John Wiley and Sons, Inc., 1986.

[356] Seppo Linnainmaa. Error linearization as an effective tool for experimental analysis of the numerical stability of algorithms. *BIT*, 23:346–359, 1983.

[357] J. L. Lions. *Optimal Control of Systems Governed by Partial Differential Equations*. Springer, 1971.

[358] Faydor L. Litvin. *Gear Geometry and Applied Theory*. Prentice Hall, Englewood Cliffs, N. J., 1994.

[359] M. D. Liu and A. L. Tits. User's guide for ADIFFSQP Version 0.9: A utility program that allows the user of the FFSQP constrained nonlinear optimization routines to conveniently invoke the computational differentiation preprocessor ADIFOR 2.0. Technical Report, University of Maryland, Systems Research Center, College Park, MD, USA, 1997.

[360] G. Löbel. Ein neues Programmkonzept zur Simulation von Dampferzeugern. Siemens-Bericht TSE3/B17, Siemens.

[361] Bernd Maar and Volker Schulz. Interior point multigrid methods for topology optimization. *Struct. Optim.*, 19(3):214–224, 2000. Also appeared as IWR, University of Heidelberg Technical Report 98–57, 1998.

[362] S. A. Mahlke, W. Y. Chen, J. C. Gyllenhaal, and W.-M. W. Hwu. Compiler code transformations for superscalar-based high performance systems. *Proceedings of the 1992 Conference on Supercomputing (Washington D.C.)*, pages 808–817, 1992.

[363] Kamel G. Mahmoud. Approximations in optimum structural design. In B. H. V. Topping and M. Papadrakakis, editors, *Advanced in Structural Optimization*, pages 57–67. Civil-Comp Press, Edinburgh, 1994.

[364] Kyoko Makino. *Rigorous Analysis of Nonlinear Motion in Particle Accelerators*. PhD thesis, Michigan State University, East Lansing, Michigan, USA, 1998. Also MSUCL-1093 and bt.nscl.msu.edu/papers-cgi/display.pl?name=makinophd.

[365] Kyoko Makino and Martin Berz. Remainder differential algebras and their applications. In Berz et al. [57], pages 63–74.

[366] Kyoko Makino and Martin Berz. Efficient control of the dependency problem based on Taylor model methods. *Reliable Computing*, 5(1):3–12, 1999.

[367] Kyoko Makino and Martin Berz. New applications of Taylor model methods. In Corliss et al. [144], chapter 43, pages 359–364.

[368] Kazimierz Malanowski. Sufficient optimality conditions for optimal control subject to state constraints. *SIAM J. Cont. Opt.*, 35:205–227, 1997.

[369] Kazimierz Malanowski and Helmut Maurer. Sensitivity analysis for parametric optimal control problems with control-state constraints. *Comp. Optim. Appl.*, 5:253–283, 1996.

[370] T. Maly and Linda Petzold. Numerical methods and software for sensitivity analysis of differential-algebraic systems. *Applied Numerical Mathematics*, 20(1-2):57–79, 1996.

[371] Marco Mancini. A hierarchical approach in automatic differentiation. In Christèle Faure, editor, *Automatic Differentiation for Adjoint Code Generation*, Technical Report no. 3555, pages 39–45. INRIA, 1998.

[372] Marco Mancini. A parallel hierarchical approach for automatic differentiation. In Corliss et al. [144], chapter 27, pages 231–236.

[373] W. Donald Marquardt. An algorithm for least squares estimation of non-linear parameters. *J. Soc. Indust. Appl. Math.*, 11(2):413–441, 1963.

[374] Georgy A. Martynov. *Fundamental Theory of Liquids: Methods of Distribution Functions*. Adam Hilger, 1992.

[375] Kurt Maute and Ekkehard Ramm. Adaptive topology optimization. *Struct. Optim.*, 10:100–112, 1995.

[376] Vladimir Mazourik. Integration of automatic differentiation into a numerical library for PC's. In Griewank and Corliss [239], pages 286–293.

[377] P. R. McHugh, Dana A. Knoll, and David E. Keyes. Application of a Newton-Krylov-Schwarz algorithm to low Mach number combustion. *AIAA J.*, 36:290–292, 1998.

[378] B. A. Megrey. Review and comparison of age-structured stock assessment models from theoretical and applied point of view. In B.A. Megrey and E.F. Edward, editors, *Mathematical Analysis of Fish Stock Dynamics*. American Fisheries Society, 1989.

[379] Scott Meyers. *Effective C++: 50 Ways to Improve Your Programs and Designs*. Addison-Wesley, 1992.

[380] Leo Michelotti. MXYZPTLK: A C++ hacker's implementation of automatic differentiation. In Griewank and Corliss [239], pages 218–227.

[381] A. Miele and J. W. Cantrell. Study on a memory gradient method for the minimization of functions. *Journal of Optimization Theory and Applications*, 3:459–470, 1969.

[382] Bijan Mohammadi, Jean-Michel Malé, and Nicole Rostaing-Schmidt. Automatic differentiation in direct and reverse modes: Application to optimum shapes design in fluid mechanics. In Berz et al. [57], pages 309–318.

[383] Michael Monagan and René R. Rodoni. Automatic differentiation: An implementation of the forward and reverse mode in Maple. In Berz et al. [57], pages 353–362.

[384] Ramon E. Moore. *Interval Analysis*. Prentice Hall, Englewood Cliffs, N.J., 1966.

[385] Ramon E. Moore. *Methods and Applications of Interval Analysis*. SIAM, Philadelphia, Penn., 1979.

[386] Ramon E. Moore, editor. *Reliability in Computing: The Role of Interval Methods in Scientific Computing*. Academic Press, San Diego, 1988.

[387] Ramon E. Moore. Private communication, 1998.

[388] Jorge J. Moré. The Levenberg-Marquardt algorithm: Implementation and theory. In G. A. Watson, editor, *Lecture Notes in Mathematics*, pages 105–116. Springer Verlag, Berlin, 1977.

[389] Jorge J. Moré. Automatic differentiation tools in optimization software. In Corliss et al. [144], chapter 2, pages 25–34.

[390] T. Morita and Kazuo Hiroike. A new approach to the theory of classical fluids. *Prog. Theor. Phys.*, 23:1003, 1960.

[391] Steven S. Muchnick. *Advanced Compiler Design Implementation*. Morgan Kaufmann Publishers, San Francisco, California, 1997.

[392] D. M. Murray and S. J. Yakowitz. Differential dynamic programming and Newton's method for discrete optimal control problems. *Journal of Optimization Theory and Applications*, 43(3):395–414, 1984.

[393] Nathan C. Myers. Traits: A new and useful template technique. *C++ Report*, 1995. See www.cantrip.org/traits.html.

[394] G. A. Mykut and N. Untersteiner. Some results from a time dependent thermodynamic model of sea-ice. *J. Geophys. Res.*, 76:1550–1575, 1971.

[395] L. A. Mysak, D. K. Manak, and R. F. Marsden. Sea-ice anomalies observed in the Greenland and Labrador seas during 1901-1984 and their relationship to an interdecadal Arctic climate cycle. *Climate Dyn.*, 5:111–133, 1990.

[396] NAG. *NAG Fortran Library Manual, Mark 18*. The Numerical Algorithms Group Limited, Oxford, 1997.

[397] NAG. Iris explorer. www.nag.co.uk, 2001.

[398] Uwe Naumann. *Efficient Calculation of Jacobian Matrices by Optimized Application of the Chain Rule to Computational Graphs*. PhD thesis, Technical University of Dresden, December 1999.

[399] Uwe Naumann. Elimination techniques for cheap Jacobians. In Corliss et al. [144], chapter 29, pages 247–253.

[400] I. Michael Navon and Xiaolei Zou. Application of the adjoint model in meteorology. In Griewank and Corliss [239], pages 202–207.

[401] NEOS Server for Optimization Problems. See www-neos.mcs.anl.gov.

[402] Arnold Neumaier. *Interval Methods for Systems of Equations*. Cambridge University Press, Cambridge, 1990.

[403] Garry N. Newsam and John D. Ramsdell. Estimation of sparse Jacobian matrices. *SIAM J. Alg. Disc. Meth.*, 4(3):404–417, 1983.

[404] Joseph Noailles and Cong Thanh Le. Contrôle en temps minimal et transfert orbital à faible poussée. *Équations aux dérivées partielles et applications*, pages 705–724, Gauthier-Villars, 1998. Articles in honour of J. L. Lions for his 70th birthday.

[405] Jorge Nocedal and Stephen J. Wright. *Numerical Optimization*. Springer Series in Operations Research. Springer-Verlag, New York, 1999.

[406] Numerical Objects AS. Diffpack World Wide Web home page. See www.nobjects.com.

[407] L. S. Ornstein and F. Zernike. *Proc. Acad. Sci. Amsterdam*, 17:793, 1914.

[408] Constantinos C. Pantelides. The consistent initialization of differential-algebraic systems. *SIAM J. Sci. Stat. Comput.*, 9(2):213–231, 1988.

[409] J. F. A. De O. Pantoja. *Algorithms for Constrained Optimization Problems*. PhD thesis, Imperial College of Science and Technology, University of London, 1983.

[410] J. F. A. De O. Pantoja. Differential dynamic programming and Newton's method. *Int. J. Control*, 47(5):1539–1553, 1988.

[411] Seon Ki Park, Kelvin K. Droegemeier, and Christian H. Bischof. Automatic differentiation as a tool for sensitivity analysis of a convective storm in a 3-D cloud model. In Berz et al. [57], pages 205–214.

[412] C. L. Parkinson and W. M. Washington. A large-scale numerical model of sea-ice. *J. Geophys. Res.*, 84:311–337, 1979.

[413] Terence Parr, John Lilly, Peter Wells, Ric Klaren, Mika Illouz, John Mitchell, Scott Stanchfield, Jim Coker, Monty Zukowski, and Chapman Flack. ANTLR Reference Manual. Technical report, Mage-Lang Institute's jGuru.com, January 2000. See www.antlr.org/doc/index.html.

[414] David A. Patterson and John L. Hennessy. *Computer Architecture: A Quantitative Approach*. Morgan Kaufmann, San Mateo, CA, 1990.

[415] Ken R. Patterson and G. D. Melvin. Integrated catch at age analysis. Version 1.2. Scottish Fisheries Research Report 58, The Scottish Office. Agriculture, Environment and Fisheries Department., HMSO Edinburgh, October 1996.

[416] H. W. Pele, R. Hempelmann, M. Prager, and M. D. Zeidler. Dynamics of 18-crown-6 ether in aqueous solution studied by quasielastic neutron scattering. *Berichte der Bunsen-Gesellschaft für physikalische Chemie*, 95(5):592–598, 1991.

[417] Michèle Pichat. Correction d'une somme en arithmétique à virgule flottante. *Numer. Math.*, 19:400–406, 1972.

[418] Niles A. Pierce and Michael B. Giles. Peconditioned multigrid methods for compressible flow calculations on stretched meshes. *J. Comput. Phys.*, 136:425–445, 1997.

[419] John R. Pilbrow. *Transition Ion Electron Paramagnetic Resonance*, chapter 5.3.1 Powder Averaged Transition Probability, Formula (5,22), page 222. Clarendon Press, Oxford, 1990.

[420] Olivier Pironneau. *Optimal Shape Design for Elliptic Problems.* Springer, 1982.

[421] Paul E. Plassmann. Sparse Jacobian estimation and factorization on a multiprocessor. In Thomas F. Coleman and Y. Li, editors, *Large-Scale Optimization*, pages 152–179. SIAM, Philadelphia, Penn., 1990.

[422] Michael J. D. Powell. A fast algorithm for nonlinear constrained optimization calculations. In G. A. Watson, editor, *Numerical Analysis*, number 630 in Lecture Notes in Mathematics. Springer, Berlin, 1978.

[423] Michael J. D. Powell and Philippe L. Toint. On the estimation of sparse Hessian matrices. *SIAM J. Numer. Anal.*, 16:1060–1074, 1979.

[424] William H. Press, Saul A. Teukolsky, William T. Vetterling, and Brian P. Flannery. *Numerical Recipies in Fortran. 2nd edn.* Cambridge University Press, Cambridge, 1992.

[425] John D. Pryce. Solving high-index DAEs by Taylor series. *Numerical Algorithms*, 19:195–211, 1998.

[426] John D. Pryce. A simple structural analysis method for DAEs. Technical Report DoIS/TR05/00, RMCS, Cranfield University, March 2000.

[427] John D. Pryce and John K. Reid. ADO1, a Fortran 90 code for automatic differentiation. Technical Report RAL-TR-1998-057, Rutherford Appleton Laboratory, Chilton, Didcot, Oxfordshire, OX11 OQX,

England, 1998. See `ftp://matisa.cc.rl.ac.uk/pub/reports/prRAL98057.ps.gz`.

[428] Gordon D. Pusch. Jet space as the geometric arena of automatic differentiation. In Berz et al. [57], pages 53–62.

[429] H. Rabitz, M. Kramer, and D. Dacol. Sensitivity analysis in chemical kinetics. *Ann. Rev. Phys. Chem.*, 34:419–461, 1983.

[430] Louis B. Rall. Convergence of the Newton process to multiple solutions. *Numer. Math.*, 9:25–37, 1966.

[431] Louis B. Rall. *Automatic Differentiation: Techniques and Applications*, volume 120 of *Lecture Notes in Computer Science*. Springer-Verlag, Berlin, 1981.

[432] Louis B. Rall. Point and interval differentiation arithmetics. In Griewank and Corliss [239], pages 17–24.

[433] Louis B. Rall and George F. Corliss. An introduction to automatic differentiation. In Berz et al. [57], pages 1–17.

[434] Ekkehard Ramm, Kai-Uwe Bletzinger, Reiner Reitinger, and Kurt Maute. The challenge of structural optimization. In B. H. V. Topping and M. Papadrakakis, editors, *Advanced in Structural Optimization*, pages 27–52. Civil-Comp Press, Edinburgh, 1994.

[435] Dietmar Ratz and Tibor Csendes. On the selection of subdivision directions in interval branch-and-bound methods for global optimization. *J. Global Optim.*, 7:183–207, 1995.

[436] Gunther Reißig, Wade S. Martinson, and Paul I. Barton. Differential-algebraic equations of index 1 may have an arbitrarily high structural index. *SIAM J. Sci. Comput.*, 21(6):1987–1990 (electronic), 2000.

[437] I. Rigor and M. Ortmeyer. Observations of ice SLP, SAT and ice motion. See `iabp.apl.washington.edu/Summary`, 2000.

[438] Klaus Röbenack and Kurt J. Reinschke. Reglerentwurf mit Hilfe des Automatischen Differenzierens. *Automatisierungstechnik*, 48(2):60–66, 2000.

[439] Klaus Röbenack and Kurt J. Reinschke. Nonlinear observer design using automatic differentiation. In Corliss et al. [144], chapter 15, pages 137–142.

[440] Philip L. Roe. Approximate Riemann solvers, parameter vectors, and difference schemes. *Journal of Computational Physics*, 43:357–372, 1981.

[441] Nicole Rostaing, Stéphane Dalmas, and André Galligo. Automatic differentiation in Odyssée. *Tellus*, 45A:558–568, 1993.

[442] Nicole Rostaing-Schmidt and Eric Hassold. Basic functional representation of programs for automatic differentiation in the Odyssée system. In François-Xavier Le Dimet, editor, *High Performance Computing in the Geosciences*. Kluwer Academic Publishers, Dordrecht, 1994.

[443] Ralf Rothfuß, Johanna Schaffner, and Michael Zeitz. Rechnergestützte Analyse und Synthese nichtlinearer Systeme. In *Nichtlineare Regelungen: Methoden, Werkzeuge, Anwendungen*, pages 267–291. VDI-Verlag, 1999.

[444] John Shipley Rowlinson. Self-consistent approximation for molecular distribution functions. *Mol. Phys.*, 9:217–227, 1965.

[445] Sirpa Saarinen, Randall Bramley, and George Cybenko. Neural networks, backpropagation, and automatic differentiation. In Griewank and Corliss [239], pages 31–42.

[446] Andrea Saltelli, Karen Chan, and Evelyn Marian Scott, editors. *Sensitivity Analysis*. Wiley, 2000.

[447] Johanna Schaffner and Michael Zeitz. Variants of nonlinear normal form observer design. In Henk Hijmeijer and Thor I. Fossen, editors, *New Direction in Nonlinear Observer Design*, Number 244 in Lecture Notes in Control and Information Science, pages 161–180. Springer, Berlin, 1999.

[448] G. A. F. Seber and C. J. Wild. *Nonlinear Regresssion*. Wiley Series in Probability and Mathematical Statistics. John Wiley and Sons, Inc., New York, 1989.

[449] P. K. Seidelmann, editor. *Explanatory Supplement to the Astronomical Almanac*. University Science Books, Mill Valley, California, 1992.

[450] A. J. Semtner, Jr. A model for the thermodyanmic growth of sea-ice in numerical investigation of climate. *J. Phys. Oceanogr.*, 6:379–389, 1976.

[451] C. Sevin. *Optimisation de formes en mécanique des fluides numérique*. PhD thesis, Université Pierre et Marie Curie, 1999.

[452] R. Seydel, F. W. Schneider, T. Küpper, and H. Troger, editors. *Proceedings of the Conference at Würzburg, Aug. 1990, Bifurcation and Chaos: Analysis, Algorithms, Applications*. Birkhäuser, Basel, 1991.

[453] Piyush Shah. Application of adjoint equations to estimation of parameters in distributed dynamic systems. In Griewank and Corliss [239], pages 181–190.

[454] Khodr Shamseddine and Martin Berz. Exception handling in derivative computation with nonarchimedean calculus. In Berz et al. [57], pages 37–51.

[455] Khodr Shamseddine and Martin Berz. Convergence on the Levi-Civita field and study of power series. In *Proc. Sixth International Conference on Nonarchimedean Analysis*, pages 283–299. Marcel Dekker, New York, 2000.

[456] L. L. Sherman, Arthur C. Taylor III, Larry L. Green, Perry A. Newman, Gene W. Hou, and Vamshi Mohan Korivi. First- and second-order aerodynamic sensitivity derivatives via automatic differentiation with incremental iterative methods. *J. Comput. Phys.*, 129(2):307–331, 1996.

[457] A. Shestakov and J. Milovich. Applications of pseudo-transient continuation and Newton-Krylov methods fo the Poisson-Boltzmann and radiation diffusion equations. Technical Report UCRL–JC–139339, Lawrence Livermore National Laboratory, 2000.

[458] G. Sheveleva. Mathematical simulation of spiral bevel gear production and meshing processes with contact and bending stresses. In *Proc. IX. World Congr. IFToMM*, volume 1, pages 509–513, 1995.

[459] Dmitri Shiriaev. ADOL–F automatic differentiation of Fortran codes. In Berz et al. [57], pages 375–384.

[460] J. J. Skrobanski. *Optimization Subject to Nonlinear Constraints*. PhD thesis, London University, 1986.

[461] Barry F. Smith, Petter Bjørstad, and William D. Gropp. *Domain Decomposition: Parallel Multilevel Methods for Elliptic Partial Differential Equations*. Cambridge University Press, Cambridge, 1996.

[462] J. E. Smith and G. Sohi. The microarchitecture of superscalar processors. *Proceedings of the IEEE*, 1995.

[463] Marc Snir, Steve W. Otto, Steven Huss-Lederman, David W. Walker, and Jack Dongarra. *MPI: The Complete Reference*. MIT Press, 1995.

[464] Edgar J. Soulié. User's experience with Fortran precompilers for least squares optimization problems. In Griewank and Corliss [239], pages 297–306.

[465] Edgar J. Soulié, Christèle Faure, Théo Berclaz, and Michel Geoffroy. Electron paramagnetic resonance, optimization and automatic differentiation. In Corliss et al. [144], chapter 10, pages 99–106.

[466] Edgar J. Soulié and Pierre C. Lesieur. Quantitative analysis of the electron paramagnetic resonance spectrum of a uranium (III) compound. *J. Chem. Soc. Faraday Trans. I*, 85(12):4053–4062, 1989.

[467] Bert Speelpenning. *Compiling Fast Partial Derivatives of Functions Given by Algorithms*. PhD thesis, Department of Computer Science, University of Illinois at Urbana-Champaign, Urbana-Champaign, Ill., January 1980.

[468] Hermann J. Stadtfeld. *Handbook of Bevel and Hypoid Gears*. Rochester Institute of Technology, 1993.

[469] Christoph Stangl. Optimal sizing for a class of nonlinearly elastic materials. *SIAM J. Optim.*, 9(2):414–443, 1999.

[470] Trond Steihaug and A.K.M. Shahadat Hossain. Graph coloring and the estimation of sparse Jacobian matrices with segmented columns. Technical Report 72, Department of Informatics, University of Bergen, 1997.

[471] Mohamed Tadjouddine, Shaun A. Forth, and John D. Pryce. AD tools and prospects for optimal AD in CFD flux Jacobian calculations. In Corliss et al. [144], chapter 30, pages 255–261.

[472] Oliver Talagrand. The use of adjoint equations in numerical modelling of the atmospheric circulation. In Griewank and Corliss [239], pages 169–180.

[473] Leigh Tesfatsion. Automatic evaluation of higher-order partial derivatives for nonlocal sensitivity analysis. In Griewank and Corliss [239], pages 157–165.

[474] William Carlisle Thacker. Automatic differentiation from an oceanographer's perspective. In Griewank and Corliss [239], pages 191–201.

[475] Joseph M. Thames. Synthetic calculus: A paradigm for mathematical program synthesis. In Griewank and Corliss [239], pages 263–272.

[476] The MathWorks. Matlab 5.3.1, 2000. See www.mathworks.com/products/matlab.

[477] Engelbert Tijskens, Herman Ramon, and Josse De Baerdemaeker. Efficient operator overloading AD for solving nonlinear PDEs. In Corliss et al. [144], chapter 19, pages 167–172.

[478] Engelbert Tijskens, Dirk Roose, Herman Ramon, and Josse De Baerdemaeker. Automatic differentiation for nonlinear partial differential equations: An efficient operator overloading approach. *Numerical Algorithms*, submitted.

[479] Engelbert Tijskens, Dirk Roose, Herman Ramon, and Josse De Baerdemaeker. Efficient solution of nonlinear partial differential equations by integrating automatic differentiation. *International Journal for Numerical Methods in Engineering*, submitted.

[480] Engelbert Tijskens, Wim Schoenmaker, and Karin De Meyer. Automatic numerical evaluation of derivatives and its use in device simulators. In *IEEE Workshop on Numerical Modelling of Processes and Devices for Integrated Circuits, NUPAD IV*, pages 251–254. IEEE, 1992.

[481] John E. Tolsma and Paul I. Barton. DAEPACK an open modeling environment for legacy models. *Industrial & Engineering Chemistry Research*, 39(6):1826–1839, 2000.

[482] Jean Utke. Efficient Newton steps without Jacobians. In Berz et al. [57], pages 253–264.

[483] J. van der Snepscheut. *What Computing Is All About.* Texts and Monographs in Computer Science, Suppl. 2. Springer Verlag, Berlin, 1993.

[484] Pascal Van Hentenryck, Laurent Michel, and Yves Deville. *Numerica: A Modeling Language for Global Optimization.* MIT Press, Cambridge, Mass., 1997.

[485] P. Van Laarhoven and E. Aarts. *Simulated Annealing: Theory and Applications.* Reidel, Dordrecht, 1988.

[486] Todd Veldhuizen. Techniques for scientific C++, 1998. See www.oonumerics.org/blitz/papers.

[487] Todd Veldhuizen and Kumaraswamy Ponnambala. Linear algebra with C++ template metaprograms. Rapid linear algebra is just one use. *Dr. Dobb's Journal*, 1996. See www.ddj.com/ddj/1996/1996.08/veld.htm.

[488] Arun Verma. *Structured Automatic Differentiation.* PhD thesis, Cornell University Department of Computer Science, Ithaca, NY, 1998.

[489] Olaf Vogel. Accurate gear tooth contact determination and sensitivity computation for hypoid bevel gears. In Corliss et al. [144], chapter 23, pages 197–204.

[490] Olaf Vogel, Andreas Griewank, and Gert Bär. Direct gear tooth contact analysis for hypoid bevel gears. *Computer Methods in Applied Mechanics and Engineering*, submitted.

[491] Olaf Vogel and Ulf Hutschenreiter. Berechnungen zur Kontaktgeometrie beim Eingriff bogenverzahnter Kegelräder und Hypoidräder. IOKOMO 03-1997, Technical University of Dresden, Institute of Scientific Computing and Institute of Geometry, 1997.

[492] A. G. Vompe and G. A. Martynov. The bridge function expansion and the self-consistency problem of the Ornstein-Zernike equation solution. *J. Chem. Phys*, 100:5249–5258, 1994.

[493] Andrea Walther. *Program Reversal Schedules for Single- and Multiprocessor Machines*. PhD thesis, Institute of Scientific Computing, Technical University Dresden, Germany, 1999.

[494] Andrea Walther. Program reversals for evolutions with non-uniform step costs. Preprint IOKOMO-2000, Technical University of Dresden, 2000.

[495] Andrea Walther and Andreas Griewank. New results on program reversals. In Corliss et al. [144], chapter 28, pages 237–243.

[496] Andrea Walther, Andreas Griewank, and Olaf Vogel. Higher derivative tensors from univariate Taylor series with comparison to Maple on an engineering problem. *Z. Angew. Math. Mech.*, 80(3):813–814, 2000. Proceedings GAMM99 Annual Meeting, Metz.

[497] John D. Weeks, David Chandler, and Hans C. Andersen. Role of repulsive forces in determining the equilibrium structure of simple liquids. *J. Chem. Phys*, 54:5237–5247, 1971.

[498] R. E. Wengert. A simple automatic derivative evaluation program. *Comm. ACM*, 7(8):463–464, 1964.

[499] Anthony S. Wexler. Automatic evaluation of derivatives. *Applied Mathematics and Computation*, 24:19–46, 1987.

[500] R. Clint Whaley and Jack J. Dongarra. Automatically tuned linear algebra software. Technical report, Department of Computer Sciences, University of Tennessee, Knoxville, 1998.

[501] R. Clint Whaley, Antoine Petitet, and Jack J. Dongarra. Automated empirical optimization of software and the ATLAS project. Technical report, Department of Computer Sciences, University of TenneSsee, September 2000. See www.netlib.org/atlas.

[502] James H. Wilkinson. Error analysis revisited. *IMA Bulletin*, 22(11/12):192–200, 1986.

[503] M. E. Wolf and M. S. Lam. A data locality optimization algorithm. *Proceedings of the SIGPLAN Conference on Programming Language Design and Implementation (PLDI)*, pages 30–44, 1991.

[504] Brian Worley. Experience with the forward and reverse mode of GRESS in contaminant transport modeling and other applications. In Griewank and Corliss [239], pages 307–315.

[505] Wenhong Yang and George Corliss. Bibliography of computational differentiation. In Berz et al. [57], pages 393–418.

[506] G. Zerah and J. P. Hansen. Self-consistent integral equations for fluid pair distribution functions: Another attempt. *J. Chem. Phys*, 84:2336, 1986.

[507] Hong Zhang and Herschel A. Rabitz. Robust control of quantum molecular systems in presence of disturbances and uncertainties. *Physical Review A*, 49(4):2241–2254, 1994.

[508] C. Zhu, R. H. Byrd, P. Lu, and Jorge Nocedal. LBFGS-B: Fortran subroutines for large-scale bound constrained optimization. Report NAM-11, EECS Northwestern Univ., Ill., 1994.

Index

abstract time stepping scheme, 161
accuracy, 79, 103, 123, 133, 156, 208
acoustic wave equation, 165
active variable, 296
AD, *see* automatic differentiation
AD_{opt}, 120
AD01, 255
ADIC, 169, 175, 226
ADIFOR, vi, 3, 8, 26, 71, 94, 146, 155, 175, 190, 192, 215, 255, 271
adjoint
    equations, 145, 161, 181, 327
    model, 78, 184, 283, 293, 299, 327
ADMAT, 46
ADMIT, vi
Adoc library, 338
Adogen, 338
ADOL-C, vi, 3, 14, 19, 83, 86, 141, 169, 185, 203, 271, 311
ADOL-F, 8
advection-diffusion, 227
aerofoil optimisation, 153, 157
AMPL, 26
application, v, 3
    acoustic wave equation, 165
    aerofoil optimisation, 153
    chemical reaction, 205
    circuit simulation, 7
    computational aerodynamics, 40
    computational fluid dynamics (CFD), 145, 153, 255
    computational quantum chemistry, 205

cooling system, 12
coupled pendulums, 347
data assimilation, 41, 78
design optimization, 41, 145, 153, 335
disordered condensed matter, 189
diurnal kinetics advection-diffusion, 227
elastic-plastic torsion, 177
electromagnetic problem, 339
electron paramagnetic resonance, 99
Euler equations, 176
gear tooth contact, 197
hypoid bevel gears, 197
low thrust orbit transfer, 110
material modeling, 131, 136
minimum time orbit transfer, 109
model fitting, 84, 95, 99
model structure validation, 85
MOS transistor model, 8
multibody dynamics, 131
Navier-Stokes solver, 303
near-earth asteroids, 364
neuron scattering, 69
nonlinear least squares, 70
nonlinear regression, 83
nonlinear solver, v
North Sea herring, 86
observer design, 137
ODE solver, v, 223, 319
optimal control, 41, 109, 117, 125, 189, 205, 327, 330
parameter identification, v, 41, 45, 70, 84, 95